Genetic Theory and Analysis

Genetic Theory and Analysis

Finding Meaning in a Genome

Second Edition

Danny E. Miller
Angela L. Miller
R. Scott Hawley

Published by John Wiley & Sons, Inc., Hoboken, New Jersey.
Published simultaneously in Canada.

For general information on our other products and services or for technical support, please contact our Customer Care Department within the United States at (800) 762–2974, outside the United States at (317) 572–3993 or fax (317) 572–4002.

Wiley also publishes its books in a variety of electronic formats. Some content that appears in print may not be available in electronic formats. For more information about Wiley products, visit our web site at www.wiley.com.

Library of Congress Cataloging-in-Publication Data applied for
Paperback ISBN: 9781118086926

Cover Design: Wiley
Cover Image: © Romy Honda Oka Romyhonda/EyeEm/Getty Images

Set in 9.5/12.5pt STIXTwoText by Straive, Pondicherry, India

SKY10051350_071823

Contents

Preface

For the geneticist there are accordingly three ways of examining anything. Through characters [s]he can examine function; through their changes, [s]he can examine mutation; through their reassortment, [s]he can examine recombination.

– Francois Jacob in The Logic of Life (p. 224)

Although the first edition of this book was intended for an advanced course in genetic analysis, we have realized that motivated undergraduates are as capable of digesting this information as graduate students. This book does assume the reader has a basic familiarity with the genetics of eukaryotes, as well as with the basic biology of prokaryotes and their viruses. We also assume a working knowledge of the three Sirens of molecular biology: transcription, translation, and replication. In cases where specialized techniques are used or concepts required, we have endeavored to provide the essential background material. For ease of use, we have added a glossary and a more comprehensive index.

The focus of this book was, and still is, on the basic principles that underlie genetic analysis: mutation, complementation (and its bridesmaids, suppression and enhancement), recombination, segregation, and regulation. Our goal is to provide insight into the biological and analytical processes that comprise each of these tools and to explain their use. Our basic objective is for you to learn just what each tool or test does and how it can lie to you. Perhaps most importantly, the book is designed to teach you just how much you can learn when nature misunderstands your question. In other words, this is a book about genetic theory.

Although a discussion of genetic analysis invariably requires the presentation of multiple examples, this is not to be considered a textbook of genetic facts. Facts can sometimes change in the blink of an eye; the basic analytical tools change rather more slowly. We have tried to be as comprehensive and catholic as possible in the choice of examples and organisms, drawing on studies from as many genetically tractable model organisms as possible, with even the occasional reference to humans. However, we cannot escape the truth that we are fly biologists by training and so, despite our best efforts, lean a little heavily on Drosophila stories. Moreover, with advancements in genome sequencing and annotation, the model organism landscape is ever growing and evolving. Beautiful and groundbreaking analyses are undertaken daily in organisms whose genomes were not available only a few years ago such as zebrafish, cave fish, planaria, and Nematostella. There are now so many organisms available for study that we must regrettably limit the model organisms described in Appendix A to only those relevant to the examples contained within these pages.

The following nine chapters cover the basic intellectual tools that comprise modern genetics. Some of these techniques, such as mutant hunting, suppressor analysis and complementation analysis, echo issues derived from even the most current journals, while others, such as the algebraic analysis of recombination data, have fallen into disuse. While we were writing this book, we were often reminded by our colleagues that some things are no longer done in the fashions we have described. This is, after all, the post-genomic era.

Living in an era of sequenced genomes is a heady business, indeed. But we are reminded of a comment by the playwright Noel Coward, when pressured by a friend as to how he was doing on his latest play, he is reported to have answered that he was half done. He had taken all the words out of the dictionary, now he just needed to put them in proper order. The sequencing of many genomes has given us our list of words or genes. Now we, like Mr. Coward, need to put our words in the right order. Unfortunately, unlike Mr. Coward, we have been given many words that we do not understand and must learn what these words (or genes) mean or do. We suspect that the best way to understand what genes do is to mutate them. And the best way to put them in order will be epistasis analysis. One may identify interacting proteins by systems such as the yeast two-hybrid assay or mass spectroscopy, but those interactions will have real meaning only when confirmed by genetic interactions as well. In that sense, then, perhaps the genetics *that was* becomes the genetics *that is*. We suspect that these tools, and the intellectual principles that created them, will have much more than historical value for generations of biologists to come.

Writing a book like this is rarely a solitary process, and we would be remiss if we didn't express our gratitude for those who graciously travelled with us at various points along the journey. We particularly want to thank the Stowers Institute, and specifically members of the Hawley lab, as well as the students in 206H and the patients and their families who keep us motivated. We also enormously appreciate the "community of geneticists"; we are beyond lucky to be part of this community.

Finally, this book contains several hundred references. Despite every attempt to be complete, we know we have missed citing many extremely well done and important examples of genetic analysis. In our defense, we can say only that there came a time to stop reading and to actually write this book. Our apologies to those authors whose work was in the huge stack of "you know, we really ought to discuss this" papers that remain sitting on the sides of our desks.

Introduction

All in due time, my pretties, all in due time.

– The Wicked Witch of the West

One begins a genetics textbook by talking either about Gregor Mendel or about James Watson, Francis Crick, and Rosalind Franklin. That is simply the way things are done. The choice an author makes reflects their basic scientific predilections. Classical geneticists start with Mendel, and thus so will we. We will get to Drs. Watson, Crick, and Franklin in due time.

To Gregor Mendel, a gene was little more than a statistical entity, defined by effects on phenotypic variation and by segregation. Each trait was determined by two copies of a given gene. Differences in the trait (phenotype) were due to differences in the "form" of a given gene. These different forms of a given gene were called alleles. In Mendel's construction, the phenotype was a direct consequence of the alleles present in a given individual. To explain the cases where the effect of one allele predominated over the other, Mendel created the concepts of dominance and recessivity. Thus, to Mendel, the primary definition of a gene was that it was a unit of hereditary information. Genes were not structures or tissues themselves, rather they provided information required to create those things.

Mendel's concept of the gene was also firmly embedded in the idea of a gene as the unit of segregation. This is to say an individual possesses two copies of each gene – one copy that was inherited from the mother and one copy that was inherited from the father. Moreover, an individual passes on only one of those two genes to their own offspring, and they do so at random. The gene pair thus becomes the unit of segregation that ultimately leads to gametes bearing a single hereditary particle (gene) for each trait. Mendel's concept of independent assortment can be thought of as a rather simple extension of this idea: because the individual gene pair is the unit of segregation, the assortment of two gene pairs will occur at random.

As we now understand it, Mendel's concept of the gene was all but Newtonian. He saw genes as small immutable particles whose movement was controlled by natural law in such a way that it could be modeled statistically. We can describe Mendel's concept of the gene in the three laws. The first law, which we will call the Purity and Constancy of the Gene, states that genes themselves are immutable; although genes produce the phenotype, they are not themselves the phenotype, nor are they affected by the phenotype. The second law, the Law of the Gene, states that an individual carries two, and only two, copies of a gene for a given trait. This Law of the Gene also states that each time a gamete (a sperm or an egg) is made, one of those two copies is chosen at random to be included in that gamete. The third law, the Law of Independent Assortment, mandates that for two pairs of gene, the choice of which copy of gene A is included in the gamete does not affect the choice of which copy of gene B was included in the same gamete.

The only real error in Mendel's analysis resulted from his lack of understanding of the gene itself. Indeed, the first major advance in genetic thinking would focus on correcting Mendel's view of the gene as an immutable object, his idea of the purity and constancy of a gene. Genes might be very constant, and indeed they are, but they would not turn out to be immutable. Mutations in many, but perhaps not all, genes can produce some phenotypic effects. Moreover, the modern geneticist is savvy enough to know that most biological processes involve the products of multiple genes. For many traits, there may be multiple genes that can mutate to a given phenotype. We are also aware that for some fraction of genes, the connection between genotype and phenotype may be influenced both by other genes and by the environment.

We have come a long way since Gregor Mendel. We have a much clearer view of the gene. Along the way, we have developed some very impressive tools to study gene function. These are the focus of this book. But before we embark on our discussion, perhaps we should ask ourselves, "Why should we bother with doing *genetics* at all?" What can we obtain from the isolation and analysis of mutants (for that is what *genetics* is) that we cannot learn by one of the "-omics" of modern biology?

We will define genetics simply: *Genetics is both the use of mutations and mutational analysis to study a given biological process and the study of the hereditary process itself.* When done right, the two halves of that description are inextricable. If someone is isolating mutants and characterizing those mutants to study flight, they are *doing genetics*. If they are simply isolating genes expressed in bird muscle, they may be doing biochemistry – but they are not doing genetics. The very core of *genetics* is *mutation*. However, the actual, doing of genetics requires more than isolating mutations. Doing genetics well also requires that investigators isolate and characterize those mutants in a fashion that (i) maximizes their chance of answering their initial questions; (ii) provides them with as many novel biological insights as possible; and (iii) facilitates a greater understanding of the structure and function of the genome they are studying? In other words, *doing genetics well* means understanding what types of mutations one can get, how to get them, and how to analyze them. The analysis of suppressor mutants, for example, can be a powerful tool indeed when done correctly.

The proper *doing of genetics* requires that a scientist understand their tools. The basic intellectual tools of genetics are: mutation, complementation, recombination, suppression, and regulation (epistasis). This book is about the proper use of those intellectual tools. Our goal is to give you ideas of what works and cautions as to what doesn't. We will discuss the biology and biochemistry of very processes, but only when we need to do so to describe the mutants. The very essence of our story is: the mutants. So, that is where we begin . . . .

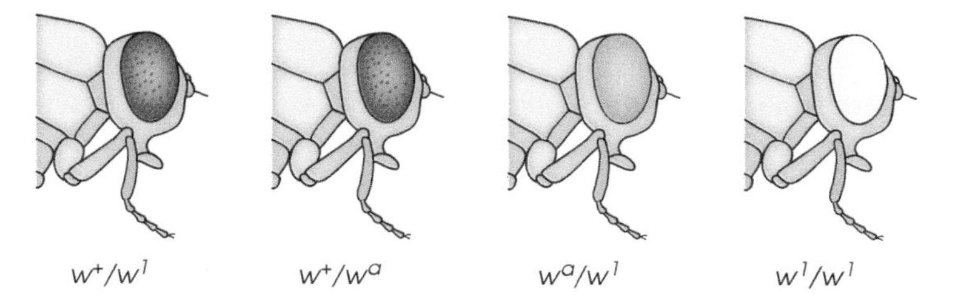

Figure 1.2 Hypomorph. In *Drosophila*, the w^a allele, a hypomorph, produces a diminished quantity of the normal red eye color pigment. Thus, flies heterozygous for w^a and a w^1 allele, which produces no pigment, have orange eyes.

protein product. One can imagine a host of genetic lesions that might produce hypomorphic mutants, ranging from mutations that decrease the level of transcription to mutations that alter messenger stability or the activity or amount of protein product. An excellent example of a hypomorphic mutation is the *white-apricot* (w^a) mutation in *Drosophila* (Figure 1.2). Wildtype (w^+) flies have red eyes, while flies mutant for the *white* gene (w^1) have white eyes. The hypomorphic allele w^a produces some of the protein product needed for red eyes, but not at wildtype levels, so w^a/w^1 flies have orange eyes. This example demonstrates the defining characteristic of a hypomorphic mutant: at least some discernable level of active product is being produced. Most hypomorphic mutants are recessive, but the same caveat about loss-of-function mutations in genes encoding proteins whose dosage is critical applies to partial loss-of-function mutants as well.

One can easily distinguish between hypomorphic and nullomorphic mutants if a deficiency (*Df*) for the gene in question is available. A true nullomorph is the genetic equivalent of a deficiency for a specific gene. Thus, in terms of phenotypic severity, a *m/Df* nullomorph is expected to be equivalent to *m/m*. For a hypomorph, however, *m/Df* is expected to be more severe in phenotype than an *m/m* homozygote. Most of the time the *m/m* homozygote would be expected to produce twice as much active product as the *m/Df* individual. By a similar algebra, an individual with three doses of a hypomorphic mutation (*m/m/m*) is expected to show a less severe phenotype than one with two (*m/m*), whereas adding more doses of a true nullomorphic allele should have no effect.

We've heard many geneticists refer to hypomorphs as the bane of their existence. True enough, the residual level of activity created by such mutations often frustrates the phenotypic or functional analysis of these genes. Nonetheless, hypomorphic mutants are often the first or only mutants to define important genes. It turns out, as we shall discuss in Chapter 2, that many mutant hunts or screens require that homozygotes for the newly induced mutations be viable. Given this restriction, a null allele of a gene required for life would not be recovered in such a screen, even if the protein product of that gene played a critical role in the process under study. In contrast, a hypomorphic mutant can, and often does, produce enough product to allow survival, but not enough to produce a normal phenotype. In such a case, the finding of a hypomorph alerts the investigator to the existence of this gene and heralds its role in the process under study.

Hypermorphs

As the name implies, **hypermorphs** produce either a harmful excess of the normal protein product or a hyperactive one. The defining characteristic of this mutant class is that *m/Df* should be less severe in phenotype than *m/+* because *m/Df* makes less overall protein product than *m/+*. Indeed, in terms of decreasing phenotypic severity, the dosage series should be $m/m > m/+ > m/Df$. Verified examples of hypermorphic mutations are few and far between. The best example of a hypermorphic mutation in *Drosophila* is a mutant called *Confluens* that affects wing vein morphology. *Confluens* is an allele of *Notch* (*N*). The phenotype of *Confluens* can be mimicked by three doses of *N*+, and *Confluens* over a nullomorphic allele of *Notch* is wildtype. All of this makes perfect sense when one realizes that the *Confluens* mutation is a **tandem duplication** of the *N*+ gene. Clearly, increases in the dose of the *N*+ gene have phenotypic consequences, and however you get to three doses, a phenotype is created. A good example from humans is achondroplasia, a genetic disorder in which individuals have short arms and legs but a normal-sized torso. This is caused by variants in fibroblast growth factor receptor 3 (*FGFR3*) that result in the protein being overactive during development.

Other types of mutations that might upregulate the transcription or translation of a given gene or its mRNA product might also produce observable phenotypes. A technique for creating hypermorphs in *Drosophila* was developed by Pernille Rørth in the late 1990s using **transposable elements** (**transposons**), or DNA segments capable of moving around within the genome. Rørth began by creating a transposon capable of driving the expression of neighboring genes in a fashion that was both tissue-specific and inducible (Rørth et al. 1998). By mobilizing that transposon within the *Drosophila* genome, Rørth and her collaborators created a collection of 2,300 lines of flies, each of which carried an independent transposon insertion. This collection of insertions was screened for the ability to suppress a hypomorphic mutation in the *slow border cells* (*slbo*) gene that confers a cell migration defect. This defect is observed in the *Drosophila* ovary and results in sterility. Because the *slbo* gene encodes a C/EBP transcription factor, Rørth and her collaborators reasoned that the high-expression suppressors "could be genes normally activated by C/EBP in border cells or genes which (in this situation) are rate-limiting for cell migration." They obtained both.

Of 2,082 insertion lines tested, 60 showed clear suppression of the mutant phenotype created by homozygosity for *slbo*. The suppressing insertions resulted in the overexpression of genes that encoded known players in actin cytoskeletal remodeling, a critical process in cell migration. They also recovered insertions in a receptor tyrosine kinase gene (*abl*) that also appears to be involved in the control of actin polymerization. The success of the Rørth suppression screen may have been largely due to the choice of a hypomorphic *slbo* allele that retains some degree of C/EBP activity. Thus, it was only necessary to provide a small increase in border cell migration to suppress the sterility caused by the *slbo* mutation.

If they can be obtained, hypermorphs can be a valuable tool for dissecting a genetic process. This may be especially true in a case where one is dealing with a group of functionally redundant genes. In such a case, a simple loss-of-function mutation in one of these genes may not produce a discernable phenotype but overexpressing one of those genes may create an observable defect. Indeed, in Section 6.7, we will consider the use of high-copy suppression libraries in yeast to mimic the creation of hypomorphic mutations by creating colonies, each of which possesses a high copy number of a plasmid carrying a given gene. Phenotypes created by such methods can also serve as the substrate for enhancer and suppressor screens aimed at identifying other genes in this process.

Antimorphs

An antimorphic mutation results in a protein product that antagonizes, or poisons, the wildtype protein. Thus, the phenotype of a true **antimorph** is expected to mimic the phenotype presented by a strong hypomorph or nullomorph. Antimorphic mutations are dominant by definition. However, increasing the dose of the wildtype allele can sometimes ameliorate the phenotype of an antimorphic mutant. For example, imagine a gene (*muct*) that encodes the protein subunit Muct, four identical copies of which are required to produce the enzyme muctinase, which is in turn required to synthesize the imaginary substance muctin. Nullomorphic mutations in *muct* should produce a muctin$^-$ phenotype when homozygous, but assuming that half the level of the enzyme is enough, loss-of-function alleles of *muct* will be recessive and +/*muct* heterozygotes should be normal. But what if a mutant allele of *muct* produces an unusual structural variant of Muct, which is incorporated into the polyprotein complex in such a way as to render the entire complex inactive? Assuming the wildtype and mutant Muct subunits are produced with equal abundance, this variant of Muct will inactivate virtually all (15/16) of the muctinase complexes, and the remaining activity may simply not be enough. However, by increasing the dose of the wildtype allele to two, approximately 20% of the muctinase complexes will be composed of normal subunits.

Sickle cell anemia is a type of antimorphic mutation. Hemoglobin A is a tetramer usually formed by two β-globin and two α-globin chains. This is the typical adult configuration. Every adult also has a low level of Hemoglobin A2, formed by two α and two δ chains, as well as Hemoglobin F, with two α chains and two fetal γ chains. A point mutation in the β-globin gene causes the hemoglobin tetramer to become structurally unstable, especially when there is no oxygen bound to the molecule. It is antimorphic in the sense that one mutant β-globin chain disrupts the entire tetramer even if the other β-globin chain is wildtype. Some treatment modalities focus on upregulating the expression of fetal hemoglobin in adults to compensate for the loss of the β-globin chain.

Another example of an antimorph is presented in Figure 1.3. The microtubules that make possible many processes of cellular movement are composed of long arrays of tubulin monomers. Each of these monomers is composed of an α-tubulin and a β-tubulin subunit. Imagine a mutant in the α-tubulin gene that produces a variant subunit that can incorporate into a growing chain but cannot support further growth. Once a mutant subunit is incorporated, chain growth freezes. However, by increasing the dosage of the normal allele, one decreases the probability of incorporating a mutant subunit from 1/2 to 1/3. One *might* see some phenotypic amelioration in such a case. Indeed, a dominant mutant allele ($\beta 2t^D$) of a testis-specific β-tubulin gene has exactly this type of effect on microtubule assembly in the *Drosophila* male germline. Both heterozygotes and homozygotes show dramatic defects in the formation of large microtubule assemblies in the testis and are sterile as males. But heterozygotes carrying an extra dose of the wildtype allele ($\beta 2t^D/+/+$) are weakly fertile (Kemphues et al. 1980, 1983).

As noted here, the vast majority of antimorphs are dominant. If we use the symbol *A* to denote the mutant, the relative phenotypic severity observed in different genotypes can be described as follows:

$$A/A \geq A/Df \geq A/+ >>> +/Df \geq +/+$$

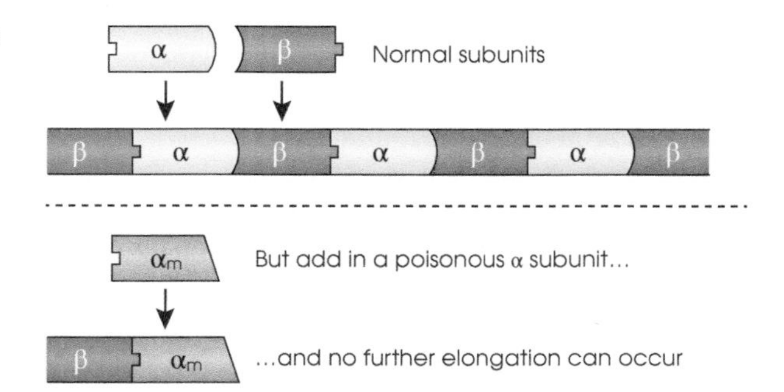

Figure 1.3 Antimorph. Incorporation of the product of an antimorphic allele of tubulin impedes further growth on the microtubule fiber.

The defining characteristic of an antimorph is that one should be able to revert (or, more precisely, pseudorevert) an antimorphic mutation to a nullomorphic mutation of the same gene. In other words, the easiest way to stop this allele from producing a poisonous product is simply to inactivate the gene by a second intragenic mutation (i.e. any mutation that blocks the production of the poisonous protein), thus creating a **pseudorevertant**. One would do this by mutagenizing *A/A* individuals and screening *A**/+ offspring (where *A** indicates a mutagenized chromosome) for revertants. The result should be a new loss-of-function, preferably nullomorphic, allele (denoted r^A). Most critically, the following should be true:

- $+/r^A$ individuals should be phenotypically similar to +/*Df* or +/+ individuals (i.e. the original antimorph should be reverted), and
- r^A/r^A individuals should be phenotypically similar to *A/A*, *A/Df*, or *A*/+ individuals

Indeed, the $\beta 2t^D$ allele generated by Kemphues can be reverted to create recessive loss-of-function alleles of the $\beta 2t^D$ gene that follow the rules above (Kemphues et al. 1983).

As a second example of the processes of mutating an antimorph into a nullomorph, consider the *Drosophila* gene *nod*. The *nod* gene is required for female meiosis in *Drosophila*, but it is also expressed in virtually all mitotically dividing cells (Zhang et al. 1990). There are many recessive loss-of-function *nod* alleles that disrupt meiotic chromosome segregation when homozygous. Curiously, none of these mutations has any demonstrable effect on mitotic cell division or mitotic cells. There is, however, one dominant allele called nod^{DTW}. Heterozygous nod^{DTW}/+ females show the same defect in chromosome segregation as do females homozygous for complete loss-of-function *nod* mutations (Rasooly et al. 1991).

The normal function of the Nod protein is to stabilize chromosomes along microtubule tracks. The nod^{DTW} mutation, which alters only a single amino acid in a critical region of the protein, poisons that process, and appears to lock the chromosomes in place. Rasooly and colleagues mutagenized males carrying the nod^{DTW} mutation on their X chromosomes and screened for pseudorevertants (mutated X chromosomes that no longer exhibited the dominant meiotic effect). They recovered four such mutants, all of which turned out to be new nullomorphic and fully recessive alleles of *nod*. When sequenced, each of these new mutants carried the original nod^{DTW} mutation as well as a second mutation that inactivated the *nod* gene.

Neomorphs

It is not clear exactly what Muller intended by this term. But over the years, **neomorph** has come to mean a mutation that causes a gene to be active in an abnormal time or place. One example is a translocation event that occurs in humans and results in a blood cancer known as Burkitt's lymphoma. As a result of a translocation between chromosomes 8 and 14, the coding sequence of the *c-myc* gene on chromosome 8, which acts to promote cell division, now lies downstream from a very powerful set of lymphocyte-specific promoter elements derived from a gene on chromosome 14. After translocation, these promoter elements inappropriately turn on the *c-myc* gene in white blood cells, resulting in uncontrolled cellular proliferation. Many, if not most, human cancers are the result of neomorphic mutations.

Alternatively, one can consider a dominant mutation in *Drosophila*, $Antp^{73b}$, that causes the antennae to be replaced by legs. (Yes, there really are two extra legs sticking out of the head just above the eyes.) The *Antennapedia* (*Antp*) gene is not normally expressed in the head, but rather in the thorax of the developing embryo where it plays a critical role in specifying the development of thoracic structures such as the leg (Struhl 1981). As diagrammed in Figure 1.4, the mutation results from an **inversion** that fuses the 3′ coding sequences of the normal *Antp* gene next to the

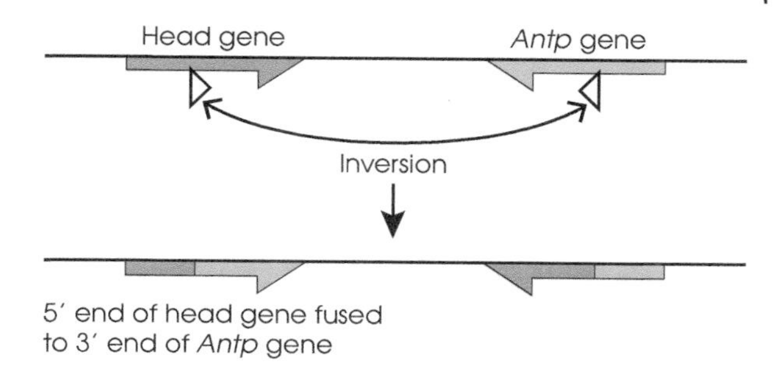

Figure 1.4 Neomorph. *Antp*^{73b} results from an inversion that fuses the 3′ coding sequences of the *Antp* gene with the 5′ coding sequences and regulatory sequences of a gene normally expressed in the head.

5′ coding region of a gene normally expressed in the head (Frischer et al. 1986). As a result of this inversion, a significant portion of the Antp protein – enough to specify leg development – is expressed in the antennae primordia of the developing head. The result is legs where there should be antennae.

Finally, a lovely set of studies by Ganetzky and colleagues has served to elucidate the identity and function of one of the most fascinating of all neomorphic mutants, the *Segregation Distorter* (*SD*) chromosome in *Drosophila* (Kusano et al. 2001). In the *Drosophila* male germline, the *SD* chromosome exhibits a process referred to as **meiotic drive** – when heterozygous with a normal chromosome, *SD* chromosomes have the endearing habit of destroying their wildtype homologs (denoted *SD*$^+$). 99% of the sperm produced by *SD/SD*$^+$ heterozygotes will carry the *SD* chromosome and less than 1% carry the wildtype *SD*$^+$ homolog. The *SD* chromosome does this by causing improper chromosome condensation in *SD*$^+$ sperm.

An *SD* chromosome is composed of several genetic units that contribute to its function. The first of these is the *Sd* mutant itself, which acts at a separate site on the chromosome called *Responder* (*Rsp*) to cause spermatid dysfunction (Ganetzky 1977). *Rsp* itself is composed of a repetitive element located in the **centric heterochromatin** (heterochromatin near the centromere) of the second chromosome (Wu et al. 1988). The sensitivity of a given chromosome to destruction by an active *Sd* element increases with the number of copies of the *Rsp* repeat. The *Sd* mutation is the result of a small tandem duplication event involving the *RanGAP* gene (Powers and Ganetzky 1991) that results in a mutant RanGAP protein, truncated by 234 amino acids at the C terminus (Merrill et al. 1999). This truncation creates a novel protein whose expression causes distortion – only the presence of this novel form of the protein gives rise to meiotic drive. Loss-of-function mutations in the wildtype (*SD*$^+$) *RanGAP* gene are not expected to create an *SD* phenotype.

There is an important lesson here. Early papers on *SD* often focused their efforts on the idea that understanding the mechanism by which *SD*-induced distortion might provide critical insights into the meiotic mechanism itself. The reality is that the actual function of the *RanGAP* gene is unrelated to the mechanism of maintaining Mendelian fairness. Rather it is a function required to mediate nuclear transport in many, if not most, cell types. But it was this mechanism that lent itself to exploitation in a wonderfully devious way. H.J. Muller was clearly prescient in putting forward the idea of a neomorph, a mutation that creates a novel phenotype unrelated to the usual function of the gene.

One major distinction between antimorphs and neomorphs is that the neomorph is not poisoning the normal function of the unmutated gene, but is instead performing the correct function at the wrong time or place. Operationally, this distinction can be made by attempting to create pseudorevertants of the dominant mutation. Reversion of the *Antp*^{73b} neomorph often results in a complete null mutation in the *Antp* gene. The phenotype of that mutation is early embryonic lethality due to malformations of the thorax. Compare this to the *nod* antimorph described earlier, where the original antimorph and its loss-of-function revertants had the same phenotype. Neomorphs also differ from antimorphs in that the effects of neomorphic alleles are independent of the presence or dosage of the wildtype allele. Thus, by definition, neomorphic alleles are dominant.

In recent years, the term neomorph has gradually been supplanted by the term gain-of-function mutation. Actually, both terms can be somewhat misleading because in most cases the mutant does not confer a truly new or gained function on the gene, but simply causes the gene to be expressed at the wrong time, in the wrong place, or at a higher rate than wildtype.[3] Nonetheless, it seems odd to us that we cannot cite even a single example that fits the true definition of a neomorph: a mutation that creates a protein with a novel function.

Modern Mutant Terminology

Following the elucidation of the central dogma of molecular biology (DNA → RNA → Protein), some geneticists felt that Muller's classically based system was too awkward to describe the alterations in gene function created by mutants at the molecular level. Thus, we must discuss terminology that is frequently used by geneticists today.

Loss-of-Function Mutants

In modern articles, mutants that reduce the level of gene product are often classified only as **loss-of-function mutants**, but this term often lumps together hypomorphic and nullomorphic mutants. More precisely, loss-of-function mutants can be further characterized as null mutants, partial loss-of-function mutants, or conditional loss-of-function mutants. **Null mutants** are complete loss-of-function mutants. Indeed, in modern parlance, the term *null mutant* is reserved for those cases where the molecular biology of the mutant gene is well enough understood to be confident that there is no functional gene product produced, and often the definition of this term is narrowed further to imply that no product is produced at all. In **partial loss-of-function mutants**, the mutant produces some degree of product activity – or product itself. Partial loss-of-function mutants include both those mutants that impair the level of product formation and those that create a partially functional product. The term **conditional loss-of-function mutant** refers to cases where the loss of activity of the gene protein is observed under one set of conditions (e.g. higher or lower temperature or treatment with a particular drug) and not under another. The canonical case of a **temperature-sensitive mutant** involves the denaturation of the mutant protein at higher temperatures. This is not, however, always the case. In some circumstances, even null mutants can be sensitive to environmental cues and produce a phenotype only in the presence of environmental or genetic stress.

Dominant Mutants

Dominant mutants are referred to by several different names, some of which seem more or less useful than their Mullerian counterparts. For example, mutants that produce poisonous products are referred to as **dominant negative** mutants. This is a well-used and often accurate term, but we fail to see its advantage over "antimorph."

---

3 For an interesting example of increased expression of a gene leading to drug resistance in cancer, see Wang et al. (2004).

Gain-of-Function Mutants

Mutants that cause a gene to be inappropriately expressed during development or that cause a gene product to be inappropriately regulated are often lumped together under the term **gain-of-function mutant**. While this term is in general use, it seems to us preferable to use the term **heterochronic mutant** to describe mutants that cause genes to be expressed at the wrong *times*. We can then reserve the term gain-of-function mutant for mutants that, for example, remove the regulatory element from some signal transduction protein and thus lock that protein in the ON state or that, by combining components from more than one gene, allow the creation of a product with novel specificities for binding partners or active sites in the cell.

Separation-of-Function Mutants

Some genes perform two functions – say, converting substrate B into C as well as facilitating the interaction of proteins X and Y. If you can recover a mutant in this gene that only converts substrate B into C but does not facilitate the interaction of proteins X and Y, then you have recovered a **separation-of-function mutant**. Separation-of-function mutants allow you to separate genes into their functional domains. How do you know if your gene is performing multiple functions? One hint may be that your gene seems to be similar to two separate genes in another closely related organism. With this knowledge, you may then decide to delete specific portions of your gene or introduce specific point mutations to gain a deeper understanding of how each region functions. Separation-of-function mutants thus allow you to make claims about how specific regions of a gene's protein product operate in your organism. Keep in mind, however, that what appears to be a separation-of-function mutant might simply be a hypomorph whose level of expression lies between two phenotypic thresholds such that there is enough protein to perform task A but not enough to also perform task B.

DNA-Level Terminology

The ultimate classification scheme for mutations requires the DNA sequence of the wildtype and mutant alleles themselves. The following are the basic classes of mutations at the DNA level.

Base-Pair-Substitution Mutants

Often called **point mutants**, these mutants result from the change of one base in the sequence to another (Figure 1.5). Changes such as A–T to G–C or C–G to T–A that replace a like nucleotide (purine or pyrimidine) on each strand are referred to as **transitions**. Changes such as A–T to C–G, in which a purine on one strand is replaced with a pyrimidine on the same strand (or vice versa), are referred to as **transversions**.

Missense mutants are a class of base-pair-substitution mutants that change the sequence of a given codon, which then directs the incorporation of an amino acid different from the one specified at the same position in the wildtype allele. They can be either conservative or nonconservative. A **nonconservative mutation** changes an amino acid to one with different properties. A **conservative mutation** results in a change that incorporates a chemically or structurally similar amino acid, which makes it less likely to disrupt the function of a protein than a nonconservative missense mutation.

Nonsense mutants are a class of base-pair substitution mutants that alter a given codon to create one of the three stop codons, UAA, UAG, or UGA. Geneticists old enough to remember the moon landing may sometimes refer to these mutants as *ochre* (UAA), *amber* (UAG), and *opal* (UGA).

Silent substitutions (or **silent mutations**) are either mutants in coding sequence that do not change the amino acid directed by that codon, as is often true for third-base substitutions, or

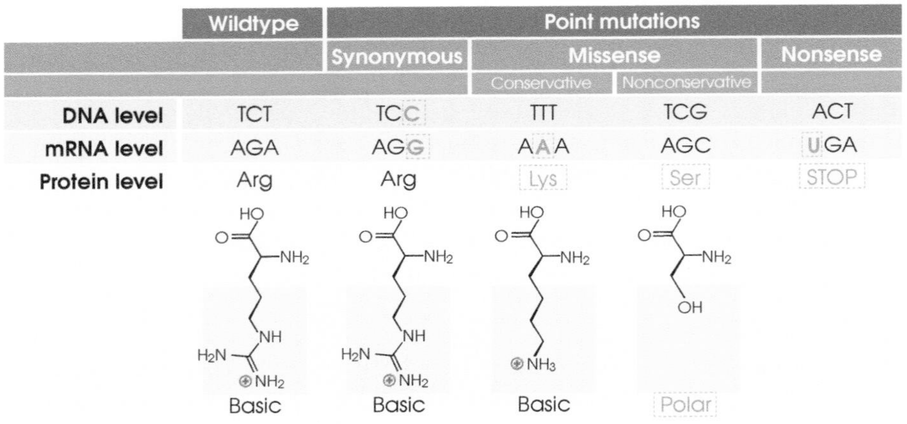

	Wildtype	Point mutations			
		Synonymous	Missense		Nonsense
			Conservative	Nonconservative	
DNA level	TCT	TCC	TTT	TCG	ACT
mRNA level	AGA	AGG	AAA	AGC	UGA
Protein level	Arg	Arg	Lys	Ser	STOP
	Basic	Basic	Basic	Polar	

Figure 1.5 Base-pair-substitution mutations. A synonymous mutation is a type of silent substitution that does not change the amino acid expressed by a codon. A conservative missense mutation changes the expressed amino acid to one with a similar structure or functionality as the original. A nonconservative missense mutation results in a different amino acid with different properties than the wildtype amino acid. A nonsense mutation results in a stop codon.

changes in noncoding regions that do not affect gene expression, such as changes in an intron that do not affect splicing or gene function. This class of mutants has taken on real value in providing markers such as **restriction fragment length polymorphisms** (RFLPs) and other single-nucleotide variants for human genetic mapping. Also note that conservative missense mutations might be phenotypically silent if the substitution involves a chemically similar amino acid. (Do not forget that silent mutations can create new splice acceptor or donor sites within a gene, or they might alter an enhancer for your gene of interest or a neighbor – so do not become too comfortable ignoring them.)

Base-Pair Insertions or Deletions

The name says it all. **Frameshift mutants** result from the insertion or deletion of one or more base pairs within the coding sequence that alters the reading frame, typically soon after the insertion or deletion. Because of their tendency to result in premature stop codons, frameshift mutants are often also classified as nullomorphs. Larger deletions are also sometimes referred to as **deficiencies**.

Chromosomal Aberrations

We should also note that a series of terms used to describe molecular events have been borrowed from the terminology used to describe chromosome rearrangement. The meanings of most of these terms, such as **inversion**, **duplication**, and **deficiency** are self-evident (see Box 1.1). The term **translocation** refers to a breakage and rejoining event involving sequences on nonhomologous chromosomes.

The changes we just described can have multiple effects on the function of genes, or they may have no effect at all. For example, a single base change at a splice site might ablate that site, causing the exon to be spliced out as if it were intronic sequence. This kind of change clearly affects the resulting protein structure. Single changes may also ablate start codons or occur close enough to the gene (in the 5′ untranslated region, for example) to change the level of expression of that gene.

1.2 Dominance and Recessivity

We use the term dominant to mean that if *A/a* individuals or cells are phenotypically similar or identical to *A/A* cells or individuals, while the *a/a* genotype confers a different phenotype, then *A* is the dominant allele and *a* is the recessive allele. If *A/a* individuals or cells are phenotypically intermediate between their *A/A* and *a/a* counterparts, then *A* and *a* are said to be **semidominant**. If, on the other hand, *A/a* individuals exhibit the phenotypes of both *A/a* and *a/a* individuals, then *A* and *a* are **codominant**. A simple example for understanding the difference between codominance and semidominance is flower color. If a flower has alleles for both red and white color but produces pink flowers, the alleles are semidominant; if both red and white stripes are produced, the alleles are codominant. These are seemingly straightforward terms that all of us learned in high school biology. Unfortunately, their usage is often rather careless. You need to think about the situation in which the term is being applied.

To a certain extent, the terms dominant and recessive are simply matters of perspective. Suppose we look at a mutant in a human that encodes an essential metabolic enzyme. Heterozygosity for a simple loss-of-function mutant in that enzyme is likely to have little effect on the metabolism in most cell types. Thus, heterozygotes for this mutation are likely to be normal and we would classify this mutation as fully recessive. But, if we refocused our interest only on the amount of active enzyme produced by a given cell type, then our loss-of-function mutant might be codominant. Indeed, in the absence of cellular controls that limit the level of enzyme production, one might expect that virtually all mutants could be shown to be codominant at the molecular level. Thus, our terminology only has meaning when put into proper perspective.

A dramatic example of the importance of perspective can be seen in the human hereditary cancer disorder retinoblastoma. Children who inherit one defective copy of the *Retinoblastoma1* (*RB1*) gene classically develop retinal tumors very early in life and have a high rate of other childhood tumors, such as osteosarcoma. The inherited form of retinoblastoma is a simple autosomal dominant ($RB1^-$) and the disorder behaves in a pedigree as a dominant mutation should. But in $RB1^+/RB1^-$ individuals, only 10 or so cells in the retina of each eye form tumors. These cells only become tumorous because they have, by subsequent somatic loss or mutation, lost the normal $RB1^+$ allele and unmasked the inherited $RB1^-$ allele. The key to this example is that at the cellular level, the defective gene, $RB1^-$, is fully recessive – one wildtype copy of *RB1* is enough for normal function. The $RB1^-$ mutation only induces tumor formation when the wildtype allele is removed. The requisite somatic mutation/loss events are rare, but because there are hundreds of millions of retinal cells in each eye, the somatic mutation events required to unmask the $RB1^-$ allele become virtually certain to occur somewhere in the eye (see Box 1.3 and Chapter 8). So, are $RB1^-$ mutants dominant or recessive? The answer depends on your perspective – in pedigrees they are dominant; in cells they are recessive.

Because it is exactly one's perspective that matters here, we will begin our discussion of dominance and recessivity at the level of the individual cell.

The Cellular Meaning of Dominance

The critical point is that dominance will result when one of three things occur:

1) a single copy of the wildtype gene is insufficient, or
2) the product of the mutant gene is poisonous to the process, or
3) the mutation causes the gene to be incorrectly expressed in a way that creates a phenotype.

Box 1.3 De Novo Mutation

Every cell division is an opportunity for a new error or a **de novo mutation**. Because of this, every multicellular organism is a **mosaic**, where the genome of every cell differs slightly from the original single-celled zygote that the organism arose from (Biesecker and Spinner 2013; Lupski 2013). Examples of visible mosaics in humans can be seen in Figure B1.1, and a detailed discussion of mosaicism is found in Chapter 8: Mosaic Analysis.

The simplest error encountered during cell division is a **single nucleotide variant (SNV)** or point mutation. Each human cell contains about 6.4×10^9 nucleotides (remember, most of your cells are diploid), and the rate of de novo single-nucleotide changes is approximately 5×10^{-11} mistakes per base copied (Drake et al. 1998).[4] This tells you that every time a cell divides, 0.32 new mutations are created, or about one new mutation every three divisions. Fascinating work by Gilissen et al. (2014) shows that de novo SNVs cause many more genetic diseases than previously thought. Studies that carefully compare individuals with a particular disease to their parents are showing us that de novo mutation should be considered in any individual who presents with a genetic disorder and a family history that doesn't seem to suggest a cause (Frank 2014; Acuna-Hidalgo et al. 2015).

The second type of de novo change you may encounter is an insertion or deletion (or **indel**) error. It was previously somewhat difficult to estimate the rate of de novo indel formation, but the falling cost of whole-genome sequencing and the advent of single-cell sequencing exposed a picture of more frequent indel formation than expected (McConnell et al. 2013; Glessner et al. 2014; Hannibal et al. 2014). A third type of error, chromosome segregation errors, are less common than either SNVs or indels simply because they are extreme events that are often not tolerated by a cell.

Individuals can also be germline mosaics due to mutations that occur during **gametogenesis**, or the creation of sperm and eggs. Germline mosaic individuals may have offspring that appear to "inherit" a mutation that the parent doesn't seem to carry. Consider human

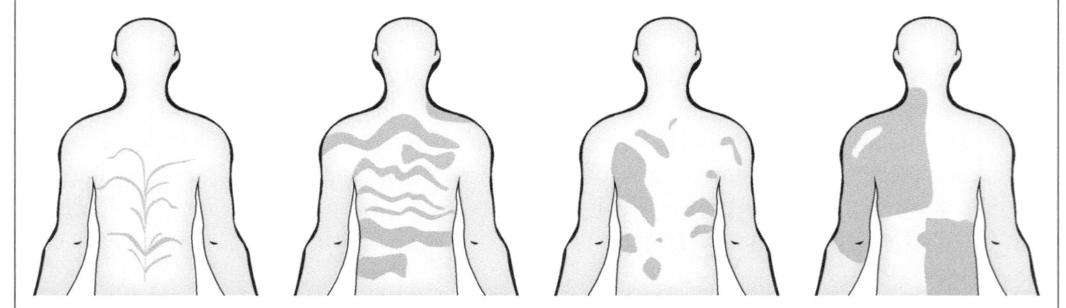

Figure B1.1 Mosaicism. The error inherent in each cell division means that every multicellular organism is a mosaic at some level, with closely related cells being different at a few nucleotides. These minor genetic differences can sometimes be visualized. For example, a variety of genetic disorders can cause patterns of skin discoloration in humans, with cells carrying the mutated gene expressing more or less pigment than normal cells. This visual presentation occurs because of the migratory nature of precursor skin epithelial cells.

4 A fantastic way to find numbers such as this and the references for them is the BioNumbers website at http://bionumbers.hms.harvard.edu.

development and how many cell divisions it takes to go from a single cell to the approximately 1×10^{14} (10 trillion!) cells it takes to make an adult human (Frank 2014). Because of mutation, you likely inherited 50–100 de novo single nucleotide mutations that neither of your parents have. Strikingly, the age of the father at conception has a lot to do with how many de novo mutations an individual inherits: the older the father, the more mutations (Genome of the Netherlands Consortium 2014). The frequency and impact of de novo germline and somatic mutations acquired during normal development are becoming more apparent as the cost of sequencing a genome becomes more affordable.

When thinking about mutations and whether a mutation is dominant or recessive at the organismal or cellular level, it's important to also keep in mind that each cell in your organism might be ever so slightly different from its neighboring cells. If a dominant somatic variant arises in the person, mouse, or fly you're studying, it may affect only a small subset of tissue, and it may or may not exert an overt phenotype. While rare, these types of mutations have great potential to mislead and should always be considered. This is especially important to keep in mind when working with complicated systems such as mammals or tumor models, or when working with cell culture where the number of chromosomes in your cells or even the sex of the cells is uncertain.

We have tried to make the point that all classes of mutants can, in the right context, be dominant. Neomorphs and antimorphs are dominant, almost by definition. Hypermorphs could conceivably be dominant, but one could easily imagine cases where they are recessive. Nullomorphs or hypomorphs in genes that encode structural or other dosage-sensitive proteins could also be considered dominant.

So, how can you better classify a dominant mutation you just recovered? The first and best question to answer is this: Is the mutation mimicked by a deficiency? If *Df/M* looks like *M/M*, then your mutant is most likely a nullomorph. If *Df/M* is more severe in phenotype than *M/M*, then what you have is a hypomorph. If a deficiency does not mimic your mutant, then consider the remaining possibilities:

- If a duplication of your gene (i.e. +/+/+) looks like *M*/+, you have a hypermorph.[5]
- If your mutant cannot be mimicked by either a deficiency or a duplication, then the odds are that you have an antimorph or a neomorph. Antimorphs can often be partially suppressed by adding extra copies of the wildtype gene, while neomorphs cannot.

The Cellular Meaning of Recessivity

There is a simple reason that most mutants are recessive. Most genes encode enzymes, and for most of the enzymatic reactions that take place in a cell, the limiting factor affecting the amount of product (or even the rate of product accumulation) is substrate concentration. So, a twofold reduction in the concentration of enzyme itself often has minimal effect on the rate of product formation, and thus on the phenotype. For that reason, it is often of little consequence to the cell whether there are one or two functional copies of a gene. However, as noted earlier, mutants in genes that produce structural proteins that are required in stochastic amounts are not expected to be recessive.

5 If you find a true hypermorph. Please email us and tell us about it! Better yet, just send it to us!

But surely there are some cases where even a 50% reduction in enzyme concentration might be sufficient to produce a phenotype. One could imagine that heterozygosity for a null mutant in an enzyme-encoding gene might place the cell (or individual) near some threshold of function that would make it more susceptible to the effects of genetic or environmental background. Indeed, there are many examples of mutants known as **enhancers** (not to be confused with transcriptional enhancers) that allow a normally recessive mutant at another gene to exert a strong phenotypic effect, even when heterozygous with a wildtype allele at that gene. Consider two genes, *A* and *B*. You have recovered loss-of-function alleles of gene *A*, denoted *a*, in a background homozygous for the *b* allele of gene *B*. In this background, which is homozygous for *b*, the *a* allele is recessive. However, you now move your *A* and *a* alleles into a genetic background carrying the *B* allele, only to discover that in this background the *a* allele is dominant. In other words:

- *Aa bb* is wildtype
- *aa bb* is mutant
- *Aa Bb* is mutant
- *AA Bb* is wildtype

Such interactions can reveal a great deal about the functions of the products of gene *A* and gene *B*. We will return to a consideration of these types of interactions in Section 3.4 and Chapter 7.

Similarly, there are examples in which recessive mutants can exert strong phenotypes at extremes of the environment, such as high or low temperatures, even when heterozygous with wildtype alleles. There is a mutant called *miniature* (*m*) in *Drosophila* that produces small wings. Under most conditions, *m* alleles are fully recessive, but raise an *m*/+ fly at 16–18 °C instead of the usual 24° temperature, and the flies will have small wings, just like an *m*/*m* fly. The reduction of protein products in *m*/+ heterozygotes falls near a critical threshold. Once an environmental stressor (cold temperature) is applied, the effect of the reduction in gene copy number is revealed. The sickle cell mutation in humans, where heterozygous carriers become more symptomatic at high altitudes, is another example of this type of genotype–environment interaction.

Such cases are not common, though. The vast majority of loss-of-function mutants are phenotypically recessive. One of the virtues of diploidy is that we apparently can tolerate such 50% reductions in protein production for many, if not most, of our genes. Still, one presumes that many 50% reductions in the normal gene dosage, as would occur in individuals heterozygous for large deficiencies, might be deleterious. Indeed, in a careful study of the effects of duplication and deletion in the *Drosophila* genome, Lindsley et al. (1972) estimated that flies could not usually tolerate heterozygosity for autosomal deficiencies that were much larger than 1% of the genome (~150–200 genes). In humans, small deletions resulting in heterozygosity for all or part of a gene product can also play a role in human disease (Tuzun et al. 2005).

Difficulties in Applying the Terms Dominant and Recessive to Sex-Linked Mutants

Many organisms tolerate hemizygosity for the genes residing on the X chromosome (e.g. sex-linked genes in XY male flies, worms, mice, or humans). Their ability to do so reflects the capacity of each of these organisms to perform a process referred to as **dosage compensation**. In the case of flies, the dosage compensation mechanism increases the activity of X chromosomal genes in XY males to match that observed in XX females, while in worms, the activity of the two X chromosomes in the hermaphrodite is reduced to equal the activity of the single X chromosome in XO males. In mammals, XX individuals (females) shut down most of one X at random in each somatic cell

Reason 1: To Identify Genes Required for a Specific Biological Process

The first reason to isolate new mutations is to identify the genes required for a specific biological process. Flight, for example, has always fascinated biologists. Anatomists might be concerned with the structure of a hawk's wings; physiologists may want to examine the mechanics of muscle contraction and wing motion; and somewhere out there, biochemists may be boiling down the wings, pouring the resulting soup over a column and trying to reconstitute flight in the cold room. But the geneticist may wonder, "Can I find mutant hawks that fly funny, or better yet, don't fly at all? And if I could get such mutants, what might they tell us about the genes, and the corresponding protein products, that are required for flight?"

Thus, your first motivation for isolating mutants might be that you are interested in some biological process – such as flight, metabolism, or the cell cycle – and wish to identify the genes whose products are essential for that process. In this case, you would probably want to identify as many important genes as possible. Your major and perhaps most daunting concern will, in fact, be to determine how many mutants need to be recovered to be assured that most, if not all, of the genes whose products are required for this process are identified. A major component of answering this last question is determining how many different genes are defined by your mutations (e.g. your collection of 20 new mutants might define 20 new genes or 20 new alleles of a single gene). Because you are at an early point in the study, you are likely not concerned with obtaining specific alleles of a given gene.

The first concerted attempt to acquire a collection of mutations that affected a defined biological process was the effort of Larry Sandler and Dan Lindsley and their collaborators to isolate mutants that impaired meiosis in *Drosophila* (Sandler et al. 1968). They set out to identify a collection of mutants defective in a specific biological process (meiosis) in *Drosophila* by screening through natural populations of *Drosophila* found near vineyards in Italy for mutants that increased the levels of meiotic chromosome missegregation. As they did so, two of Sandler's graduate students were performing a similar screen using chemically mutagenized strains of *Drosophila* (Baker and Carpenter 1972). Both of these screens identified a large set of new mutants whose subsequent analysis would be critical to increasing our understanding of the meiotic process (Hawley 1993). These efforts were followed by seminal screens by Hartwell for cell-cycle mutants in yeast (Hartwell et al. 1970, 1973; Hartwell 1991) and by Nüsslein-Volhard and Wieschaus (1980) for segmentation-defective mutants in *Drosophila*. Each of these screens recovered a large number of mutants whose analysis provided critical insights into the process being studied.

The screen by Nüsslein-Volhard and Wieschaus (1980) is particularly instructive. Their goal was to identify genes required for early embryonic development in *Drosophila*. They began by searching for a very specific class of mutants (those that confer recessive embryonic lethality) and then devised a set of rapid subscreens to identify the mutations of interest. That screen is discussed in detail in Box 2.1. Out of 10,000 mutagenized chromosomes tested, they recovered 4,217 chromosomes carrying at least one new lethal mutation. Each of these lines was then tested to determine at which stage of development the embryo died. Of these, approximately two-thirds (2,843) caused death in early embryos. Of those 2,843 lines, 321 carried mutations that were both embryonic lethal and exhibited clearly abnormal morphology among the homozygous embryos. Nüsslein-Volhard and Wieschaus went on to show that these 321 mutants defined some 60 genes whose products were critical for early events in *Drosophila* development. A variant of this scheme, designed to isolate lethal mutants on the X chromosome, is presented in Box 2.2. Structurally aberrant chromosomes, known as **balancer chromosomes** (see Box 2.3), were critical to the success of these screens, and of most mutant screens in flies.

Box 2.1 A Screen for Embryonic Lethal Mutations in *Drosophila*

Nüsslein-Volhard and Wieschaus (1980) used the mating scheme below to collect recessive lethal mutations. They began by treating males homozygous for the mutants *cn* (*cinnabar*) and *bw* (*brown*) (both on the 2nd chromosome of *Drosophila melanogaster*) with the mutagen EMS. When doubly homozygous, the *bw* and *cn* mutations produce a white-eye phenotype. These treated males were then crossed to *DTS91/CyO* females. *CyO* is a *Drosophila* 2nd chromosome balancer (see Box 2.3), and *DTS91* denotes a dominant temperature-sensitive mutant that kills developing (but not adult) flies at 29 °C. The asterisk (*) indicates the mutagenized chromosome.

$$\text{G0} \rightarrow DTS91/CyO \text{ females} \times cn\ bw \text{ males (treated with EMS)}$$

$$\text{F1} \rightarrow DTS91/CyO \text{ females} \times (cn\ bw)^*/CyO \text{ males (one male per vial)}$$

10,000 F1s were set up, with one male and several females in each vial. The flies were placed at the restrictive temperature (29 °C) for the first four days, killing all F2 progeny that received the *DTS91* chromosome. This leaves only $(cn\ bw)^*/CyO$ F2 progeny (*CyO/CyO* is lethal). F2 siblings were then mated:

$$\text{F2} \rightarrow (cn\ bw)^*/CyO \text{ females} \times (cn\ bw)^*/CyO \text{ males}$$

And two classes of F3 progeny were recovered:

$$\text{F3} \rightarrow (cn\ bw)^*/(cn\ bw)^* \text{ and } (cn\ bw)^*/CyO$$

In vials where the $(cn\ bw)^*$ chromosome carries a new recessive lethal mutation, no $(cn\ bw)^*/(cn\ bw)^*$ progeny will be produced. Thus, no white-eyed progeny will be seen. However, $(cn\ bw)^*/CyO$ progeny of both sexes will survive and can be used to create a **balanced stock**. Because the *CyO/CyO* progeny die, the only progeny produced by successive brother–sister mating will be $(cn\ bw)^*/CyO$. Nüsslein-Volhard and her collaborators sorted through several 1,000 lethal-bearing stocks to find those in which 25% of the embryos (the lethal homozygotes) showed unusual morphologies.

Box 2.2 A Screen for Sex-Linked Lethal Mutations in *Drosophila*

Objective

To isolate recessive (usually loss-of-function) mutations in vital genes on the X chromosome. Embryos homozygous for such mutants might then be examined visually in search of any number of defects.

Basic Stocks

- **Stock 1: *m/Y*.** This stock consists of males carrying the recessive marker *miniature wings* (*m*) on the X chromosome and females homozygous for *m*.
- **Stock 2: *FM7*.** The males in this stock carry the X-chromosome balancer *FM7* and a normal Y chromosome. Females are homozygous for *FM7*. (See Box 2.3 for a review of balancer chromosomes.)

The balancer *FM7a* carries three overlapping inversions and does an excellent job of suppressing crossing over. It is marked with *yellow* (*y*), *white-apricot* (w^a), and *Bar* (*B*). Females in *FM7a* can become homozygous because *FM7a* does not carry a recessive sterile. *FM7c* carries the female-sterile allele singed (sn^{X2}). The *y* mutant in *FM7* is fully recessive, and *y/Y* males or *y/y* females display yellow bodies and yellow wings (normal flies are darker). The *w* mutant is fully recessive and *w/Y* males or *w/w* females display white eyes. The *B* mutant is dominant and both *B/Y* males and homozygous *B/B* females display abnormal slit-like eyes. Heterozygous *B/+* females display kidney bean-shaped eyes.

Two features of the *FM7* chromosome make it an excellent balancer. First, the three inversions suppress crossing over. Second, the dominant *B* mutation allows you to follow this chromosome in stocks. Also, keep in mind that recombination does not occur in *D. melanogaster* males.

The Screen Itself

- **Generation 0 (G0).** Mutagenize *m/Y* males from Stock 1 by feeding them EMS, and then cross them to virgin *FM7a/FM7a* females from Stock 2.
- **Generation 1 (F1).** Pick up the *FM7/m** daughters (the symbol * denotes the mutagenized X chromosome). Mate these females to *FM7/Y* males (either their brothers or more males from Stock 1). (You need not collect virgin females at this step, since females mating with their brothers is exactly the cross that we want to perform, but be aware that some of your offspring may be homozygous for mutagenized autosomes.) Place each female and 3–5 *FM7/Y* males (usually brothers) in a single vial.
- **Generation 2 (F2).** Examine the progeny of each vial. Look for vials that do not produce *m* B^+ (miniature-winged, non-Bar) sons. Such vials must have been started with mothers heterozygous for a new recessive lethal mutant (denoted $lethal^1$). That is to say that the sperm from the mutagenized G0 male may have carried a new lethal mutation.
- **Generation 3 (F3).** Take the *FM7/m* $lethal^1$ daughters that did survive and cross them to their *FM7/Y* brothers. You have now established a stock of the new lethal mutant. You can use a variety of tricks to map the new mutant, to determine the time of death of the *m* $lethal^1$ males, etc. (It is our experience that even though *FM7a* does not carry a female sterile, the stock will remain heterozygous for some time. That said, the wise researcher would cross *FM7a/m* $lethal^1$ females to *FM7c* males, pickup virgin *FM7c/m* $lethal^1$ females, and start a new stock by crossing again to *FM7c* males.)

A Complication

If you used a chemical mutagen and sample progeny derived from mature sperm, there is a good chance that many of your F1 females are, in fact, mosaics. If EMS got into a mature sperm and alkylated a given G, denoted G*, there are no subsequent replications to resolve the G*/C base pair. The alkylated base will remain opposite the C until the sperm fertilizes the egg and the first embryonic S phase commences. The result of that replication will give a G*–T daughter strand *and* a normal G–C daughter strand. When these two daughter chromatids segregate at the first embryonic mitosis, you have two genetically different populations of daughter cells.

If you imagine that the GC→AT transition indeed produced a recessive lethal, then you can see that this female has some cells bearing a new lethal mutation and some cells with a normal X chromosome. If that mosaicism extends to her germline, then the cells without the new lethal mutation will allow the female to produce miniature-winged, non-Bar progeny. You will fail to recover this mutation.

There is a way to get around this problem:

- **Generation 0 (G0).** Mutagenize *m/Y* males from Stock 1 by feeding them EMS and cross them to virgin *FM7a* females.
- **Generation 1 (F1).** Pick up the *FM7a/m** females. Mate these females to *FM7a/Y* males. (Assure yourself that you need not collect virgin females at this step.). Place each female and 3–5 males in a single vial.
- **Generation 2 (F2).** Individually mate 5–10 *FM7a/m** daughters from each female to *FM7a/Y* males. Place each female and 3–5 males in a single vial (you will have 5–10 vials with individual F2 females and 3–5 males). This allows you to sample a number of progeny females from the original F1 female. If she is mosaic for a lethal, it is likely that some of the vials will produce no $m\ B^+$ male progeny but others would.
- **Generation 3 (F3).** If none of the 5–10 vials produced $m\ B^+$ male progeny, then such vials must have been started with mothers heterozygous for a new recessive lethal mutant (denoted $lethal^2$). Mate the *FM7a/m* $lethal^2$ females to *FM7a/Y* males. Again, it would be wise to replace the *FM7a* chromosome with *FM7c*.

In practice, many geneticists do not worry about mosaicism. First off, the best data suggest that only 20–25% of the F1s will be mosaic in the germline. Secondly, the cure described here is just too much work. (It involves increasing the number of single-pair matings by 5–10times.) Given that most screens look at 10,000 to 20,000 single-pair matings in the F1, setting up 100,000 crosses in the F2 generation is a serious impediment to success and a really easy way to burn out your lab during the cross. We are willing to accept the fact that we lose some fraction of newly induced mutations in this manner, so we just set up a larger number of crosses in the original screen.

Box 2.3 The Balancer Chromosome

For *Drosophila* genetics, the single most critical tool for the analysis of mutants is the balancer chromosome. A good balancer consists of three elements:

1) A set of overlapping inversions that suppress both the occurrence and recovery of crossovers.
2) An easily recognized dominant mutation that allows the balancer to be easily followed in crosses.
3) A recessive lethal or sterile that prevents balancer homozygotes from surviving. (Note that X chromosome balancers in *Drosophila* do not usually carry a recessive lethal because males are hemizygous. Instead, they carry a recessive female sterile mutation.)

Inversions suppress crossing over for a variety of reasons. Small **paracentric inversions** (those that do not include the centromere) suppress the occurrence of recombination, especially in the regions surrounding their breakpoints (Novitski and Braver 1954; Theurkauf and Hawley 1992). Larger paracentric inversions prevent the recovery of crossover chromatids, thus they appear to suppress crossing over (Beadle and Sturtevant 1935; Sturtevant and Beadle 1936). As shown in Figure B2.1, a crossover event within the inversion creates a **dicentric bridge** and an **acentric fragment**. Dicentric bridges are ripped apart during chromosome

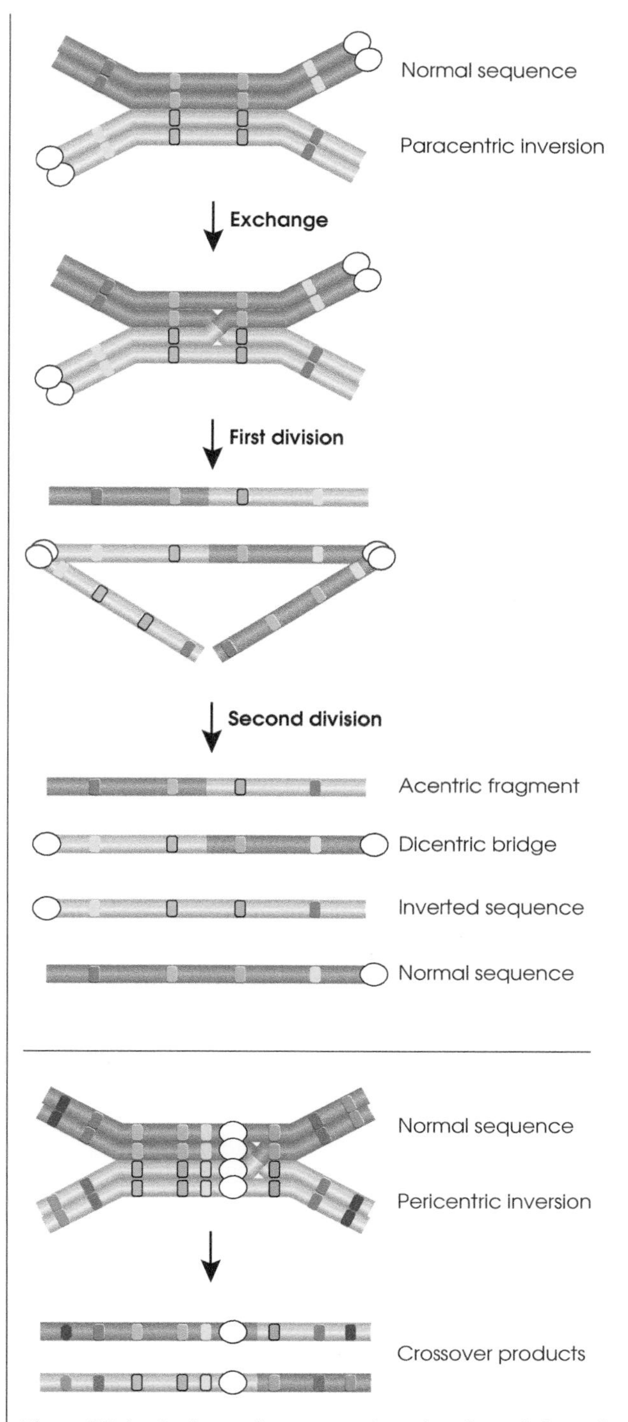

Figure B2.1 Exchange in paracentric and pericentric inversions. A crossover event within a paracentric inversion creates a dicentric bridge and an acentric fragment. The diagram is schematic, for instead of an inversion loop formed when the chromosomes are paired throughout their length, synapsis is indicated only for the inverted segment. This makes the consequences of the single crossover – the formation of a dicentric bridge and acentric fragment – readily apparent. Exchange within pericentric inversions creates large duplications and deficiencies that are virtually always lethal to the zygote.

segregation and acentric chromosomes cannot be properly segregated. Exchange within **pericentric inversions** (those that span the centromere) creates large duplications and deficiencies that are virtually always lethal to the zygote.

Most good balancers combine both pericentric and paracentric inversions, and most balancers truly do have enough breakpoints that they do indeed suppress the occurrence of exchange. But some autosomal balancers are less effective at blocking crossovers than others, especially near the telomeres. So, be careful to choose a good balancer and pay attention to the markers on your balancer, if you lose one, something is likely to have happened in your stock.

All balancers in *Drosophila* carry a strong, easily recognized dominant mutation – or claim to. But some "easily recognized" dominants are easier to recognize than others. All autosomal dominants also carry at least one recessive lethal mutation to prevent homozygosity. Readers interested in choosing a balancer for a given purpose in *Drosophila* are referred to Lindsley and Zimm (1992), papers by Miller and colleagues (Miller et al. 2016a, b, 2018a, 2019), or Flybase (http://flybase.org). Balancers are also available in other organisms, such as mice (Zheng et al. 1999; Hentges and Justice 2004; Klysik et al. 2004) and *C. elegans* (Edgley and Riddle 2001), although not for every chromosome, and they do not always balance the majority of the chromosome as they do in *Drosophila*.

Lee Hartwell began his search for mutants in genes that controlled the yeast cell cycle by casting an equally wide net for any mutant that was temperature sensitive for growth. He then screened a collection of 1,500 temperature-sensitive mutants for new mutants. He did this by screening through cultures of each mutant, raised at the restrictive temperature, looking for the "rather uniform and sometimes unusual cellular and nuclear morphologies that the mutant cells assume after an interval of growth at the restrictive temperature" (Hartwell et al. 1973). The assumption was that those colonies bearing a temperature-sensitive mutant in a common housekeeping function would simply die, but mutants in genes required for progression through the cell cycle would arrest at that control point. He recovered 148 such cultures that defined 32 different genes.

As evidenced by the success of those and subsequent larger screens, the right collection of mutations can provide a strong basis for the analysis of a given biological process. (A brief list of references for mutant hunting in a variety of organisms is provided in Table 2.1.) One just needs to be able to isolate the mutants and then know what to do with those mutants once they are obtained. Indeed, RSH's graduate advisor, Larry Sandler, often got quite aggravated when asked by a genetically naive colleague whether he thought it was "possible" to generate mutants of some given phenotype. "Of course, you can! You can get any mutant you want," Larry would bellow from his office. "I could mutate *E. coli* into an elephant, if given enough time."[1] That may be stretching it a bit, but if you think enough about a given biological process, you can devise a screen for obtaining mutants in the process. It doesn't matter if such mutations are expected to be lethal, sterile, or whatever, as homozygotes. We can cope with such things; we do it all the time. But there is a second, more difficult question. Can we make those mutants tell us what we want to know? Yes, but it is certain to require some serious thought. This will be the subject of the rest of this book.

1 We presume he also meant to add, ". . . and with enough graduate students."

Table 2.1 Resources for mutant isolation schemes and techniques in various organisms.

Prokaryotes and their viruses

Phage lambda

- Davis RW, Botstein D, Roth J. 1980. *Advanced Bacterial Genetics: A Manual for Genetic Engineering*. Cold Spring Harbor (NY): Cold Spring Harbor Laboratory Press.
- Hendrix RW, Roberts JW, Stahl FW, Weisberg RA (eds.). 1983. *Lambda II*. Cold Spring Harbor (NY): Cold Spring Harbor Laboratory Press. [See especially the chapters by Arber (A beginner's guide to lambda biology) and by Arber et al. (Experimental methods for use with lambda).]

Phage T4

- The Bacteriophage Ecology Group website: http://www.phage.org
- Calendar R. 1988. *The Bacteriophages*, Volumes 1–2. New York (NY): Plenum Press.
- Mathews CK, Kutter EM, Mosig G, Berget P. 1983. *Bacteriophage T4*. Washington (DC): ASM Press.

Phage P22

- Susskind M, Botstein D. 1978. Molecular genetics of bacteriophage P22. *Microbiol. Rev*. 42:385–413.

Escherichia coli

- The Coli Genetic Stock Center: http://cgsc.biology.yale.edu
- Davis RW, Botstein D, Roth J. 1980. *Advanced Bacterial Genetics Laboratory Manual*. Cold Spring Harbor (NY): Cold Spring Harbor Laboratory Press.
- Miller JH. 1992. *A Short Course in Bacterial Genetics: A Laboratory Manual and Handbook for Escherichia coli and Related Bacteria*. Cold Spring Harbor (NY): Cold Spring Harbor Laboratory Press.
- Zhang L, Foxman B, Manning SD, Tallman P, Marrs CF. 2000. Molecular epidemiologic approaches to virulence gene discovery in uropathogenic *Escherichia coli*. *Infect. Immun*. 68(4):2009–2015.

Salmonella Typhimurium

http://www.salmonella.biocyc.org

- The Salmonella Genetic Stock Centre: http://people.ucalgary.ca/~kesander
- Davis RW, Botstein D, Roth J. 1980. *Advanced Bacterial Genetics Laboratory Manual*. Cold Spring Harbor (NY): Cold Spring Harbor Laboratory Press.
- KE Sanderson, Hessel A, Rudd KE. 1995. Genetic map of *Salmonella typhimurium*, edition VIII. *Microbiol. Rev*. 59(2):241–303.

Caulobacter

- Ely B. 1991. Genetics of *Caulobacter crescentus*. *Methods Enzymol*. 204:372–84.

More articles dealing with approaches to genetic analysis in a variety of bacteria

- Akerley BJ, Rubin EJ, Camilli A, Lampe DJ, Robertson HM, Mekalanos JJ. 1998. Systematic identification of essential genes by *in vitro mariner* mutagenesis. *Proc. Natl. Acad. Sci. USA* 95(15):8927–8932.
- Biery MC, Stewart FJ, Stellwagen AE, Raleigh EA, Craig NL. 2000. A simple *in vitro* Tn7-based transposition system with low target site selectivity for genome and gene analysis. *Nucleic Acids Res*. 28(5):1067–1077.
- Braunstein M, Griffin TJ IV, Kriakov JI, Friedman ST, Grindley ND, Jacobs WR Jr. 2000. Identification of genes encoding exported *Mycobacterium tuberculosis* proteins using a Tn*552'phoA in vitro* transposition system. *J. Bacteriol*. 182(10):2732–2740.
- Conner CP, Heithoff DM, Mahan MJ. 1998. *In vivo* gene expression: contributions to infection, virulence, and pathogenesis. *Curr. Top. Microbiol. Immunol*. 225:1–12.
- Shea JE, Holden DW. 2000. Signature-tagged mutagenesis helps identify virulence genes. *ASM News* 66(1):15–20.

(*Continued*)

Table 2.1 (Continued)

Unicellular eukaryotes: yeasts, other fungi, algae
Saccharomyces cerevisiae
• Saccharomyces Genome Database: http://www.yeastgenome.org
• DeRisi JL, Iyer VR, Brown PO. 1997. Exploring the metabolic and genetic control of gene expression on a genomic scale. *Science* 278:680–686.
• Forsburg SL. 2001. The art and design of genetic screens: yeast. *Nat. Rev. Genet.* 2(9):659–68.
• Guthrie C, Fink GR (eds.). 1991. *Methods in Enzymology, Vol. 194: Guide to Yeast Genetics and Molecular Biology*. New York (NY): Academic Press.
• Jones EW, Pringle JR, Broach JR (eds.). 1992. *The Molecular and Cellular Biology of the Yeast* Saccharomyces, *Vol. 2: Gene Expression*. Cold Spring Harbor (NY): Cold Spring Harbor Laboratory Press.
• Sherman F. 2002. Getting started with yeast. *Methods Enzymol.* 350:3–41. [A modified and updated PDF of this paper can be found at https://instruct.uwo.ca/biology/3596a/startedyeast.pdf (accessed 2021-12-16)].
Schizosaccharomyces pombe
• Forsburg Lab website (PombeNet): http://www-bcf.usc.edu/~forsburg
• PomBase: www.pombase.org
• Forsburg SL. 2001. The art and design of genetic screens: yeast. *Nat. Rev. Genet.* 2(9):659–668.
• Hagan IM, Carr AM, Grallert A, Nurse P, editors. 2016. *Fission Yeast: A Laboratory Manual*. Cold Spring Harbor (NY): Cold Spring Harbor Laboratory Press.
• Hayles J, Nurse P. 1992. Genetics of the fission yeast *Schizosaccharomyces pombe*. *Annu. Rev. Genet.* 26:373– 402.
• Moreno S, Klar A, Nurse P. 1991. Molecular genetic analysis of fission yeast *Schizosaccharomyces pombe*. *Methods Enzymol.* 194:795–823.
Aspergillus nidulans
• The Aspergillus Website: www.aspergillus.org.uk
• Fungal Genetics Stock Center: http://www.fgsc.net
• Kaminskyj SGW. Fundamentals of growth, storage, genetics and microscopy of Aspergillus nidulans: http://www.fgsc.net/fgn48/Kaminskyj.htm
Chlamydomonas reinhardtii
• Chlamydomonas Resource Center: http://www.chlamycollection.org
• Dutcher SK. 1995. Mating and tetrad analysis in *Chlamydomonas reinhardtii*. *Methods Cell Biol.* 47:531–40.
• Harris EH. 1989. *The* Chlamydomonas *Sourcebook: A Comprehensive Guide to Biology and Laboratory Use*. San Diego (CA): Academic Press.
Dictyostelium
• dictyBase: http://dictybase.org
• Kessin RH. 2001. Dictyostelium*: Evolution, Cell, Biology, and the Development of Multicellularity*. Cambridge (UK): Cambridge University Press.
• Kuspa A, Loomis WF. 1994. REMI-RFLP mapping in the *Dictyostelium* genome. *Genetics*. 138:665–674.
• Kuspa A, Loomis WF. 2006. The genome of *Dictyostelium discoideum*. *Methods Mol. Biol.* 346:15–30.
• Scherczinger CA, Kenetch DA. 1993. Co-suppression of *Dictyostelium discoideum* myosin II heavy-chain gene expression by a sense orientation transcript. *Antisense Res. Dev.* 3(2):207–217

Table 2.1 (Continued)

Neurospora crassa

- Fungal Genetics Stock Center: http://www.fgsc.net
- Davis RH. 2000. Neurospora: *Contributions of a Model Organism*. New York (NY): Oxford University Press. [See especially, Chapter 14: Genetic, Biochemical, and Molecular Techniques.]
- Davis RH, de Serres FJ. 1970. Genetic and microbiological research techniques for *Neurospora crassa*. *Methods Enzymol.* 27A:79–143.
- Perkins DD. 1997. Chromosome rearrangements in *Neurospora* and other filamentous fungi. *Adv. Genet.* 36:239–398.
- Perkins DD, Barry EG. 1977. The cytogenetics of *Neurospora*. *Adv. Genet.* 19:133–285.
- Perkins DD, Radford A, Sachs MS. 2000. *The* Neurospora *Compendium: Chromosomal Loci*. San Diego (CA): Academic Press.

Sordaria

- Fungal Genetics Stock Center:
- Fields WG. 1970. An introduction to the genus *Sordaria*. Neurospora *Newsletter*. 16:14–17. [Available online here: http://www.fgsc.net/fgn/nn16/16fields.pdf (accessed 2021-12-16).]
- Glase JC. 1995. A study of gene linkage and mapping using tetrad analysis in the fungus *Sordaria fimicola*. Pages 1–24, *in* Tested studies for laboratory teaching, Volume 16 (CA Goldman, editor). Proceedings of the 16th Workshop/Conference of the Association for Biology Laboratory Education (ABLE), 273 pages. [Available online here: http://www.ableweb.org/biologylabs/wp-content/uploads/volumes/vol-16/1-glase.pdf (accessed 2019-07-15).]

Invertebrates

Caenorhabditis ***elegans***

- WormBase: http://www.wormbase.org
- Anderson P. 1995. Mutagenesis. *Methods Cell Biol.* 48:31–58.
- Epstein HF, Shakes DC, editors. 1995. *Caenorhabditis elegans: Modern Biological Analysis of an Organism*. (Methods in Cell Biology, vol. 48)
- Riddle DL, Blumenthal T, Meyer BJ, Priess JR, editors. 1997. C. elegans II, 2nd ed. Cold Spring Harbor (NY): Cold Spring Harbor Laboratory Press. [Available at: https://www.ncbi.nlm.nih.gov/books/NBK19997]

Drosophila melanogaster

- FlyBase: http://flybase.org
- Ashburner M, Golic KG, Hawley RS. 2011. Drosophila*: a laboratory handbook*, 2nd ed. Cold Spring Harbor (NY): Cold Spring Harbor Laboratory Press.
- Greenspan RJ. 2004. *Fly-Pushing*, 2nd ed. Cold Spring Harbor (NY): Cold Spring Harbor Laboratory Press.
- Lindsley DL, Grell EH. 1968. *Genetic Variations of Drosophila melanogaster*. Carnegie Institute of Washington Publication 624.
- Lindsley DL, Zimm GC. 1992. *The Genome of Drosophila melanogaster*. San Diego (CA): Academic Press.
- Wolfner MF, Goldberg ML. 1994. Harnessing the power of Drosophila genetics. *Methods Cell Biol.* 44:33–80.
- Adams MD, Sekelsky JJ. 2002. From sequence to phenotype: reverse genetics in *Drosophila melanogaster. Nat Rev Genet.* Mar;3(3):189–98.

(*Continued*)

Table 2.1 (Continued)

Anopheles

- Severson DW, Brown SE, Knudson DL. 2001. Genetic and physical mapping in mosquitoes: molecular approaches. *Annu. Rev. Entomol.* 46:183-219.
- Holt RA and others. 2002. The genome sequence of the malaria mosquito *Anopheles gambiae*. Science. Oct 4;298(5591):129-49.
- Neafsey DE and others. 2015. Highly evolvable malaria vectors: The genomes of 16 *Anopheles* mosquitoes. 347(6217):1258522.

Vertebrates

Zebrafish

- The Zebrafish website: http://zfin.org
- Patton EE, Zon, LI. 2001. The art and design of genetic screens: zebrafish. *Nat. Rev. Genet.* 2(12):956–966.

Goldfish

- Smartt J. *Goldfish Varieties and Genetics: Handbook for Breeders*. 2001. Oxford: Blackwell Science.

Mus musculus

- Mouse Genome Informatics: http://www.informatics.jax.org
- Copeland NG, Jenkins NA, Court DL. 2001. Recombineering: a powerful new tool for mouse functional genomics. *Nat. Rev. Genet.* 2(10):769–779.
- Justice MJ. 2000. Capitalizing on large-scale mouse mutagenesis screens. *Nat. Rev. Genet.* 1:109–115.
- Justice MJ, Noveroske JK, Weber JS, Zheng B, Bradley A. 1999. Mouse ENU mutagenesis. *Hum. Mol. Genet.* 8(10):1955–1963.
- Stanford WL, Cohn, JB, Cordes, SP. 2001. Gene-trap mutagenesis: past, present and beyond. *Nat. Rev. Genet.* 2:756–768.
- Yu Y, Bradley A. 2001. Engineering chromosomal rearrangements in mice. *Nat. Rev. Genet.* 2:780–790.

Medaka

- Wittbrodt J, Shima A, Schartl M. 2002. Medaka—a model organism from the far East. *Nat. Rev. Genet.* 3:53–64.

Rattus rattus

- Rat Genome Database: http://rgd.mcw.edu
- Jacob HJ, Kwitek. 2002. Rat genetics: attaching physiology and pharmacology to the genome. *Nat. Rev. Genet.* 3:33–42.

Homo sapiens

- Online Mendelian Inheritance in Man: http://www.omim.org
- Collins FS. 1995. Positional cloning moves from perditional to traditional. *Nat. Genet.* 9(4):347–350.
- Ghosh S, Collins FS. 1996. The geneticist's approach to complex disease. *Annu. Rev. Med.* 47:333–353.
- International Human Genome Sequencing Consortium. 2001. Initial sequencing and analysis of the human genome. *Nature.* 409:860–921.
- Ott J. 1999. *Analysis of Human Genetic Linkage*, 3rd ed. Baltimore (MD): Johns Hopkins University Press.
- Scriver CR, Beaudet AL, Sly WS, Valle D, Childs B, Kinzler KW, Vogelstein B. *The Metabolic and Molecular Bases of Inherited Disease,* Vol. I–IV, 8th ed. New York (NY): McGraw-Hill.
- Venter JC and others. 2001. The sequence of the human genome. *Science.* 291(5507):1304-1351.

Table 2.1 (Continued)

- Weiss KM. 1995. *Genetic Variation and Human Disease Principles and Evolutionary Approaches*. Cambridge (UK): Cambridge University Press.

Tyrannosaurus rex

- Crichton M. 1999. Jurassic Park. New York (NY): Ballantine Books.
- Organ CL, Schweitzer MH, Zheng W, Freimark LM, Cantley LC, Asara JM. 2008. Molecular phylogenetics of mastodon and *Tyrannosaurus rex*. *Science* 320(5875):499.

Plants

Arabidopsis thaliana

- The Arabidopsis Information Resource https://www.arabidopsis.org
- Meyerowitz EM, Somerville CR, editors. 1994. *Arabidopsis*. Cold Spring Harbor Monograph 27, 1300 pp
- Weigel D, Glazebrook J. 2002. *Arabidopsis: A Laboratory Manual*. Cold Spring Harbor (NY): Cold Spring Harbor Laboratory Press.

Maize

- Maize Genetics and Genomics Database: https://www.maizegdb.org
- Neuffer MG, Coe EH, Wessler SR. 1997. *Mutants of Maize*. Cold Spring Harbor (NY): Cold Spring Harbor Laboratory Press.

Wheat

- Wheat Genetics Resource Center: http://www.k-state.edu/wgrc

Reason 2: To Isolate more Mutations in a Specific Gene of Interest

As your analysis continues, you will likely focus your efforts on only a few of the genes defined by the mutations recovered in your general screen. In some cases, a specific gene might be important enough that you require multiple alleles of that gene, or you might require specific kinds of mutations in that gene, such as conditional alleles (e.g. temperature-sensitive mutants), separation-of-function alleles, or null alleles. Multiple alleles of a single gene also help you validate that the gene you think you've found is really the gene you've found. Imagine you isolate a single mutant through a mutagenesis screen. You then perform whole-genome sequencing of the line bearing your new mutant and find, to your horror, candidate deleterious mutations in several genes within the region you believe your gene lies. If you had multiple alleles of the same mutant, you could sequence those stocks (or you could simply PCR and Sanger sequence specific genes in your allelic stocks) and compare them all. If you find different deleterious changes in a single gene in multiple stocks, then you've probably got your mutant.[2]

Another reason to generate multiple alleles of the same gene is that you're not always going to get single point mutations. Perhaps an inversion or rearrangement caused your initial mutation, which may be difficult to identify computationally. Even relatively small deletions can be difficult to confidently nail down, and you may need to assay gene expression to identify the gene being affected (see Box 1.2). All of these complications are made easier by generating multiple alleles of the same gene. This may seem daunting at first but remember that genetics has moved out of the dark ages of using chemical mutagenesis to generate alleles and into the Jetson-like future of

2 Although, sometimes this does not work just like you think it would.

targeted mutagenesis, which makes generating alleles of specific genes much easier than going through an entirely new mutagenesis screen.

Reason 3: To Obtain Mutants for a Structure-Function Analysis

There will come a point in your analysis when you know the sequence of the gene and its protein product. Perhaps the structure of that protein provides clues to its function. The next step is a high-resolution **structure-function study** of this protein to determine what functions are carried out by which amino acid domains. A large collection of missense mutations now becomes an invaluable tool in your analysis, precisely because it allows you to examine the effects of changing the protein sequence one amino acid at a time. One example of such a dissection, which involves the use of the *nanos* gene in *Drosophila* (Arrizabalaga and Lehmann 1999), will be discussed in Section 2.2.

Reason 4: To Isolate Mutations in a Gene So Far Identified only by Computational Approaches

Say, for example, that you have just discovered a Zebrafish gene that, based on sequence similarity, appears to be the homolog of an interesting human gene. To make this a useful result, you need mutants in this homolog, because until you demonstrate that those Zebrafish mutants have a phenotype similar to the human mutant or a phenotype consistent with a defect in the predicted protein, all you have is sequence homology. You need to "knock out" that gene. There are several tools available for knocking out a specific gene (historically known as targeted gene disruption), such as RNAi, TALENs, or CRISPR/Cas9. These are discussed in Section 2.2.

Regardless of your motive for making mutants, you do want to be efficient about it. There are many means by which one can induce new mutations, and the method you use will determine the sort of mutations you get. It is thus worth discussing the various tools for making mutants.

2.2 Mutagenesis and Mutational Mechanisms

Prior to the pioneering work of Hermann Muller in *Drosophila*, mutations were simply found as serendipitous accidents. They occurred spontaneously and became the property of those individuals sharp-eyed enough to notice them and far-sighted enough to breed them. Unfortunately, the spontaneous mutation rate for most genes is low, with approximately one new mutation per gene in every 1,000,000 or so gametes. Geneticists thus became dependent either on finding existing mutations within natural populations or on the rare recovery of new mutations in their laboratory strains. For example, one of the greatest of all *Drosophila* geneticists, Calvin Bridges, made his first contribution to genetics by noticing a fly with an unusual eye color, vermillion, among the flies buzzing in a bottle he was about to wash.

One of the fundamental advances of twentieth-century genetics was the finding that a variety of physical, chemical, or biological agents could induce mutations. Muller won the Nobel Prize in 1946 for demonstrating that X-rays induced mutations, and studies carried out later in the century would reveal the variety of chemicals that are also potent mutagens. Transposable elements, long known as **mutator elements** by corn geneticists, were recognized as powerful mutators in a variety of organisms by the 1980s, and the technology of in vitro DNA manipulation quickly allowed transposable elements to be modified in such a way as to facilitate cloning of the mutant genes.

that are unknown. Furthermore, this level of saturation may allow the identification and study of genes that, although lethal as a null, may have just enough residual activity to survive with a transposon nearby while still demonstrating a mutant phenotype – a major advantage over targeted deletions. With so many transposon insertions, Guo and colleagues were able to ask which open reading frames in *S. pombe* are essential by simply counting the transposon density in any particular window of the genome. Nonessential open reading frames showed a 10-time higher transposon density than essential open reading frames, allowing the authors to refine the list of essential and nonessential open reading frames in *S. pombe*.

Another advantage of transposon mutagenesis is that it exploits the penchant of many transposons to move to new sites that are close to the original insertion. Thus, if one already has a transposon inserted near to, but not within, the gene of interest, one can often easily mobilize that transposon into the gene of interest. By doing so, one can subsequently obtain flanking DNA sequences corresponding to that gene. A second value of transposon mutagenesis is that excision is often, or can be made to be, an imprecise process that leads to the deletion of flanking DNA sequences. Thus, once you have a transposon insertion in your gene, you can produce deletions within that gene by **imprecise excision** events (Figure 2.2).

Identifying Where Your Transposon Landed

Target P elements can carry marker genes that confer a visible phenotype, allowing these elements to be followed in crosses. These elements also carry a bacterial origin of replication, unique restriction sites within the transposon, and a selectable antibiotic-resistant gene. One could thus re-clone the transposon and flanking DNA sequences by a technique known as **transformation rescue** or **plasmid rescue** (Figure 2.3). Briefly, genomic DNA is isolated from individuals carrying the new insertion and digested with a restriction enzyme that cuts only once in the transposon. The DNA is then ligated at a low-enough concentration so that the ends of each DNA molecule are ligated to form circles. The ligated mix is transformed into *E. coli*. Following antibiotic selection, the resulting colonies carry plasmids bearing the transposon and the DNA sequences of your gene that flank the site of P element insertion. Thus, if you can mobilize a P element into your gene, you have at least part of your gene cloned.

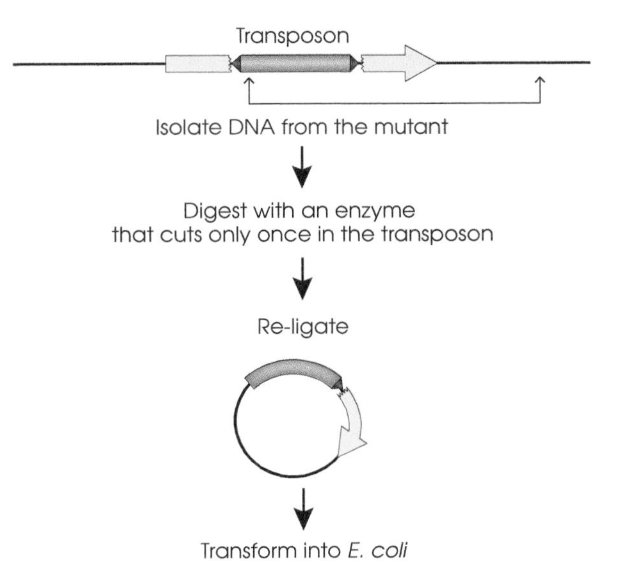

Figure 2.3 Transformation rescue. Provided a transposon carries sequences that allow replication and selection in bacteria, you can reclone the transposon and flanking DNA sequences by a technique known as transformation rescue or plasmid rescue.

In the absence of a visual marker or easily recovered DNA, you could turn to whole-genome sequencing to determine where your transposon ended up, but if you do short-read whole-genome sequencing, you need to be careful with this method. Your transposon won't align properly to your reference genome, masking its insertion site from you. It is likely to align to some poorly assembled heterochromatic region nowhere near the gene it is actually in. To identify the location of your transposon by short-read whole-genome sequencing, you'll need a copy of the original stock you started your transposon mutagenesis with, along with the affected stock. Sequence both and ask where you see evidence of a new transposon that isn't present in your original stock. Alternatively, you can use newer long-read sequencing methods to recover reads that contain the entire transposon and the flanking sequences (see Box 2.5 for a more detailed discussion about finding new TEs).

Box 2.4 De Novo Genome and Transcriptome Assembly

Genome assemblies are available for nearly all commonly used laboratory organisms, but perhaps you are interested in working on a species for which no genome assembly is available. Maybe you desire to understand the unique adaption a species has made to a specific environmental challenge. What do you do? There are two common approaches you could take in this situation. The first, and now likely easier approach, is to perform long-read sequencing, using either PacBio or Oxford Nanopore, and assembly. Combining these two technologies allowed for the first complete, telomere-to-telomere assemblies of a human genome recently (Nurk et al. 2021). This is a somewhat remarkable statement given the history of genome assembly and that the first genome assembly projects were tedious, using Sanger sequencing to basically "walk" the entire length of the genome. This was expensive and time-consuming. Because one fragment was used to make the next fragment, you could determine how the fragments aligned to one another. In the late 2000s, the advent of massively parallel sequencing facilitated the collection of multiple fragments of sequence from an entire genome at once. But this also presented a difficult puzzle: these short fragments had to be matched together to make longer fragments, ideally representing entire chromosomes. This puzzle is not too complicated for smaller organisms like viruses or bacteria, but it becomes challenging for large eukaryotes whose genomes are filled with low-complexity or repetitive sequence that may appear hundreds or thousands of times in the genome. Using short-read data, genome assemblers can reconstruct the nonrepetitive parts of genomes, but they are typically unable to assemble across repetitive segments.

To address the problems caused by repetitive and low-complexity sequence, long-read sequencing was introduced that generates reads longer than these problematic areas. As of this writing, there are primarily two technologies available that produce large amounts of long DNA reads. PacBio uses a fixed polymerase and fluorescently tagged nucleotides to "watch" the sequencing process, while Nanopore uses an electrochemical sensor to observe changes in a DNA or RNA strand as it passes through a small pore, thus generating the sequence. Both technologies can yield individual reads that are tens of thousands of nucleotides long. Either technology can be used to create high-quality de novo assemblies (meaning they have few gaps) of new genomes and to update existing assemblies of commonly used organisms whose genomes have been available for some time (Miller et al. 2018b). If starting a genome project today, there seems to be little reason to complete an assembly using short reads.

If you aren't able to do a DNA-based assembly you could assemble your species' **transcriptome**, which is generally considered to be all the mRNA molecules that an organism produced or could produce. To do this, isolate mRNA from your new species, sequence multiple biological replicates (see Box 1.2), and then de novo assemble the transcripts into genes. You can then compare them to other closely related organisms and identify if any key genes are missing or are surprisingly different from their close ancestors. Keep in mind that with long-read sequencing of cDNA or RNA molecules, the need to assemble short-sequencing fragments back into full-length genes goes away. Sequencing of full mRNA molecules provides the huge advantage of giving you the full isoform of the transcript, and not a reassembled isoform.

Once a genome has been sequenced and assembled, it will need to be annotated. Genome annotation involves identifying and labeling coding, regulatory, and repetitive regions (among other things), using tools such as MAKER (Cantarel et al. 2007) or the Comparative Annotation Toolkit (Fiddes et al. 2018). It is usually advantageous to include RNA-seq data, which will increase the quality of your annotation by helping the annotation software correctly identify different features of the genome. It is a good idea to consider what you expect your annotated genome to look like and not blindly trust what the software has given you. For example, are the number of protein-coding genes appropriate? Are highly conserved genes, or those that you expect to see in all species, present? Asking some very simple questions about your genome can save you a lot of headache later in your analysis and experiments.

A list of current genome assemblies, along with the actual data, can be found at the UCSC Genome Browser Gateway (http://genome.ucsc.edu/cgi-bin/hgGateway) or at Ensembl (http://www.ensembl.org/info/about/species.html). Both sites regularly update their hosted genomes and provide a browser you can use to analyze the genome.

Box 2.5 Identifying New Transposon Insertion Sites

If you suspect your new mutation is caused by the insertion of a transposon but you don't know which gene is affected, then you have a complicated (but not impossible) problem on your hands. The key to finding your new insertion is to compare your mutagenized stock to the original, nonmutagenized stock. You can address this using whole-genome sequencing and RNA-seq.

Using the short-read whole-genome sequencing approach, you will need to sequence your genomes using at least 100-base-pair paired-end reads. Then isolate the reads that align partially to the reference genome and partially to some repetitive part of the genome. These are known as either **split reads** (reads in which two halves of one read can be aligned to multiple locations in the genome) or **discordant reads** (when one read pair aligns to one position in the genome and the other read pair aligns to a different position in the genome). Typically, we identify split or discordant reads at unique sites in the genome; then **de novo assemble** the original fastq reads (see Box 2.4), looking for assembled fragments that partially map back to a position in the genome and partially to a transposable element. The positions to which these fragments map back can be identified using BLAST. One problem with this approach is that **repetitive** and **low-complexity regions** are spread throughout most eukaryotic genomes, which can make searching difficult.

Alternatively, expression analysis on the whole genome using RNA-seq (see Box 1.2) or a microarray first may offer clues about where your TE landed. A **microarray** is a surface with a large number of known single-stranded oligonucleotides attached to it. Both DNA (or DNA prepared from reverse transcribing RNA) from the organism or cell preparation under study and control DNA are labeled with a fluorescent molecule and placed on the surface. The experimental and control DNA compete with one another to anneal to complementary sites on the surface. The quantity of experimental DNA that annealed is then quantified by fluorescence, allowing one to determine whether certain genes are up- or downregulated. DNA polymorphisms can be detected by the presence or absence of specific sequences in the experimental sample. Microarrays have the advantages of being relatively cheap to use, having a well-developed suite of available analysis tools, and being fairly quick to analyze. Alternatively, RNA-seq can be used to compare expression data from your mutant stock and your original, nonmutagenized control. This should be done in multiple replicates. If you're lucky, you'll see one or two genes up- or downregulated in your mutant, suggesting the presence of the TE. You can then confirm this with either whole-genome sequencing or PCR.

Identifying transposon insertion sites is made simpler by newer long-read-based methods. If you perform long-read whole-genome sequencing of your line of interest, you're likely to recover reads that contain flanking DNA sequence as well as the entire transposon. The trick then becomes identifying these reads by some method such as alignment of reads to a transposon database or de novo assembly of your data. This space is changing quite quickly, and new tools using long-read data are frequently released.

Why not Always Screen With TEs?

The reader might wonder, "Why would I mess with X-rays or deadly chemicals? Who wants to finger paint with carcinogens when you get the mutant *and* have a reasonable chance of finding the gene using transposon mutagenesis?" That reader might have a point. But there are reasons to be just a bit shy of these screens. The primary difficulty with doing transposon mutagenesis is that most transposons show some degree of target-site specificity. Thus, they insert efficiently into some genes while ignoring others. There are genes that are very mutable with some transposons, and others that are hit rarely, if at all. Additionally, some transposons prefer regulatory regions or introns as their target sites, and insertion into these regions produces only a partial decrease in gene activity. You can get around these difficulties, but it requires additional effort.

Method 4: Targeted Gene Disruption

It is sometimes possible to use computational methods to identify novel genes in your organism that are homologs of known genes in another organism (see Chapter 5). Or perhaps you have only one mutant allele of a gene of interest and you want to study the effects of additional alleles. In cases like this, it is often possible to create specific mutants using **targeted gene disruption**. There are a few techniques available for this purpose, such as RNAi, CRISPR/Cas9, and TALENs. While RNAi can be used to disrupt the translation of a particular gene, the CRISPR/Cas9 and TALEN systems allow precise heritable or tissue-specific changes to be made in the genome of your organism of interest. These latter two methods are similar in that they allow you to make a directed break in DNA yet differ in the tools they use to make that break. In addition to targeting one specific gene for knockdown or mutation, all three methods also provide some interesting opportunities for mutant screening.

RNA Interference

One method for targeting a specific gene for further study is to use **RNA interference (RNAi)**. RNAi involves the introduction of double-stranded RNA (dsRNA) by either feeding it to an animal, injecting it, or expressing it from a transgenic construct. Depending on the reagents used, either the introduced dsRNA or smaller versions of it, such as small interfering RNAs (siRNA), bind complementary mRNA sequences to inhibit translation or degrade the mRNA. This dsRNA therefore "phenocopies" the effect of a strong hypomorphic or nullomorphic mutant.

RNAi does come with its drawbacks – particularly off-target effects and incomplete knockdown of protein products. You also must know something about the genome of the organism you're working in – great for the *Caenorhabditis elegans* and human studies out there, not so great if you're working in an unstudied organism. Nonetheless, RNAi certainly has its advantages.

First, RNAi has been in use for many years now and many organisms have libraries of RNAi constructs available that target a specific gene of interest. RNAi also allows you to study the tissue-specific effects of knockdowns. By driving RNAi expression with a tissue-specific promoter, you can study a mutation that may otherwise be lethal or have some global deleterious effect on your organism of interest (see Section 8.4). For example, some *Drosophila* genes that function in the ovary during early meiosis are lethal when completely knocked out. To study their role in oogenesis, one could target those genes by driving the expression of RNAi only in the female germline. RNAi also allows you to target two or more genes at the same time. If you're looking for a mutant that rescues the lethal phenotype of the loss of gene *X*, you can knock down gene *X* in every cell or animal in your screen, looking for the rare survivors as candidate rescue genes.

Additionally, RNAi can be used to screen for interactors of a gene of interest. Typically, a screening library is used, which targets thousands of known genes within an organism. For example, Dopie et al. (2015) used the *Drosophila* RNAi Screening Center to screen cells in 384-well plates for RNAi that interfered with nuclear actin formation. Using automated image processing, the group was able to detect cells with increased or decreased nuclear actin formation and identified 19 regulators of nuclear actin. RNAi libraries similar to this exist for most model organisms and high-throughput screening techniques have been developed that make assaying your mutants easier.

CRISPR/Cas9

The **CRISPR/Cas9 system** is a bacterial adaptive immune defense system that has been isolated and modified for use in a number of prokaryotic and eukaryotic systems, including humans. Clustered regularly interspaced short palindromic repeats (CRISPR) is the name given to short sequences of DNA found in the bacterial genome that are used to target invading viral DNA, while Cas9 is simply an enzyme that makes a double-strand break in DNA (Figure 2.4). A guide RNA is included that tells the Cas9 nuclease where to make the break, so you can guide the system to a specific location in the genome. Once a break has been made in the genome, the cell must repair that break either by **non-homologous end-joining** (NHEJ), which repairs DNA by directly ligating it, or by **homology-directed repair** (HDR), which repairs DNA by copying sequence from a provided DNA template. Repair of breaks by NHEJ tends to result in the deletion of one or more nucleotides around the break site. This is useful if you're trying to knock out a gene because a break like this within the coding sequence could result in a frame-shifting mutation or the removal of enough amino acids to yield a nonfunctional protein of interest. HDR, on the other hand, allows you to replace a segment of the organism's genome with new sequence that you have designed yourself. This would allow you to make significant modifications to a protein of interest, such as lengthening a structural protein or introducing new sequence from a closely related species to see how it functions in your organism.

Another use for CRISPR/Cas9 might be in screening for enhancers and suppressors of mutants you already have in hand. For example, a library of 1-kb deletions 10 kb upstream and downstream of a gene could be used to study how disrupting those regions affects the regulation and expression of your gene (see Section 6.3). Indeed, projects are underway in *Drosophila* to create knockouts of every gene in the genome using CRISPR/Cas9 so that a mutant is always available for

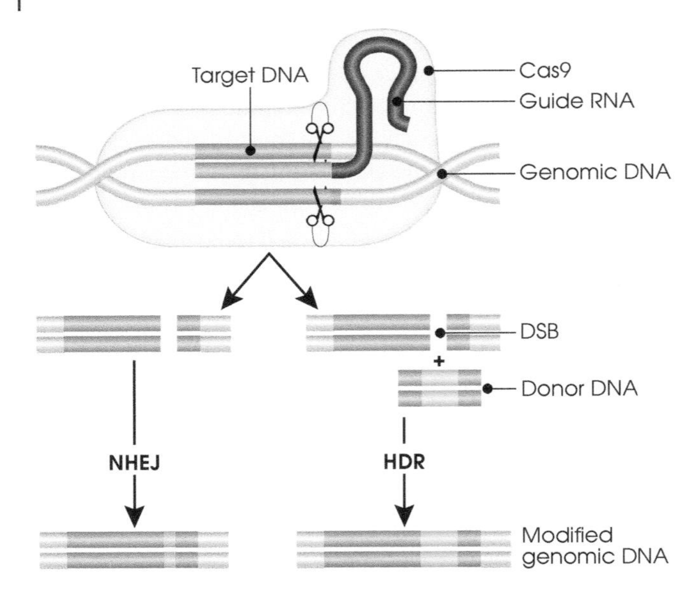

Figure 2.4 The CRISPR/Cas9 system. A guide RNA directs the Cas9 endonuclease to a target site, where a double-strand break (DSB) is induced. This break is then repaired by nonhomologous end joining (NHEJ), an error-prone process that typically results in small deletions of DNA, making this an efficient way to knock out a specific gene of interest. Alternatively, homologous donor DNA may be added to encourage repair of the DSB from a modified template. Recovery of homologous repair products is low, making this process promising, but challenging.

complementation testing when needed (see Chapter 3). CRISPR/Cas9 also provides a way to study genes that in the past have seemed immutable by X-ray or chemical mutagenesis (of course, RNAi may be able to do the same).

Despite the exciting possibilities of CRISPR/Cas9 system, it does have a couple of limitations. First, when designing a guide, it must contain the sequence 5′-NGG-3′ at the end, which means only those regions of the genome that contain the NGG sequence can be targeted. Second, the targeting mechanism is not perfect. Other regions of the genome that are similar, but not identical, to the sequence you are targeting may also be cut by the Cas9 enzyme. These are known as **off-target effects**. Most software used to design guide RNAs takes these off-targets into account and attempts to minimize them by ranking sequences with fewer possible off-target effects higher than those with multiple off-target sites.

TALENs

Like CRISPR/Cas9, **transcription activator-like effector nucleases (TALENs)** are used to create DSBs at specific sites in a genome. TALENs are simply restriction enzymes that are guided by two DNA-binding domains, with each domain targeting either strand of DNA (Figure 2.5). The DNA-binding domain (also called the TAL effector) is actually a series of ~34-amino-acid proteins that are each identical except for their 12th and 13th amino acids, which can be modified to target different bases. The TAL effector is attached to the DNA cleavage domain of a FokI endonuclease, which, when paired with another FokI domain, creates the DSB. Doing an unbiased genetic screen using TALENs is rarely, if ever done. A more likely use of TALENs in screening is directed knockout of a handful of genes known or thought to be part of a biological process of interest, and then assaying the results of those knockouts (Beumer et al. 2013; Beumer and Carroll 2014).

A benefit of TALENs is that any site in the genome can be targeted – there is no requirement for a specific sequence at or near the DSB site. Additionally, targeting a break from two different sides decreases the likelihood of off-target effects – a possible, but still low, risk when using CRISPR/Cas9. Furthermore, while CRISPR/Cas9 matches only 20 or so nucleotides, TALENs match 40 nucleotides, likely reducing the probability that a cut will be made in an undesirable position.

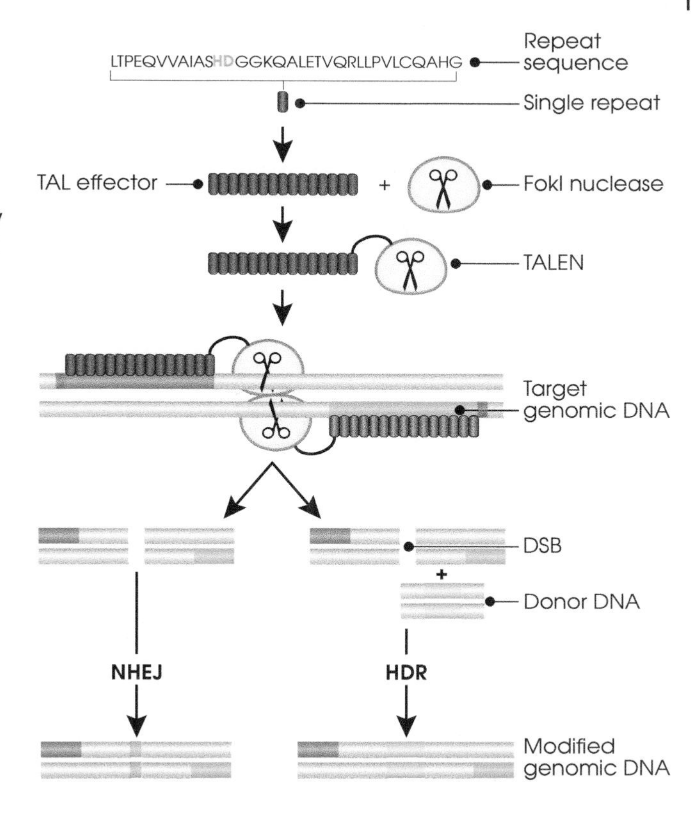

Figure 2.5 TALENs. Specific DNA sequences can be cut with transcription activator-like effector nucleases (TALENs). These are restriction enzymes that can be targeted at a specific sequence of DNA. Advantages of TALENs over the CRISPR/Cas9 system are that they may have lower off-target activity and that they can be used in regions that are difficult to target with CRISPR.

TALENs are also rather affordable. Once you identify the region of the genome you want to target, you simply synthesize DNA that will code for the entire TALEN. This sequence can be ordered from a commercial supplier and introduced into whatever plasmid is necessary for your experiment.

Making DSBs using TALENs generally occurs by one of two mechanisms. First, a plasmid can be inserted into a cell, from which mRNA can be made and the TALENs themselves created. This technique works well for cell culture or other organisms in which it's easy to insert a plasmid, but is more challenging for larger, multicellular organisms. A second method involves introducing mRNA directly into a cell. This bypasses the transcription step but has the same problems as using a plasmid – namely getting the material into the cell or cells of interest.

As with RNAi, by including a tissue-specific promoter in your plasmid, you can drive the expression of the TALEN sequence specifically in a target tissue. For example, Treen and colleagues used promoters that function only in epidermal or neural tissue to knock out *Hox12* and *Fgf3*, respectively, in those tissues in the ascidian chordate *Ciona intestinalis* (Treen et al. 2014). A high rate of knockout of those genes was facilitated by the fact that in this organism, plasmids can be ubiquitously introduced using electroporation. Thus, TALENs could be used for screening if you have a fairly reasonable panel of genes you are looking to study.

So Which Mutagen Should You Use?

This depends on where you start and what you are trying to achieve. In general, our advice is to start with EMS. An EMS mutagenesis will tell you what genes are out there and give you some idea of what kind of phenotypes you can expect. If an EMS screen did not generate enough mutants, then you should first make sure your screen was set up in such a way to identify the class of mutants you were interested in (see Section 2.3). You can then consider using ionizing radiation or transposon mutagenesis. All three methods work, and unless you are luckier than most of us, you will probably need to use all three. Alternatively, if you have a new organism that you're eager to generate mutants in and you've done a reasonably good genome assembly and annotation, then you may wish to use targeted gene disruption such as RNAi, CRISPR/Cas9, or TALENs to target specific genes in the pathway or cellular process that most interests you.

2.3 What Phenotype Should You Screen (or Select) for?

The success of this effort will depend, as we will reiterate constantly, on how cleverly you set up your initial screening criteria. The phenotype of the mutants you select will determine both the genes you identify and the types of mutations you recover. No aspect of a mutant hunt is as critical as determining the phenotype on which the screen will be based.

Suppose – to follow up the flight example presented here – we search for mutants in *Drosophila* based on their inability to fly. As long as our screen is based on testing *live* animals for an inability to fly, we will obtain only those mutants that are not themselves lethal. Thus, we are likely to miss those genes whose protein products are essential to the organism's normal development or survival *in addition* to being required for flight. This may be fine, perhaps even desirable in some circumstances, but it may severely limit the value of the mutant collection that you obtain. Similarly, there may be other genes that encode subtle components of flight (e.g. navigation), whose effects might be missed by a simple "Does it fly, or not?" screen.

Focus on too narrow or demanding a phenotype and you may miss a great many genes. Cast your net too widely (e.g. trying to identify *any* mutation that is homozygous lethal or sterile) and you will be swamped by mutants of genes that may be of little interest to you. On this matter, we can offer only three bits of advice.

First, start by looking for a phenotype that is specific to the process in which you are interested. In other words, if you are interested in flight, look for flightless mutants. Do not expand your screen to "any gene that affects flight and wing development" just because flight may require some vital genes. Do not worry initially about missing some genes; no screen is perfect. Once you obtain several good mutants, they can usually serve as tools to help you find mutants in the sorts of genes that your initial screen may have missed. This is not to say that there is not an enormous value to huge screens designed to recover lethal mutations in every vital gene or sterile mutants in every gene required for fertility. Such screens, when done cooperatively and made available to the entire community, are of enormous value. But they are usually not a good way for an individual scientist to invest her or his time, especially if they are interested in a specific biological process.

Second, to the best of your ability, choose a clean, discrete, and easy-to-score phenotype. Remember that there may be phenotypes that are "scorable" in any given experiment that may not be useful as an endpoint in a large-scale screen. Because mutant hunts are messy and time-consuming adventures, you need to be able to take a quick look at each individual and reliably determine its phenotype. Selecting for a simple, visible phenotype is usually best. As difficult as it might be to imagine, it is often possible to devise screens for a given biological process that are based on an externally visible phenotypic change. For example, Gerry Rubin dissected the process of signal transduction in the developing eye of the fruit fly by using screens in which the endpoint was the texture of the adult eye (Simon et al. 1991; Rebay et al. 2000). A mutant in a signal transduction protein was found that changed the eye texture of the fly. By screening for mutants that enhanced or suppressed this phenotype, Rubin obtained mutants in other genes whose protein products participate in this pathway.

In some cases, the observable phenotype, while process-specific, need not be obviously related to the process under study. Gary Karpen obtained a fascinating collection of mutants that affect the function of centromeres in *Drosophila*, isolated entirely by screens in which the sought-after change in phenotype was a change in the eye and body color of the fly (Murphy and Karpen 1995). Karpen created small "marker" chromosomes in which the genes responsible for eye and body color were moved close to the centromeric regions. He then screened for deletions of these markers, some of which extended into the centromeric regions. Just as in the case discussed earlier, Karpen developed a genetic system in which changes in chromosome structure or the signal transduction apparatus resulted in changes in an easily observed visible phenotype. Simply put, the biological system you are studying may be as complex as you wish, but your screens need to be as simple as possible.

Third, devise a scheme in which organisms displaying the mutant phenotype are selected for you in some fashion, rather than using a screen where one just searches through all the progeny looking for mutants. If the mutants you want are the only progeny that get to live, then such selections are much less work than most screens. We refer to such mutant selection schemes as mutate-my-way-or-die experiments. The drawback to such screens is that the tighter you make your selection scheme, the fewer types of mutations you recover.

2.4 Actually Getting Started

Your Starting Material

A colleague of ours tells a sad story about doing a huge screen for behavioral mutants in flies only to recover the same odd mutation 20 or more times. It turns out that this unusual allele preexisted, albeit at a low frequency, in the starting stocks. We cannot urge you strongly enough to check multiple isolates of your starting stock for the phenotype you are about to assay. Better yet, take the time to build an **isogenic** starting stock (with individuals homozygous for all the loci on a designated chromosome or set of chromosomes) from single pairs of individuals you know to be wildtype for the phenotype. The old maxim "garbage in, garbage out" truly applies here. Also be sure to isolate genomic DNA from your starting stock, as you will want to compare later mutant stocks to this initial "reference" genome. It may not be a bad idea to isolate mRNA in order to also have baseline expression and isoform data from your starting stock.

Pilot Screen

There are many screens that work just great on whiteboards and fail miserably in the lab. Trust us, however you design your first screen, it probably won't work exactly the way you thought – some genotype won't survive, some class of females won't be fertile enough, etc. And the more cute tricks you designed into your screen, the more things can go wrong. So, do a small pilot screen to make sure your ideas work in real life.

What to Keep?

There is a tendency in the first phase of a large screen to keep everything that looks odd, no matter how subtle the difference from wildtype. Remember what we told you before. Choose a strong, clean phenotype and use it. This is especially true if your phenotype is to any degree quantitative. Choose your "this is interesting" cutoff point rigidly before you start, or after your pilot screen, and stick to it.

How many Mutants is Enough?

Okay, so you've been through 20,000 or 200,000 mutagen-treated chromosomes to screen for new mutants in process X. You have 40 or 400 new mutants. Should you stop? The critical issue here, unfortunately, is not how many mutants you have, but how many mutants you don't have. You are interested in estimating the fraction of the total number of genes that can mutate to the phenotype you seek, which are represented by your current collection. If, for example, you have 100 mutants with 20 alleles defining each of five genes, then for heaven's sake stop. You can keep this up for years and not identify any additional new genes. A plot of the numbers of new genes identified versus the number of mutants recovered was generated by Nüsslein-Volhard et al. and is displayed in Figure 2.6. You can see that the identification of more and more genes just as a consequence of finding more and more new mutants is by no means a sure thing. The law of diminishing returns certainly applies here. On the other hand, if you have 40 mutants, with 20 alleles of one gene and only one or two alleles of some 10–15 other genes, then no, you have probably not yet reached saturation. There are still likely to be genes that aren't represented by mutants in your collection. You *might* want to keep going.

Estimating the Number of Genes not Represented by Mutants in Your New Collection

To cope with this problem, geneticists have agreed by convention to lie to ourselves. If we assume that all loci are equally mutable at some low frequency, we can use the **Poisson distribution** to estimate the fraction of presumably mutable genes for which no alleles have yet been recovered ($f[0]$). For a screen in which the mean number of alleles per locus already identified is m, then $f[0] = e^{-m}$.

Our rule of thumb is that saturation is achieved when $m = 5$ and thus $f[0] = 0.007$. That's fine. It is a total delusion for two reasons, but it is fine. First, we all know that some loci are far more mutable than others, and thus one should not use the Poisson distribution. Indeed, the fact that some loci are far more mutable than others goes back to (cited in Lefevre and Watkins 1986). Secondly,

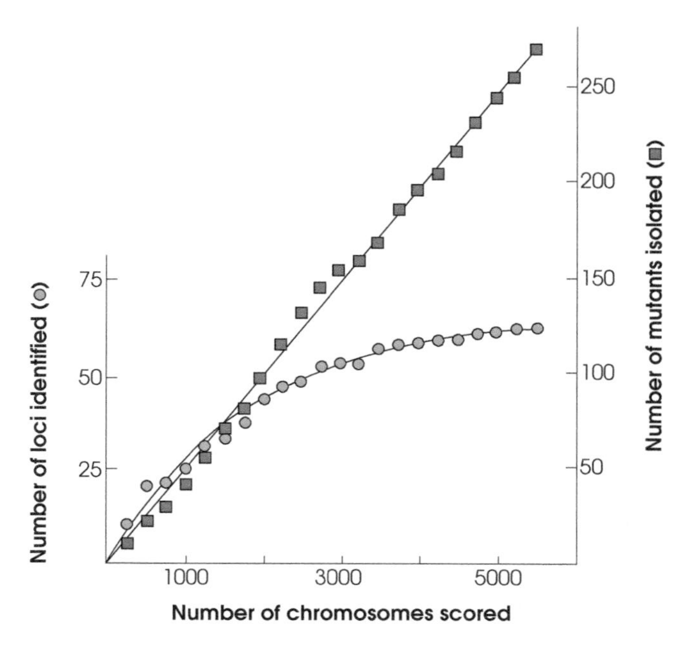

Figure 2.6 Saturation of mutant screens. A plot of the number of new genes identified (circles) and the number of mutants recovered (squares) versus the number of mutagen-treated lines tested in the screen performed by Nüsslein-Volhard and her collaborators.

and of equal concern, is the fact that the mean value (the essence of using the Poisson distribution) is actually being miscalculated in this equation. A true estimation of the mean would require knowing the number of loci represented by zero alleles, which is of course exactly what one is trying to calculate. This error will greatly inflate the estimation of the mean (m), and thus underestimate the number of loci predicted to be represented by smaller numbers of alleles, especially the number of loci for which no alleles were obtained.

Figure 2.7 presents the observed distribution of alleles per locus for the Hartwell et al. (1973) screen. It is important to consider here because the fit to the Poisson distribution is clearly relatively poor. There are both too many loci defined by only one allele and too many defined by large numbers of alleles. These are not unusual results. We once did, in collaboration with Dr. Ken Burtis, a large screen for DNA repair-defective mutants in flies. After screening approximately 6,000 EMS-treated third chromosomes, we recovered 69 new mutants. But 25 of these mutants were shown to be alleles of the same (very mutable) gene. The remaining 44 mutants defined genes that were represented by either 1 (22 cases), 2 (3 cases), 4 (1 case), 5 (1 case), or 7 (1 case) new alleles. If we just take the data at face value, then the average number of alleles per gene is approximately 69/29, or 2.38, and thus $f[0] = 0.093$. Wow, if we believe those data, then we have identified more than 90% of the genes that can be identified in such a screen. That really is more than enough.

But seriously, these data don't fit a Poisson distribution. Clearly, that gene with 25 alleles is a lot more mutable than the other genes. Either the mutation that was recovered 25 times preexisted in the stock prior to mutagenesis (a rude possibility that most folks won't consider, but a possibility nonetheless) or that gene is hypermutable under these conditions. Either way, these alleles cannot be used to assess the degree of saturation. So, what if we repeat the calculation leaving out the overrepresented gene? In this case, the average number of alleles per gene is approximately 44/28 or 1.57, and thus $f[0] = 0.208$. This calculation still suggests that we have close to 80% of the genes one can get by this method. Nonetheless, the math argues that the screen is by no means saturated. Moreover, had we screened another 2,000 chromosomes (and not identified a new

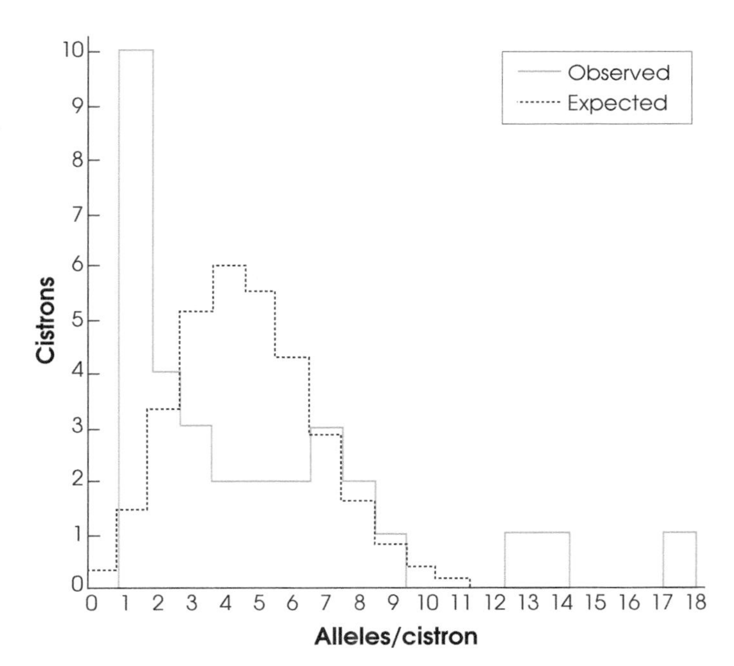

Figure 2.7 Distribution of alleles per locus. The observed and predicted distributions of *cdc* alleles per cistron for the Hartwell et al. (1973) screen, based on an average of 4.625 alleles per cistron from the Poisson distribution. This demonstrates that the observed distribution does not follow the Poisson distribution.

complementation group), we would still only be confident that we had 90% of the genes. Hardly a worthwhile effort. What does one do? The problem is that to continue turning the crank and screening more mutants may not be worth the effort. In our case, we decided that enough was enough, and we stopped screening.

In general, despite the flaws heralded above, this misapplication of the Poisson distribution can be considered to be a reasonable approximation as long as the mean number of alleles per locus is large. We should note that other statistical approaches for estimating the degree of saturation (including using truncated binomial expansions and gamma distributions) have been proposed and tested (for review, see Lefevre and Watkins 1986). The distributions predicted by these methods produce better, though far from perfect, fits to the observed distributions.

But, pragmatically, if you already have an average of three mutants for each gene, then the effort to find a much less represented gene is probably just not worth the work. The answer then is: you can stop at $m = 5$ and no one will fault you. You can also stop at $m = 3$, you just cannot say that your screen was saturating. In reality, you will stop at the point that you simply cannot – or the people in your lab *will* not – keep on going. At some point, they will decide to invest their time in characterizing the mutants in hands, rather than making more mutants.

2.5 Summary

This chapter described three key issues: why you might wish to screen for mutants, how you might choose to induce mutations, and how to actually execute the screen. We looked at several examples of screens and selections, but our focus was always the same: to make it clear that the kind of mutants you will recover will be dictated by the type of screen you do. Each type of screen or selection has limitations, and no screen will get every mutant. Thus, it is important to carefully choose the initial strain to mutagenize, the mutagen to use, and the phenotype to be hunted for or selected. The more carefully designed – and tested – a screen is, the more likely it is to work. Still, not even the most careful of designs can predict all of the outcomes. Sometimes the most interesting mutants are the ones you never expected to get.

References

Arrizabalaga G, Lehmann R. 1999. A selective screen reveals discrete functional domains in *Drosophila* Nanos. *Genetics* 153:1825–1838.

Ashburner M, Golic K, Hawley RS. 2005. *Drosophila – A Laboratory Handbook*, 2nd ed. Cold Spring Harbor (NY): Cold Spring Harbor Laboratory Press.

Baker BS, Carpenter AT. 1972. Genetic analysis of sex chromosomal meiotic mutants in *Drosophila melanogaster*. *Genetics* 71:255–286.

Beadle GW, Sturtevant AH. 1935. X chromosome inversions and meiosis in *Drosophila melanogaster*. *Proc. Natl. Acad. Sci. U.S.A.* 21:384–390.

Beumer KJ, Carroll D. 2014. Targeted genome engineering techniques in *Drosophila*. *Methods* 68:29–37.

Beumer KJ, Trautman JK, Christian M, Dahlem TJ, Lake CM, *et al.* 2013. Comparing zinc finger nucleases and transcription activator-like effector nucleases for gene targeting in *Drosophila*. *G3 Genes Genomes Genetics* 3:1717–1725.

Cantarel BL, Korf I, Robb SMC, Parra G, Ross E, *et al.* 2007. MAKER: an easy-to-use annotation pipeline designed for emerging model organism genomes. *Genome Res.* 18:188–196.

are doubly heterozygous at both genes. These progeny carry one wildtype (or normal-functioning) allele of each gene (*B* and *C*). In this case, the wings of these flies have normal length because the wildtype alleles of the two genes provide normal copies of each protein. An example of the complementation test as used in yeast is provided in Box 3.2.

Transformation rescue experiments may also be considered permutations of complementation tests (see Box 3.3). If the introduced construct can "rescue" the function ablated by the mutant

Box 3.2 An Example of Using the Complementation Test in Yeast

The baker's yeast, *Saccharomyces cerevisiae*, can live as either a haploid or a diploid. Haploids can be mated to produce diploids, and the resulting diploids can be induced to undergo meiosis, producing four haploid spores. This latter process is called **sporulation.** Haploid spores can be grown up on petri dishes to form haploid colonies. Diploid cells can produce diploid colonies.

Imagine that you have two newly induced mutants. The first mutant destroys a gene required for the cell to make the amino acid leucine. These cells cannot grow unless leucine is added to the media. This mutation is referred to as a leucine-minus (*leu*$^-$) **auxotroph.** The other mutant is a tryptophan-minus (*trp*$^-$) auxotroph (i.e. this mutation inactivates a gene whose protein product is required for the cell to synthesize tryptophan). Both mutations are recessive, therefore a diploid carrying one mutant and one normal copy of either gene can still synthesize the necessary amino acid. Both types of haploids will die if plated separately on minimal media containing neither leucine nor tryptophan. However, if the two haploids are mated to form a diploid, the *leu*$^-$ cell brings a functional tryptophan synthesis gene (denoted *TRP*$^+$) to this union, and the *trp*$^-$ haploid brings in a functional leucine synthesis gene (*LEU*$^+$). The genotype of the resulting diploid is *TRP*$^+$/*trp*$^-$; *LEU*$^+$/*leu*$^-$. This diploid will be able to survive on minimal media because it carries functional genes for both leucine and tryptophan biosynthesis. In this sense, the two mutant-bearing genomes complement each other.

Now suppose that you have two independently arising mutants that were both *leu*$^-$. Leucine biosynthesis requires a succession of enzymatic steps, therefore one could easily imagine that the two mutants define different genes and thus different enzymatic defects. Consider the hypothetical pathway drawn here:

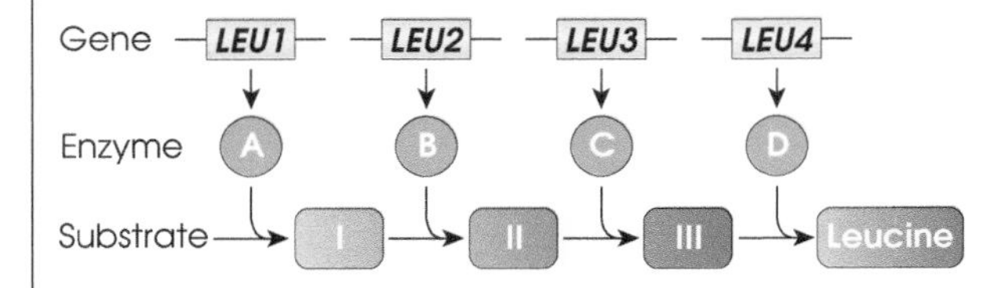

If the first mutant (named *leu*$^{-m1}$) is in the gene that encodes enzyme A and the second mutant (*leu*$^{-m2}$) is in the gene for enzyme C, then the genotype of a *leu*$^{-m1}$/*leu*$^{-m2}$ diploid should in fact be rewritten as *LEU1*/*leu1*$^{-m1}$; *LEU3*/*leu3*$^{-m2}$. You can easily see that in such a diploid, the genes exist to make all four necessary enzymes.

But now suppose you find a third mutant (*leu*$^{-m3}$) that also defines the *LEU1* gene. The genotype of a diploid created by mating a *leu*$^{-m1}$ haploid and *leu*$^{-m3}$ haploid is best rewritten as *leu1*$^{-m1}$/*leu1*$^{-m3}$. There is no functional copy of the *LEU1* gene. This diploid will remain a leucine auxotroph.

Thus, by creating diploids for various pairwise combinations of a large number of independently isolated leucine auxotrophs, one can quickly identify those mutants that define the same gene, or complementation group. You should realize that sporulating diploids with noncomplementing mutants should only rarely produce *LEU*$^+$ haploid spores, while such wildtype spores should be common when complementing diploids are sporulated. Why? How might the rare wildtypes derived from the noncomplementing diploids have arisen? The answer is recombination events that occur within a gene to create a normal allele (see Section 4.5).

Box 3.3 Transformation Rescue is a Variant of the Complementation Test

In the matter of gene finding, **transformation rescue** is considered the gold standard of proof. If you can show that a homozygote for the mutant of interest can be "rescued" by the addition of a wildtype copy of your candidate gene, then you have cloned the gene defined by that mutant. If you are careful, you will double-check this conclusion by sequencing that gene in lines homozygous for one or more mutant alleles. In reality, the transformation rescue test is really just a permutation of the complementation test – can a wildtype allele rescue the function of the mutant allele?

Could the transformation rescue test ever lie? It doesn't happen very often, but it does happen. The simplest and by far the most common error is a consequence of a phenomenon called **multicopy suppression**. Multicopy suppression is described in detail in Section 6.7. For our purposes here, we need only note that if the rescue is done in yeast using a high-copy-number plasmid, then increasing the dosage of gene *X* can sometimes suppress a mutation in gene *Y*, perhaps by allowing the cell to bypass the defect. Lesson: never do transformation rescue using a multicopy plasmid. Or if you do, first sequence the supposedly mutant alleles.

But even single-copy rescue tests can still get you into trouble, albeit rarely. Beall and Rio (1996) reported that mutants in the *Drosophila mus309* gene were rescued by a single-copy insertion of a gene encoding the DNA repair protein Ku (now known in *Drosophila* as Irbp). Indeed, the *Ku* gene had been mapped to a small region that contained the *mus309* mutations. Unfortunately, it turns out that *mus309* mutations are not alleles of the *Ku* gene. Rather, they were shown to be alleles of the nearby gene *blm* – a homolog of the human Bloom syndrome gene, which encodes a helicase of the RECQ family (Kusano et al. 2001a, b). The *Ku* and *blm* genes are separated by about 330 kb and both happen to function in the repair of double-strand breaks. Increasing the dose of *Ku*, even by 50%, can suppress the effects of homozygosity for loss-of-function mutations at the *blm* gene. This misunderstanding could have been avoided had the original workers sequenced the *Ku* gene in *mus309* homozygotes. Be careful out there: sometimes Nature has a nasty little mean streak.

allele(s), then DNA carried by that construct defines the same gene. If it cannot rescue, then it carries a different gene.

The complementation test is only a test of gene function and provides no information regarding the nature or position of the mutants. Two mutants that alter the same base pair in the same gene will fail to complement, just as will two very different mutants in the same gene. To determine whether two noncomplementing mutants occur at the same or different sites in a given gene, one can simply sequence the gene in both stocks, or in cases where the nature or position of the aberration makes sequencing challenging, one can turn to intragenic recombination (see Appendix B). (The confounding of the concepts of complementation and intragenic recombination has a noble history in genetics. Nonetheless, it should not be perpetuated.)

3.2 Rules for Using the Complementation Test

The Complementation Test Can be Done Only When Both Mutants are Fully Recessive

Because the complementation test works by revealing the presence or absence of a normal allele, it will work only for recessive mutants. (In Box 3.4, we present an alternative method, based on pseudoreversion, for determining whether a given dominant mutation is allelic to one or more recessive alleles of a given gene.)

Box 3.4 A Method for Determining Whether a Dominant Mutation is an Allele of a Given Gene

This section might also be called "How to make dominants into recessives by pseudoreversion." If you need to determine whether a given dominant mutation is allelic to an existing tightly linked recessive or to another closely linked and phenotypically similar dominant, your only hope is to induce revertants of the dominant allele. If you are lucky, some of these "revertants" won't be true revertants at all, but rather new loss-of-function alleles exhibiting a phenotype when homozygous. Such mutants are referred to as **pseudorevertants**.

This approach depends on the dominant allele in question being a gain-of-function mutation, either an antimorph or a neomorph. The basic concept is simple: if the mutation is dominant because the gene now produces a poisonous product or produces the correct product at the wrong place or time, then the addition of a second knockout mutation that ablates the ability of that allele to make *any* product should destroy the dominance of that allele. Indeed, the result should be a loss-of-function allele. Such mutants are referred to as pseudorevertants because, although they appear to revert the "dominant" effects of the mutation, they are not true reversions of the original dominant mutation to wildtype. They are simply second mutational events in the same gene that inactivate the poisonous or misexpressed allele.

Recall the case of the *nod* gene in *Drosophila*, discussed in Section 1.1. The *nod* gene is required for female meiosis in *Drosophila*, but it is expressed in virtually all mitotically dividing cells. There are a number of recessive loss-of-function *nod* alleles that disrupt meiotic chromosome segregation when homozygous. However, there was also a mutation, originally called *l(1) TW-6cs*, that mapped very close to the *nod* gene and that exhibited a dominant meiotic phenotype identical to the phenotype exhibited by loss-of-function *nod* mutations.

Rasooly et al. (1991) obtained three pseudorevertants of *l(1)TW-6cs*, all of which no longer caused a meiotic defect when heterozygous. However, homozygotes for these pseudorevertants displayed a meiotic phenotype identical to that observed for loss-of-function alleles. A series of complementation tests using the pseudorevertants and existing loss-of-function alleles of *nod* revealed that all three of the pseudorevertants failed to complement existing *nod* alleles or each other. By this test, it was concluded that *l(1)TW-6cs* was, in fact, a dominant allele of *nod* and was renamed nod^{DTW}. Thus, in this case, allelism of a dominant antimorphic mutation could be demonstrated only by reverting the dominant to a recessive loss-of-function mutation. The point is that you can do a complementation test only with recessive mutations.

The reversion of dominant antimorphic or neomorphic alleles is well documented in the literature. Indeed, an excellent example of the reversion of a dominant allele of *nanos* (Arrizabalaga and Lehmann 1999) was presented in Section 2.2. Other examples involve the reversion of dominant mutations at the *doublesex* locus (Denell 1972; Duncan and Kaufman 1975; Belote et al. 1985), the *Sex lethal* locus (Cline 1978, 1984), the *Antennapedia* gene (Hazelrigg and Kaufman 1983), *Dichaete* (Russell 2000), *Enhancer of split* (Nagel et al. 1999), and *Aberrant X segregation* (Whyte et al. 1993).

The Complementation Test Does Not Require that the Two Mutants Have Exactly the Same Phenotype

As noted previously, different mutations in the same gene can produce rather different phenotypes. Sometimes a mutant that alters but does not destroy function will have a weaker effect on the organism's phenotype than does a null, or knockout, mutation. In these cases, the compound heterozygote usually exhibits the phenotype characteristic of the weaker of the two alleles.

As an example, consider two different, independently isolated mouse mutants. When homozygous in females, the first mutant, denoted *fs(a)*, results in sterility. The second mutant, denoted *l(b)*, is a recessive lethal. These two are tightly linked, as evidenced by the tight linkage of both *fs(a)* and *l(b)* to some other mutation, but the exact position of the mutants is unknown. The compound heterozygote is sterile, exhibiting the weaker of the two phenotypes. The finding of sterility in *fs(a)/l(b)* heterozygotes suggests that the two mutations define the same gene; they fail to complement each other with respect to sterility. One explanation for this observation is that the *fs(a)* mutation is a hypomorphic allele, while *l(b)* is a nullomorphic allele. According to this explanation, the quantity of protein required for viability is less than the quantity required for fertility. As long as some protein is produced, the individual will survive, but a normal or near-normal level of the protein is required for fertility. A second explanation argues that the wildtype protein has both a general function required for viability and a second, rather more specialized, function for fertility. In the case of a lethal mutation, the gene is disrupted to produce a null allele, while the *fs(a)* mutation, which is recessive female-sterile, specifically disrupts the site on the protein required for fertility.

The Phenotype of a Compound Heterozygote Can be More Extreme than that of Either Homozygote

There are several well-described instances in which the phenotype of a compound heterozygote for two mutants within a given gene is considerably more extreme than that of either of the two homozygotes. This phenomenon is referred to as **negative complementation** (Fincham 1966). Negative complementation presumably reflects the ability of the two abnormal protein products to form a dimer or multimer that is not only nonfunctional (as are homodimers of the two mutant proteins), but poisonous as well. Although this phenomenon is best described in *Drosophila* (Raz et al. 1991; Bickel et al. 1997), it has also been described in several other instances (Fincham 1966).

3.3 How the Complementation Test Might Lie to You

The complementation test is generally an excellent and reliable tool, but there are well-documented examples of the test yielding erroneous results. We present the ways in which the test can give a misleading result not to scare you from doing one, but because catching the complementation test in a lie can be incredibly informative.

Two Mutations in the Same Gene Complement Each Other

The fact that proteins have different functional domains, which can be separately mutated, raises an intriguing question. Suppose a given gene encodes a protein with several separate functional domains, and that mutants in a given domain sometimes only inactivate that specific function. Could two mutations exist in the same gene, but map in separate functional domains, and thus complement each other? There is a gene in *Drosophila* called *rudimentary* that encodes a protein that exhibits three spatially and functionally distinct domains. Missense mutants that alter the first domain of this gene can be complemented by mutants that specifically alter the sequence of the second or third domain. All that the cell, or fly, seems to require is that at least one functional copy of all three domains is present in the cell. The difficulties inherent in the analysis of *rudimentary* will be discussed in detail in Appendix B.3. Suffice it to say that working out structure–function relationships between genes and their protein products using **intragenic complementation** is a tricky business.

Another example of intragenic complementation is provided by cases in which one of the two mutants lies in an upstream regulatory element and the other lies within the coding region. In most Dipteran insects, including *Drosophila*, there is a sufficient degree of somatic pairing of homologous chromosomes to allow the functional copy of the regulatory region to act properly on the gene carried by the homologous chromosome (Bosco 2012). As shown in Figure 3.2, this allows the functional regulatory element on the chromosome with the coding region mutant to "reach over" and activate the normal-sequence promoter region on its homolog – the regulatory element may be thought of as acting in *trans*. This complementation is dependent on proper somatic chromosome pairing and is referred to as **transvection** (see Box 3.5).[2]

A Mutation in One Gene Silences Expression of a Nearby Gene

In our own laboratory, we came across a particularly confusing example of how the complementation test might lie to you. We performed an EMS screen for new meiotic mutants on the *Drosophila* third chromosome and recovered several stocks that gave high levels of meiotic

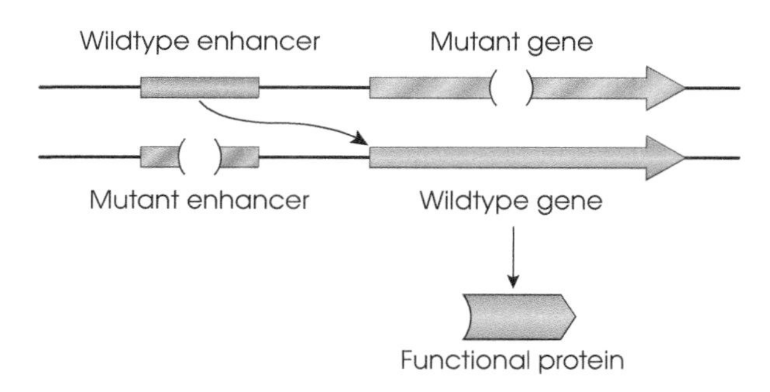

Figure 3.2 Pairing-dependent complementation. Transvection can result in intragenic complementation in which the nonmutant enhancer functions in *trans* to activate transcription of the nonmutant gene on the homologous chromosome.

2 A detailed consideration of the pairing-dependent complementation patterns observed for the *yellow* and *BX-C* genes in Drosophila is found in Appendix B.5.

Box 3.5 Pairing-Dependent Complementation: Transvection

There are many examples in *Drosophila* in which intragenic complementation can be disrupted by heterozygosity for chromosomal rearrangements [for review, see Wu and Morris (1999)]. These effects appear to reflect the ability of various enhancer-like elements to function in *trans* in an organism like *Drosophila*, which has ubiquitous somatic pairing. That is to say, an *m1/m2* heterozygote (in which *m1* defines an enhancer element and *m2* defines the coding sequence) may exert a wildtype appearance because the wildtype copy of the enhancer borne by the *m2* chromosome can activate the wildtype coding sequence borne by the *m1* chromosome (Figure 3.2). However, this *trans*-activation will be suppressed by genetic rearrangements that suppress pairing.

Transvection was first described and demonstrated at the bithorax complex (BX-C) in *Drosophila* by Ed Lewis (1954). Using aboveground nuclear detonations to produce rearrangements,[3] Lewis demonstrated that even breakpoints located quite far away on the same chromosomal arm from the gene in question could generate transvection. Homozygosity for those breakpoints does not usually cause transvection. Subsequently, transvection was demonstrated at multiple loci in *Drosophila* [for review, see Ashburner (1989) but c.f. Gelbart (1982)]. Perhaps the most interesting of these cases involves the pairing-dependent interaction of the *zeste* and *white* genes. Homozygotes for loss-of-function mutations at the *zeste* locus can modify the expression of the *white* locus, but only when two copies of the *white* locus are paired (Gelbart 1982; Green 1977, Jack and Judd 1979). An elegant comparison of transvection at several loci in *Drosophila* is provided by the work of Smolik-Utlaut and Gelbart (1987), and a mechanistic basis for the observed differences is provided by Golic and Golic (1996). There is some evidence that transvection may also occur in Neurospora (Aramayo and Metzenberg 1996; Perkins 1997). Indeed, similar effects have been noted in a variety of organisms (Tartof and Henikoff 1991; Matzke et al. 2001).

chromosome missegregation. We then complementation tested each allele with known meiotic mutants on the third (because you shouldn't spend a lot of time discovering genes that have already been described). Four stocks failed to complement the gene *ald* (the fly homolog of *Mps1*), a kinetochore-associated kinase that functions in meiotic and mitotic spindle assembly checkpoints (Gilliland et al. 2005, 2007). Notably, two of the alleles exhibited only a meiotic phenotype, but not the mitotic segregation errors typically associated with *ald* mutants. At first glance, this is quite interesting, because we love separation-of-function mutants! Recall from Chapter 1 that separation-of-function mutants help you dissect protein function. Furthermore, the phenotype of both unusual stocks could be rescued by an *ald* transgene (remember, transformation rescue is just another form of the complementation test), so we were *sure* we were looking at the right gene. But surprisingly, DNA sequencing revealed no alterations in the *ald* sequence of either stock (the other alleles of *ald* recovered in this screen did have nonsense or frameshift mutations in the gene itself, and they also had both the meiotic and mitotic phenotypes).

So, what was going on? We finally discovered that a transposon had inserted itself into *alt*, the gene immediately 3′, or downstream, of *ald*. The transposon was being silenced by

3 Wouldn't *that* be a fun methods section to write?

Piwi-interacting RNAs (piRNAs), which function in the germline to suppress the movement of transposons (Chambeyron et al. 2008). This silencing also affected the expression of the surrounding genes, including *ald*. Because piRNAs do not function in the soma, these stocks had the meiotic phenotype, but not the mitotic phenotype typically associated with *ald* (Hawley and Gilliland 2006). Interestingly, we were able to revert the phenotype with a second EMS screen, which led to a second nearby transposon insertion that piRNAs preferentially bound to, which then allowed expression of *ald* in the germline (Miller et al. 2020).[4]

Mutations in Regulatory Elements

Imagine a mutant that contains a deletion in an intron of gene *A*. It so happens that this intron contains an enhancer for gene *B*, which lies adjacent to gene *A*, and that deletion of the enhancer causes gene *B* to be expressed at such low levels that it is essentially undetectable. Because the deletion is spliced out during transcription, it has no effect on the function of gene *A*. If you complementation test the mutant with a deficiency covering both genes, it will fail to complement your mutant and you'll think your mutation is in either gene *A* or gene *B*. You then PCR and Sanger sequence the mutant and note the deletion in the intron of *A*, perhaps telling yourself it disrupts splicing, thus rendering gene *A* inactive. You wouldn't know you were being misled until you checked the expression of gene *A* in your mutant and noticed that it was expressed at wildtype levels.

In this example, failure to measure expression doesn't have to doom you to report the wrong gene. Presumably, you'll do transformation rescue to *prove* that gene *A* is the culprit. Of course, this will not rescue your phenotype (because your phenotype is caused by low expression of gene *B*), and you'll wonder why the complementation test lied to you. If you're thorough, you'll go on to check expression levels of gene *B* (because it was also covered by the deficiency) and hopefully figure out the problem. Although the complementation test didn't give you a clear result in this situation, it did still help narrow down the list of potential hits.

3.4 Second-Site Noncomplementation (Nonallelic Noncomplementation)

Could two mutations in different genes ever fail to complement each other? The answer to this question is *yes*. In rare cases, two mutants in different genes, which by themselves are fully recessive, can create a mutant phenotype in double heterozygotes. In other words, although individuals heterozygous for either *m1* or *m2* alone (*m1*/+ or *m2*/+) are wildtype, double heterozygotes (*m1*/+; *m2*/+) exhibit a mutant phenotype. This phenomenon was first noted by Calvin Bridges in the 1930s and was referred to as dominant enhancement or unlinked noncomplementation (Lewis 2003). It is now referred to as **nonallelic noncomplementation** or **second-site noncomplementation** (SSNC). We divide this SSNC into three types. In type 1 SSNC, mutant forms of the two different proteins interact to produce a poisonous product. In type 2 SSNC, the mutant form of one protein sequesters the wildtype form of the other protein into an inactive complex. In type 3, the simultaneous reduction of the two proteins results in a phenotypic abnormality. Only type 1 and type 2 SSNC suggest physical interaction between the two proteins.

4 If you think this story is as interesting as we do . . . then welcome to genetics.

Type 1 SSNC (Poisonous Interactions): The Interaction is Allele Specific at Both Loci

This type of SSNC is both the rarest and the most interesting. By definition, **poisonous interactions** require allele specificity at both genes, and neither allele can be a null mutant. In this sense, the double mutant combination may be thought of as a **synthetic antimorph** because the combination of the two mutant proteins produces a poisonous gene product. This type of interaction is often explained by asserting that the two mutant proteins physically interact to produce a poisonous protein dimer or complex (Figure 3.3). However, the proof of such a hypothesis requires biochemical evidence for the physical association of the two mutant proteins and elucidation of the mechanism by which they create a poisonous effect. Moreover, as our second example will show, there are cases where this type of interaction is not mediated by the direct physical interaction of the two gene products.

An Example of Type 1 SSNC Involving the Alpha- and Beta-Tubulin Genes in Yeast

There are two genes in yeast for α-tubulin and one for β-tubulin. The TUB1 and TUB3 genes encode variants of α-tubulin, while the TUB2 gene encodes β-tubulin. To identify genes whose products interacted with β-tubulin, Stearns and Botstein (1988) screened for mutants that displayed cold-sensitive SSNC in the presence of a recessive cold-sensitive allele of β-tubulin (tub2cs) (this screen is diagrammed in Figure 3.4). In other words, they looked for recessive mutants that produced a cold-sensitive lethal phenotype when doubly heterozygous with a recessive mutant in the β-tubulin gene. The focus on cold sensitivity here reflects a prejudice arising from phage genetics that cold sensitivity is a hallmark of mutations that affect protein assembly or protein–protein interactions.

Stearns and Botstein created a large collection of mutagenized haploid yeast colonies. They grew each colony from a single mutagenized cell and required that these cells be viable at the permissive temperature. Each of these lines was then separately mated to a haploid strain carrying the recessive *tub2cs* mutant. Thus, the resulting diploids were heterozygous for both the new mutants from the mutagenized haploid line and for the recessive *tub2cs* mutant. Diploids in which the two mutations show second-site noncomplementation should be expected to show the cold-sensitive phenotype. In the actual experiment, cells from each of the haploid colonies were mated to haploids carrying the recessive *tub2cs* mutant, and other cells from the same colonies were mated to wildtype haploids. All of the resulting diploids were then plated at the restrictive (nonpermissive) temperature.

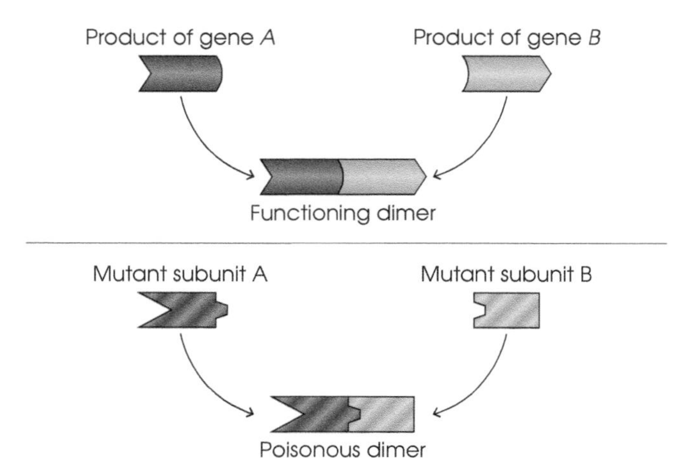

Figure 3.3 Type 1 SSNC: Poisonous interactions. A functioning dimer requires two normal protein subunits. While neither mutant subunit alone would be able to interact with a normal subunit to create a functional dimer, combining two different mutant protein subunits can sometimes create a poisonous dimer molecule.

Figure 3.4 A screen for second-site noncomplementing mutants involving the tubulin genes in yeast. *TUB2* is the gene for β-tubulin and *TNC* is a tubulin noncomplementer gene. A mutant that forms a Cs^- diploid with a $tub2^{cs}$ mutant, but a Cs^+ diploid with wildtype is a noncomplementer. The mutation responsible for the noncomplementation is then tested for linkage to the *TUB2* locus. (An excellent summary of the basic techniques of yeast mutant hunting can be found in Forsburg 2001.) *Source:* Adapted from Forsburg (2001).

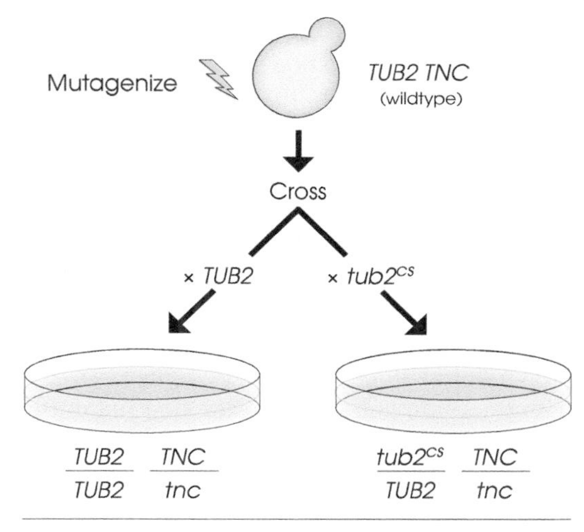

There are four types of new mutations that could cause the diploid with *tub2cs* to fail to grow at the restrictive temperature. These include mutants that block mating (diploid formation), new dominant cold-sensitive mutations, new mutants in the *TUB2* gene, and lastly the sought-after second-site noncomplementers. Fortunately, the first two types of mutants are easily recognized because they will fail to produce colonies when mated to the wildtype cells. The last two classes can be distinguished by a simple segregational assay.

20,000 colonies grown from mutagenized cells were screened for failure to complement either of two *tub2cs* alleles (*tub2cs-104* and *tub2cs-401*), and only three noncomplementers were found. Two of these noncomplementers turned out to be new alleles of *TUB2*. However, one of these mutants, designated *tnc1*, turned out to be a mutation at another gene that fails to complement the *tub2cs* gene. Thus, *tnc1/+ tub2cs/+* double heterozygotes show the same cold-sensitive lethality that is observed in *tub2cs* homozygotes.

The *tnc1* mutant fails to complement *tub2cs-401* at 16 °C, but weakly complements *tub2cs-104* at the same temperature. That is, the interaction of *tnc1* with the *TUB2* gene is allele-specific. Some allelic combinations were strongly cold-sensitive, others weakly cold-sensitive, and still others were not at all cold-sensitive (see table 6 in Stearns and Botstein 1988). The *tnc1* mutant is also cold sensitive by itself and exhibits a defect in spindle assembly even at the permissive temperature. Both the cold sensitivity of *tnc1* and its inability to complement *tub2cs-401* segregate as simple single gene traits, and the two traits are not separable by recombination.

Perhaps not surprisingly, *tnc1* turned out to be an allele of *TUB1*, the first gene for α-tubulin. (The mutation was then renamed *tub1-1*.) This was an important finding because no viable mutant alleles of the *TUB1* gene were available prior to this study. Null mutants at this gene, which can be created by various techniques, are lethal in haploids. Indeed, even as heterozygotes, such mutants have dominant effects on viability, and the diploid heterozygotes rapidly become aneuploid. While we love the idea of creating null mutants via directed methods, such as CRISPR/Cas9 (see Section 2.2), this example shows why we need other methods as well. Sometimes, creating a simple null doesn't work. Moreover, Stearns and Botstein continued their efforts by screening for second-site noncomplementers of *tub1-1* (*tnc1*). This screen yielded both a new cold-sensitive allele of

TUB2 (*tub2-501*) and, more critically, the first recovered point mutation in the *TUB3* gene (*tub3-1*). The only previous existing alleles of *TUB3* were null mutations generated by gene disruption.

It is curious, though, that both the point mutant and the null allele of *TUB3* failed to complement the *tub1-1* allele. To quote Stearns and Botstein, "This means that the failure to complement in this case is not due to a specific protein–protein interaction. Instead, it seems that a reduction in the total level of functional α-tubulin in the (*TUB1/tub1; TUB3/tub3*) double heterozygote causes the failure to grow at 14°." Indeed, this case of combined haploinsufficiency is a model for the third type of SSNC, which will be discussed later in this chapter.

The Stearns and Botstein (1988) paper is widely viewed as the hallmark on SSNC research. First, it was the first such report to appear and the first concrete example that such screens can work. The screen did identify a mutant in a gene (*TUB1*), whose product (α-tubulin) physically interacted with β-tubulin and provided the first mutant in the second α-tubulin gene (*TUB3*). (Mutants in *TUB1* and *TUB3* did not exist prior to this screen.) Second, the study was truly elegant. Unfortunately, it would be one of very few such screens to be successful. We now consider a similar screen for actin noncomplementers in yeast that was less fruitful in identifying interacting proteins.

An Example of Type 1 SSNC Involving the Actin Genes in Yeast

Following the studies of Stearns and Botstein (1988), David Drubin and collaborators set out to determine whether a screen for second-site noncomplementers in yeast could be used to identify genes whose protein products interacted with actin (Vinh et al. 1993; Welch et al. 1993). These studies followed an extensive analysis by Drubin of mutants that can suppress a temperature-sensitive allele of actin (*act1-1*). That study is discussed in Section 6.6.

Drubin and collaborators began with the observation that null mutants in two genes known to encode actin-binding protein (*SAC6* and *ABP1*), failed to complement the temperature-sensitive phenotype caused by a mutant in the *ACT1* gene (Figure 3.5). This observation emboldened them to attempt to identify novel genes whose protein products interact with actin. They did this by screening for second-site mutants that failed to complement *act1-1* or *act1-4*, two temperature-sensitive alleles of *ACT1*. The screening of more than 55,000 mutagenized colonies yielded a total of 14 extragenic noncomplementing mutants and 12 new alleles of *ACT1*.

The 14 SSNC mutations they isolated were shown to be alleles of at least 4 different genes, *ANC1*, *ANC2*, *ANC3*, and *ANC4* (actin-noncomplementing). These mutations exhibited several properties

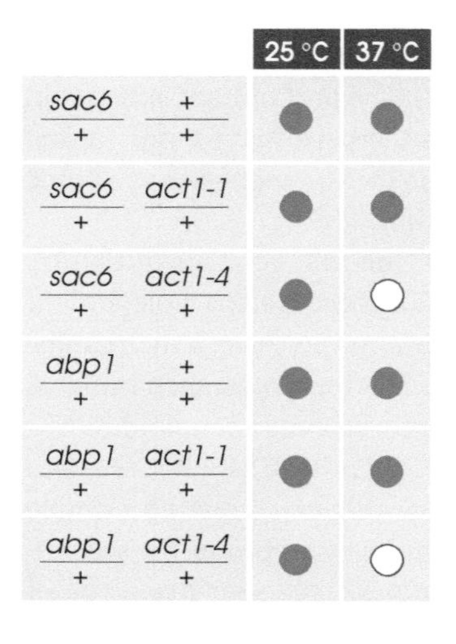

Figure 3.5 An example of type 1 SSNC. Mutants in genes whose protein products interact with actin (*sac6* and *abp1*) fail to complement the temperature-sensitive *act1-4* mutation but fully complement the temperature-sensitive *act1-1* mutant at 37 °C (i.e. they show actin allele-specific noncomplementation). Filled circles indicate growth of the genotype on petri dishes incubated at the indicated temperatures; open circles indicate little or no growth. *Source:* Figure courtesy of Diana Hiebert.

that made it likely that they defined genes of interest (i.e. genes encoding actin-interacting proteins). First, mutations in the *ANC1* gene were shown to cause defects in actin organization. Indeed, the phenotypes observed in the presence of *anc1* mutants alone were similar to those caused by *act1* mutations (see also Welch and Drubin 1994). Second, the observed noncomplementation was allele-specific. Third, when mutant alleles of four *ANC* genes (*ANC1*, *ANC2*, *ANC3*, and *ANC4*) were tested for genetic interactions with null alleles of known actin-binding protein genes, an *anc1* mutant allele failed to complement null alleles of *SAC6* and *TPM1* (genes that encode the yeast actin-binding proteins fimbrin and tropomyosin, respectively). Fourth, synthetic lethality between *anc3* and *sac6* mutants, and between *anc4* and *tpm1* mutants was observed. These rather complex genetic interactions are displayed in Figure 3.6. (Synthetic lethality is another tool for identifying genes whose products might interact either physically or in the same pathway or process. A discussion of this technique is found in Box 3.6.) Yet, the question remained: did these four genes really encode actin-interacting proteins?

Unfortunately, this search for novel genes whose protein products interact with actin didn't go quite as expected. The ANC1 protein turns out not to be a physical partner of actin. Rather, it is a transcription factor that is also known as TFG3 or TAF30 (Cairns et al. 1996). Specifically, it is a yeast-specific subunit of the transcription factor TFIIF. Indeed, the ANC1 protein is a component of the so-called "mediator complex," whose interaction with the carboxy-terminal repeat domain of RNA polymerase II enables transcriptional activation. Deletion of *ANC1/TAF30/TFG3* results in diminished transcription. Thus, *anc1* mutants appear to interact with *act1* mutants, not by physical interaction of mutant gene products, but instead by global effects on the transcription process.

This actin SSNC example points to a caveat in the use of noncomplementation screens to find interacting proteins: many of the ways in which a cell might create a given defect will not involve physical interaction with the mutant protein of interest, and, thus, may not be of real interest to you. And no matter how clever your secondary screens are, you still might get mutants in genes you weren't looking for or interested in. Of course, sometimes accidental observations such as this turn into fascinating research questions, so don't be completely discouraged if your screen turns up a gene or process that you're not immediately interested in.

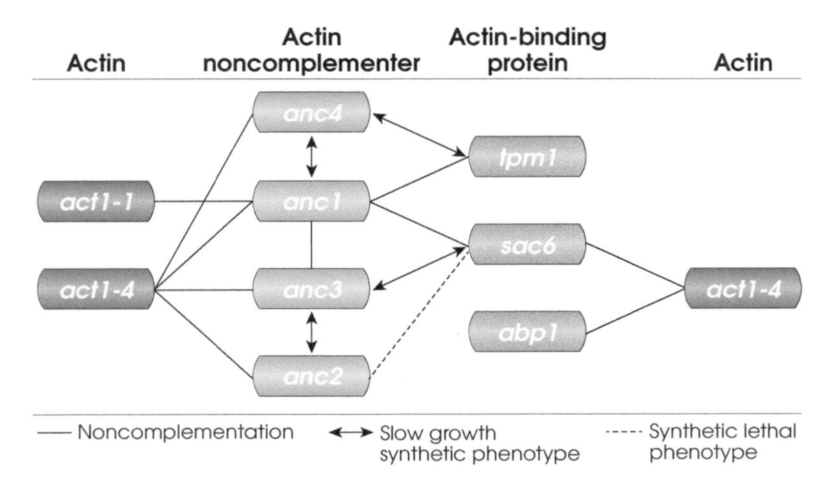

Figure 3.6 Genetic interactions among actin mutants in yeast. Interactions between *anc* mutations, between *anc* and *act1* mutations, and between *anc* and actin-binding protein mutations are illustrated. *Source:* Adapted from Welch et al. (1993).

Box 3.6 Synthetic Lethality and Genetic Buffering

As first noted by Bender and Pringle (1991), one technique for identifying genes whose products function in related or parallel pathways is to look for cases of **synthetic lethality**. Following Hartman et al. (2001), "mutations in two different genes are said to be 'synthetically lethal' if either mutation is viable in an otherwise wildtype background, but the combination of both alleles prevents growth." This definition works well in an organism like yeast, which can be grown in a haploid phase, but may become somewhat more awkward for higher eukaryotes. In such cases, the synthetic lethality may define mutations that are recessive in such a way that a heterozygote for a loss of function at either gene is viable, but the double heterozygote is lethal. (This latter case is clearly an example of second-site noncomplementation.) Alternatively, two dominant mutations may be synthetically lethal, or a recessive mutation at one locus may be synthetically lethal with some dominant allele of another gene even as a heterozygote. The simple test here is that if both alleles can survive on their own but kill the organism in combination, they are said to be synthetically lethal. Again, we quote Hartman et al. (2001): "Synthetic lethality defines a relationship where the presence of one gene (*A*) allows the organism to tolerate genetic variation (*b*) in another gene (*B*) that would be lethal in the absence of the first gene (*a*)."

We have discussed the use of semi-lethality to describe the relationships between the various actin noncomplementing mutants. A more detailed exploration of interactions using semi-lethality involves the study of the genes whose proteins mediate the secretory pathway in yeast (Figure B3.1). Synthetic lethal interactions have been found for many genes in this pathway, which can be divided into 10 discrete steps or components. Indeed, more than half of these interactions involve genes whose products act at the same step (e.g. translocation to the Golgi apparatus). As noted by Hartman et al., of 173 synthetic lethal interactions observed in this process, 116 involve the same component of the process, 68 involve proteins acting in different components of the process (a few genes act in two or more components), and 53 involve a gene that is not involved in secretion.

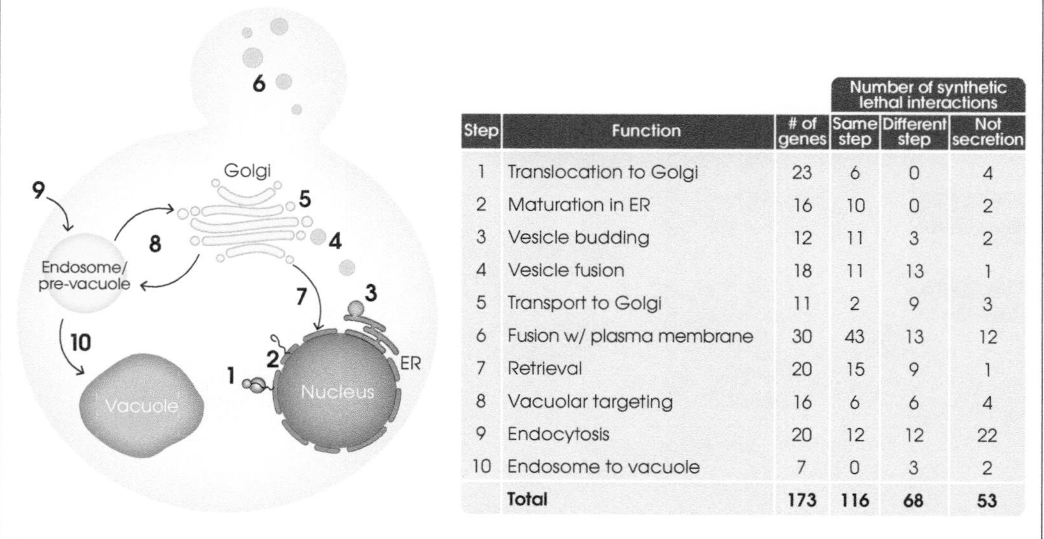

Step	Function	# of genes	Number of synthetic lethal interactions: Same step	Different step	Not secretion
1	Translocation to Golgi	23	6	0	4
2	Maturation in ER	16	10	0	2
3	Vesicle budding	12	11	3	2
4	Vesicle fusion	18	11	13	1
5	Transport to Golgi	11	2	9	3
6	Fusion w/ plasma membrane	30	43	13	12
7	Retrieval	20	15	9	1
8	Vacuolar targeting	16	6	6	4
9	Endocytosis	20	12	12	22
10	Endosome to vacuole	7	0	3	2
	Total	**173**	**116**	**68**	**53**

Figure B3.1 The secretory pathway in yeast. *Source:* Adapted from Hartman et al. (2001).

sought to find, but it is exactly what she asked for. Realize that her screen was not for genes whose protein products interact with β-tubulin, but rather for mutants that made a normally recessive *B2t* null mutation behave as a dominant.

We might also add that Calvin Bridges noted a century ago that mutants known as *Minutes* act as dominant enhancers of many recessive mutants in *Drosophila*. This is to say that double heterozygotes for the recessive mutant (*a*) and for the *Minute*, denoted *A/a; Minute/+*, often displayed the phenotype characteristic of homozygotes for the recessive mutant. *Minute* mutations turn out to define ribosomal protein genes in *Drosophila* and, as heterozygotes, reduce the overall level of translation. In such heterozygotes, the reduced level of wildtype gene product produced by the normal (*A*) allele is insufficient to produce a normal phenotype.

Thus, a screen for second-site noncomplementers can be considered to be a screen for dominant enhancers, especially in the case where the interaction is not allele-specific at both loci. Although screens for dominant enhancers can be powerful (discussed in Section 3.5), such a screen can also dredge up a host of undesired interacting loci. The maxim goes as follows: *Nature is both eminently fair and eminently cruel. Nature is fair because it will answer any question you ask, and cruel because it will answer the question you actually ask, not the question you thought you were asking.*

But surely, some of the *B2t*-interacting genes must encode the types of microtubule-associated proteins that Fuller and her collaborators were seeking? To date, none of the interacting genes yet identified by Fuller and her colleagues has been shown to encode a protein that physically interacts with microtubules. However, the genetic analysis of the *whirligig* locus is encouraging (Green et al. 1990). The interaction of *whirligig* mutants with the *B2t* null is allele-specific, and *whirligig* mutants have an interesting phenotype on their own. Two copies of the *whirligig* locus are necessary for male fertility. Both a deficiency of *whirligig* and loss-of-function alleles are dominant male-sterile mutations even in a genetic background wildtype for tubulin. This dominant male sterility is suppressed if the flies are also heterozygous for the null allele $B2t^n$, for a deficiency of α-tubulin, or for the hay^{nc2} allele. These results suggest that it is not the absolute level of *whirligig* gene product that matters, but rather its level relative to tubulin that is important for normal spermatogenesis. The phenotype of homozygous *whirligig* mutants suggests that the *whirligig* product plays a role in organizing the microtubules of the sperm flagellar axoneme. The flagellar axonemes show multiple anomalies in microtubule organization. It would certainly not surprise us if the *whirligig* gene product was eventually shown to encode exactly the type of protein that Fuller and her colleagues set out to find – namely a protein that physically interacts with tubulin.

An Example of Type 2 SSNC in *Drosophila* that Does *Not* Involve the Tubulin Genes

Dan Kiehart and his students subsequently applied SSNC to identify a group of genes whose protein products interact with non-muscle myosin in *Drosophila* (Halsell and Kiehart 1998; Halsell et al. 2000a). In *Drosophila*, the zipper (zip) gene encodes the non-muscle myosin protein. Mutants in this gene produce malformed legs in the adult, a phenotype that is easy to assay. One can thus test for second-site noncomplementation of zipper gene mutants by screening for mutants that behave as dominant enhancers of appropriate zipper mutants. Although null alleles of zipper are not known to display SSNC, a missense allele (zipEbr) proved useful as bait in a screen for dosage-sensitive second-site noncomplementers. Indeed, the zipEbr allele was specifically chosen for this study because it had already been shown to display SSNC with certain alleles of a set of X-chromosomal genes (the Broad-Complex genes).

Kiehart and colleagues scanned through a large collection of deficiencies for their ability to display SSNC with zip^{Ebr}. They tested 158 such deficiency heterozygotes for their ability to produce malformed legs in flies that were simultaneously heterozygous for zip^{Ebr}, and, in doing so, surveyed over 70% of the fly genome. The technique that Halsell and Kiehart (1998) used is conceptually rather different from the studies we have seen here. Rather than creating new noncomplementing mutants, they used an available collection of chromosomal deletions (each of which removes, on average, 50 to 100 genes). The question then becomes: which genes or chromosome regions strongly enhance the (*zip*^{Ebr}) mutation when present in only one dose? (Remember that all of these chromosomal regions, when present in two doses, are sufficient for a normal phenotype in *zip*^{Ebr} heterozygotes.) The benefit of this technique is its relative ease. The majority of the genome can be scanned in less than 200 crosses. The potential weakness is that the interaction you seek has to be dosage-sensitive; one is not testing mutant against mutant, but rather asking: for which genes does reducing the dosage by one copy become a problem in the presence of the *zip*^{Ebr} mutation? (If you think about it, it is really the reverse of screening for point mutants that interact with a null mutant in the *B2t* gene.)

Halsell and Kiehart (1998) identified two chromosomal deficiencies that strongly enhanced the phenotype of the *zip*^{Ebr} mutant in flies heterozygous for the *zip*^{Ebr} mutant (i.e. the *zip*^{Ebr}/+; *Df*/+ double heterozygotes displayed an obvious malformed leg defect) and 17 weakly interacting loci. From among these 19 deficiencies, they have been able to identify three whose interaction with *zipper* can be explained by individual genes located in these deficiencies. These include genes encoding cytoplasmic myosin, collagen IV, and the signal transduction protein RhoA (also known as Rho1 in *Drosophila*). In the cases of cytoplasmic myosin and collagen IV, the identity of the interacting loci was determined by testing overlapping deficiencies and by and by testing known mutants at these loci for their ability to interact with *zipper*.

The route to the discovery of the RhoA interaction was rather more complex and requires a more detailed discussion. This gene was initially identified by two independently isolated EMS-induced recessive lethal mutants (known as *E3.10* and *J3.8*) that mapped within one of the interacting deficiencies. Both mutants failed to complement the *zip*^{Ebr} mutant, and they failed to complement each other with respect to recessive lethality. However, it was not at all clear from these studies which gene these mutants defined.

The answer came by the serendipitous finding of another, but related, interacting gene. Halsell et al. (2000b) had performed a second screen for SSNC by testing 268 different *P* element transposon insertions for their ability to interact with *zipper*. In each case, these insertions produced recessive lethal mutants (i.e. insertion of the *P* element disrupted the function of a vital gene). Fourteen of these insertions failed to complement the *zip*^{Ebr} mutant, although in 11 of these cases the phenotype, and presumably the interaction, was weak. Two of these insertions were new alleles of *zipper*, but the third defined a new gene, *RhoGEF2*. Moreover, two previously isolated EMS-induced alleles of the *RhoGEF2* gene also failed to complement the *zip*^{Ebr} mutant. Thus, the *RhoGEF2* gene clearly behaved as an interacting gene. This result was perhaps not surprising given that the role of Rho proteins in remodeling the actin cytoskeleton is firmly established (for review, see Schmidt and Hall 1998). Rho-like proteins have been shown to affect both actin assembly and the organization of those actin filaments into various actin superstructures.

The finding that mutants in the *RhoGEF2* gene interacted with the *zipper* mutant raised the possibility that mutants in genes encoding other components of the Rho signaling pathway might also interact with *zipper*. Indeed, the two EMS-induced noncomplementers (*E3.10* and *J3.8*) recovered in the original deficiency search mapped in the same interval as the *RhoA* gene. Sequencing of the *RhoA* gene in flies carrying the *E3.10* and *J3.8* mutants revealed that these mutants were indeed lesions in the *RhoA* gene.

Several years after these initial screens, Patch and colleagues (2009) performed a similar SSNC screen for modifiers of RhoA signaling in *Drosophila*. Using an amorphic allele of *RhoA* that gave a weak, malformed leg phenotype, the authors were able to screen a large collection of new deficiencies with the goal of identifying genes involved in leg morphogenesis. This screen identified six interacting genes, three of which were known from the Halsell and Kiehart screen (1998) and three that were novel RhoA-interacting genes.

The screens performed by Halsell and Kiehart (1998), Halsell et al. (2000b), and Patch et al. (2009) comprise one of the more successful applications of second-site noncomplementation. From our point of view, the success of these screens reflected the use of a deficiency collection that allowed the studies to cover much of the genome in a straightforward and facile manner. The coupling of the deficiency screen to a screen for interacting *P*-insertion mutants provided this approach with even greater power.

An Example of Type 2 SSNC in the Nematode *Caenorhabditis elegans*

Yook and her collaborators (2001) also studied SSNC in genes whose protein products function at the synaptic junctions of the nervous system. These proteins regulate the release of neurotransmitters into the synapse by exocytosis, a process required for coordinated movements. They observed that mutations in the genes encoding the physically interacting synaptic proteins UNC-13 and syntaxin/UNC-64 failed to complement one another. In other words, mutants in both genes, which were fully recessive, produced an uncoordinated phenotype as double heterozygotes. The intriguing component of this study was that a clever drug-resistance assay allowed these workers to quantify the degree of noncomplementation rather than depending only on the discrete uncoordinated-versus-wildtype phenotypic assay.

Noncomplementation was observed only when at least one mutant encoded a partially functional, but weakly poisonous, gene product. Although these mutants were recessive in terms of the uncoordinated assay, the more sensitive drug-resistance assay detected a defect even in heterozygotes. This defect was not due simply to a 50% decrease in the amount of wildtype protein because no defect in the drug resistance assay was observed in heterozygotes for null alleles. Noncomplementation did not require partially functional, or poisonous, alleles at both loci. One such allele, when combined with a null allele of the second gene, also produced an uncoordinated phenotype. However, noncomplementation was not observed between null alleles of these two genes, and thus this genetic interaction does not occur with a simple decrease in dosage at the two loci. This genetic interaction requires a poisonous gene product to sensitize the genetic background.

To quote the authors, "Hypomorphic mutations . . . which are recessive in behavioral assays, can act as weak poisons as heterozygotes in quantitative drug sensitivity assays. These poisons sensitize the process of neurotransmission to perturbations at other synaptic loci, resulting in nonallelic noncomplementation. In addition, it is the presence of these poisons rather than a simple decrease in the dosage of the gene product that is essential for nonallelic noncomplementation . . ."

These researchers further demonstrated that nonallelic noncomplementation was not limited to interacting proteins. Although the strongest effects were observed between loci encoding gene products that bind to one another, interactions were also observed between proteins that do not directly interact but are members of the same complex. The authors explain such long-range effects by noting, "In processes requiring the correct assembly of protein complexes, a single faulty subunit can render a large number of gene products inactive by participating in and poisoning protein complexes."

Yook et al. (2001) also observed noncomplementation between genes that function at distant points in the same pathway, implying that physical interactions are not required for nonallelic noncomplementation. Of course, mutations in genes that function in different processes, such as neurotransmitter synthesis or synaptic development, do complement one another. Thus, this genetic interaction was specific for genes acting in the same pathway – for genes acting in synaptic vesicle trafficking.

Type 3 SSNC (Combined Haploinsufficiency): The Interaction is Allele-Independent at Both Loci

This type of SSNC neither requires nor implies the physical interaction of the two proteins. Rather, it suggests only that reducing the dosage of the product of gene *A* is a survivable event for the cell unless it is further crippled by a reduction in the dosage of gene *B*. **Combined haploinsufficiency** does not require allele specificity at either gene, and it is sometimes created by using null alleles or deficiencies at one or both genes. This type of SSNC is probably the most common, and sometimes the least interesting. One could imagine, for example, a case where the mutation in gene *B* simply depressed the rate of transcription of gene *A*, thus decreasing the level of A protein production below some threshold of function. However, one could also imagine examples where the two proteins do act in functionally related processes. This latter case can be informative in terms of understanding the specific biological process in question.

An Example of Type 3 SSNC Involving Two Motor Protein Genes in Flies

The *nod* and *ncd* genes of *Drosophila* both encode proteins that control the interaction of chromosomes with their microtubule tracks. When homozygous, loss-of-function mutations of both *nod* and *ncd* cause high frequencies of meiotic chromosome misbehavior in females (i.e. homologous chromosomes often fail to segregate from each other). Both *nod* and *ncd* mutants are fully recessive; no effect on meiotic chromosome behavior is seen in *nod/+* or *ncd/+* females. However, elevated frequencies of meiotic failure are observed in doubly heterozygous *nod/+, ncd/+* females (Knowles and Hawley 1991).

The types and frequencies of meiotic failures seen in double heterozygotes are more similar to those observed in *nod* homozygotes than they are to those seen in *ncd* homozygotes. On this basis, heterozygosity for mutants at the *ncd* gene enhances the effect of reducing the dosage of the nod^+ gene and thus creates a *nod*-like phenotype. This study went on to show that the same levels of meiotic failure could be observed when either one or both of the two mutant alleles was a deficiency. There is therefore no allele specificity for either gene. Further studies showed that these two proteins are not colocalized on the meiotic spindle; the Nod protein binds to chromosomes while Ncd is bound to the spindle. Nor is there any evidence that the proteins physically interact. Rather, it appears that reducing the copy number of the *ncd* gene creates a spindle that is less tolerant of reduced dosage of the *nod* gene.

Other examples of combined haploinsufficiency in *Drosophila* can be found in the work of Jackson and Berg (1999) and Kidd et al. (1999). An example of nonallelic noncomplementation involving null alleles of the genes encoding the transcription factors MF1 and MFH1 with respect to cardiovascular development in the mouse has been reported by Winnier et al. (1995). A second mouse example involving *Hoxb5* and *Hoxb6* genes has been reported by Rancourt et al. (1995). The *Hoxb5* and *Hoxb6* genes are adjacent in the mouse HoxB locus and are members of the homeotic transcription factor complex that governs establishment of the mammalian body plan. Although loss-of-function mutants at the *Hoxb5* and *Hoxb6* genes are fully recessive, *Hoxb5*, *Hoxb6* transheterozygotes ($Hoxb5^-$ $Hoxb6^+$/$Hoxb5^+$ $Hoxb6^-$) display a mutant phenotype.

Summary of SSNC in Model Organisms

We have listed here three basic types of SSNC and offered a simple set of rules for distinguishing the three types. In type 1 SSNC, the two mutant proteins physically interact to produce a poisonous protein dimer or complex. This type of genetic interaction is heralded by allele specificity for both genes and cannot be observed when examining a null allele at either gene. In type 2 SSNC, the mutant protein produced by a specific allele at one of the two loci acts to sequester the reduced level of normal protein produced by the other gene. This type of genetic interaction is heralded by allele specificity for one of the two genes and can be observed when examining a null allele at the other gene. In type 3 SSNC, the phenotype in double heterozygotes results not from any interaction of wildtype or mutant gene products, but rather from combined haploinsufficiency. This type of genetic interaction is not allele-specific for either gene, and it can be observed using null alleles at either or both genes.

SSNC in Humans (Digenic Inheritance)

There are perhaps several examples of **digenic inheritance** in humans. For example, digenic retinitis pigmentosa appears to be the result of double heterozygosity for recessive mutations at two unlinked genes: *peripherin2/RDS* and *ROM1* (Kajiwara et al. 1994). Mutations in both *peripherin2/RDS* and *ROM1* are fully recessive when present alone, but double heterozygosity has been shown to produce the disease (Böhm et al. 2017; Lewis et al. 2020). A possible molecular explanation for this interaction was first provided by the work of Goldberg and Molday (1996). They showed that the missense *peripherin2/RDS* mutant is conditionally defective with respect to its subunit assembly. Unlike wildtype peripherin/RDS protein, the mutant protein cannot properly assemble into homotetramers on its own. However, the authors were able to show that the mutant peripherin/RDS protein could form a structurally normal heterotetrameric complex in the presence of wildtype ROM1. Presumably then, the cells of the retina can utilize the mutant peripherin/RDS protein as long as sufficient ROM1 is available to facilitate assembly. But the loss of one functional copy of ROM1 appears to disrupt this process, and thus creates an insufficiency of assembled peripherin/RDS complexes.

A similar example of this phenomenon in humans is provided by the unusual genetics of a complex human disorder called Bardet–Biedl syndrome (BBS). Initially, mutations at 6 genes were implicated in the causation of this disorder, although now at least 26 genes have been associated with the syndrome (Niederlova et al. 2019). The genetics of the disorder was exceedingly complex, but the data led to the model that this syndrome was inherited as an autosomal recessive disorder and that mutations at any of these six other loci might be able to induce the disorder. Nonetheless, the pedigrees were not fully consistent with simple models of autosomal inheritance. There were cases of unaffected individuals who appeared to be homozygous at least at one of the six loci, while other similarly homozygous individuals were affected. Katsanis et al. (2001) later demonstrated that affected individuals are homozygous for a loss-of-function mutation at one locus and heterozygous for a loss-of-function mutation at one of the other five loci. Katsanis et al. (2001) wished to refer to this phenomenon as triallelic inheritance. However, we rather agree with Burghes et al. (2001) in suggesting that this phenomenon is best referred to as "recessive inheritance with a [dominant] modifier of penetrance" ("dominant" was inserted by us). As Burghes et al. (2001) point out, there are other examples of this phenomenon in flies and mammals. In *Drosophila*, homozygosity for a loss-of-function mutation in the *fascilin* gene is without phenotype unless this genotype is combined with heterozygosity for a mutation in the *Abl* gene (Elkins et al. 1990). While interesting, recent studies have identified second "missing variants" in some (but not all) cases of BBS thought to be due to variants in two different genes (Hirano et al. 2020) – a good reminder that although it is important to consider complex possibilities, such as the example here, one should always keep an eye out for simpler explanations.

A final interesting example of digenic inheritance comes from work by Lemmers et al. (2012) on facioscapulohumeral muscular dystrophy (FSHD). Individuals with FSHD have facial and upper-extremity muscle weakness caused by the inappropriate somatic expression of the transcription factor *DUX4*, which is typically expressed only in germline tissues. *DUX4* sits in the D4Z4 **microsatellite array** (short, tandem DNA repeats), and the number of copies of this satellite is inherited, varying from family to family. There are two types of FSHD. While type 1 is due to the contraction of the D4Z4 satellite array, this does not appear to be the case for type 2. Rather, Lemmers et al. (2012) showed that FSHD type 2 occurs in individuals who inherit a normal-sized D4Z4 satellite along with a mutation in the *SMCHD1* gene that reduces SMCHD1 protein levels. The reduced levels of SMCHD1 affect the silencing of *DUX4* through inappropriate heterochromatic packaging, allowing the somatic expression of *DUX4*. In other words, having less SMCHD1 results in open chromatin and expression of *DUX4*, conferring the phenotype of facial and upper-extremity muscle weakness. We find studies such as this fascinating, especially as the falling cost and increased use of genome sequencing leads to additional examples of digenic inheritance in human disease.

Pushing the Limits: Third-Site Noncomplementation

If SSNC works with two genes, why not try it with three? In other words, suppose that heterozygosity for mutants at two genes created a sensitized background, without creating a phenotype. Perhaps then, the addition of a third mutant might be enough to create that phenotype. In their studies of the epistatic interactions of genes controlling sex determination in *Drosophila*, Baker and Ridge (1980) discovered that *XX* female flies that were simultaneously heterozygous for mutants in the *tra* and *tra2* genes were wildtype. But the addition of a third sex-determining mutant in the heterozygous conditions created an intermediate sexual phenotype (intersex). Indeed, every known sex-determination mutant was shown to exert a dominant phenotype in the *tra2/+, tra/+* background.

These observations suggested a screen for new mutations in the sex determination pathway searching for dominant mutants that create an intersex phenotype in the *tra2/+, tra/+* background. Such a screen is described by Belote et al. (1985). The authors note that "One advantage of such a screen is that it can potentially allow the detection of pleiotropic mutations that affect not only sex determination, but also some vital process and thus are homozygous lethal." In a fashion similar to that described for the Halsell and Kiehart screen, the authors began by screening a large collection of deficiencies that together encompassed some 30 of the genome. Of 70 deficiencies assessed, 12 tested positive in this assay. However, seven of these deficiencies included a *Minute* gene. As noted, *Minute* genes encode ribosomal proteins and, as heterozygotes, have the capacity to enhance many heterozygous mutations. This is simply the consequence of combining a decrease in gene dosage with a further decrease in protein synthetic capability. However, one of the remaining five deficiencies, which removed a small region on the X chromosome, had the strongest effect on their assay. Several independent lines of evidence indicate that this region did indeed contain an important sex determination gene. We are not aware of subsequent uses of this type of screen. That is probably a pity. It is perhaps unfortunate that a technique with such significant promise has not been more widely used.

3.5 An Extension of SSNC: Dominant Enhancers

We have noted here that a screen for second-site noncomplementation is really a rather specialized form of a screen for dominant enhancers. The only difference is that in a more general screen for dominant enhancers, the target mutation need not be heterozygous. One can use a hypomorphic mutation as a homozygote, as a hemizygote on the X chromosome, or even as a transgenic insertion into the genome (with the wildtype copies knocked out) as your target for enhancement. If the choice of the target allele is good and the screen is well defined, this approach can be extremely effective. An example of one such screen is presented here.

A Successful Screen for Dominant Enhancers

The technique of screening for dominant enhancers of hypomorphic mutants is now well established in many organisms. Simon et al. (1991) performed an elegant screen for mutations in genes whose products interacted with the protein tyrosine kinase encoded by the X-chromosomal *sevenless* (*sev*) gene in *Drosophila*. The product of the *sevenless* gene specifies a membrane receptor tyrosine kinase whose function is required for the proper differentiation of one type of photoreceptor cell (R7) that is present in each of the 800 repeating units (**ommatidia**) that make up the fly's compound eye. The *sevenless* gene product appears to be required only for R7 differentiation, and in its absence, the cell that would normally become R7 takes on another fate. The result is a fly in which the organization of each of the ommatidia is disrupted, resulting in a compound eye with a rough or disorganized appearance. Thus, it is reasonable to propose, as did the authors, that the Sevenless protein "act[s] as a receptor for a signal that determines whether the presumptive R7 cell becomes a photoreceptor . . ."

Simon et al. (1991) noted that the most straightforward way to isolate new mutants in the process of R7 differentiation would be to isolate new recessive mutations in other genes that result in the specific loss of the R7 cell. Indeed, such screens, done by others, had already isolated two such genes: the *bride of sevenless* gene, which encodes the ligand for Sevenless (Reinke and Zipursky 1988), and *seven in absentia* (Carthew and Rubin 1990). Nonetheless, Simon et al. (1991) reckoned that many steps in the *sevenless* regulatory pathway might not be specific to the R7 cell, but rather might be shared with other signal transduction screens. Mutants in such genes would not specifically affect R7, and indeed they might be lethal as homozygotes. To obviate these difficulties, they designed "a more sensitive genetic screen that allows the identification of mutations that *reduce* rather than abolish the activity of downstream signaling proteins" (our italics).

Using a well-studied viral protein tyrosine kinase (v-Src) as an example, Simon et al. (1991) used site-directed mutagenesis to create two putative temperature-sensitive mutant alleles of *sevenless* and transformed these constructs back into *Drosophila*. In the absence of a wildtype *sevenless* gene, these two mutant constructs did indeed produce a temperature-sensitive loss of the R7 photoreceptor cell, such that when flies carrying one of these mutant alleles were reared at 22.7 °C the eyes were wildtype, but when flies of the same genotype were reared at 24.3 °C the eyes displayed the mutant phenotype.

The authors then screened for EMS mutations that, as heterozygotes, created a mutant *sevenless* phenotype at the permissive temperature of 22.7 °C (i.e. mutations that behaved as dominant enhancers of the temperature-sensitive *sevenless* allele). Think about this – it is really a straightforward F1 screen for "dominant" mutations that produce an easily scored phenotype: rough eyes. From over 30,000 treated chromosomes screened, 20 such *Enhancer of sevenless* [*E(sev)*] mutations were recovered. These mutants defined seven lethal complementation groups such that transheterozygotes within each group did not survive. The recessive lethality here is important; such mutants would not have been recovered in a direct screen for recessive visible mutations. More critically, none of these mutations produced a rough-eye phenotype in the presence of a wildtype allele of the *sevenless* gene. Thus, for each of the *E(sev)* mutants, the ability to produce a phenotype depends on the partially compromised Sevenless protein that was produced by the temperature-sensitive allele.

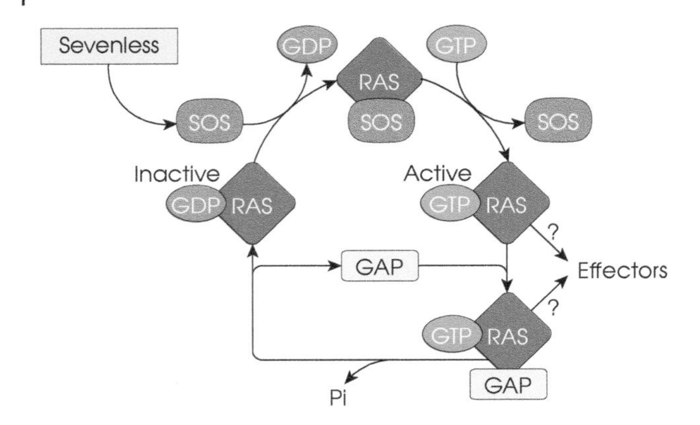

Figure 3.9 Components of the Ras signaling pathway in *Drosophila*. *Source:* Adapted from Simon et al. (1991).

Two lines of evidence indicated that the genes defined by these mutants do indeed encode shared components of a number of signal transduction pathways. First, by using a clever system to produce mitotic clones, these authors bypassed the recessive lethality of these mutations and showed that mutations in at least four of these genes appear to reduce the ability of an eye cell to develop as an R7 cell or as any photoreceptor. Thus, these genes appear to participate in some process that is common to all of the photoreceptor cells in the developing eye. Second, mutants in these four genes also attenuated the signaling by another protein tyrosine kinase (ellipse).

Perhaps not surprisingly, these mutants did turn out to define shared components of the signaling process, including the *Ras* gene (which is a component of many signal transduction pathways) and a putative *CDC25* homolog. The latter gene had previously been identified both by recessive lethal mutations and by a dominant mutation called *Son of sevenless* (*Sos*). The *Sos* mutation had been identified by its ability to suppress a specific allele of *sevenless*. Evidence in other organisms suggested that CDC25 may act to activate the Ras protein, and so the Sos protein may also act by catalyzing the activation of Ras. A drawing of the regulatory pathway elucidated from these studies is presented in Figure 3.9. The finding of these and other components of the pathway has led to a thorough understanding of the Sevenless signal transduction pathway.

The success of this screen lay in the creation of a threshold temperature-sensitive allele that was capable of creating an easily scored phenotype. By excluding enhancers that did not depend on the presence of the *sevenless* temperature-sensitive mutation to create a phenotype, the authors excluded any unrelated simple dominant mutations. Using mitotic clone analysis to verify a defect of the enhancer mutations alone in R7 cell formation further strengthened the specificity of the screen. The search for dominants allowed the recovery of loss-of-function alleles in essential genes whose products might function in shared pathways.

3.6 Summary

When the complementation test works (and it usually does), it is a common and very reliable tool. However, it *can* lie to you. Such "lies" are rare, and most of the time complementation tests can and should be taken at face value. But if the complementation data are at odds with mapping or sequencing, or the transheterozygote phenotype is too weak, or the pattern of complementation among multiple alleles is unduly complex, a few lie-detector tests to identify intragenic or nonallelic complementation effects are clearly in order. Should that become necessary, this chapter provides informative examples of such investigations that can help you think about your next steps.

References

Appling DR. 1999. Genetic approaches to the study of protein–protein interactions. *Methods* 19:338–349.

Aramayo R, Metzenberg RL. 1996. Meiotic transvection in fungi. *Cell* 86:103–113.

Arrizabalaga G, Lehmann R. 1999. A selective screen reveals discrete functional domains in *Drosophila* Nanos. *Genetics* 153:1825–1838.

Ashburner M. 1989. *Drosophila: A Laboratory Handbook*. Cold Spring Harbor Laboratory Press, New York.

Baker BS, Ridge KA. 1980. Sex and the single cell. I. On the action of major loci affecting sex determination in *Drosophila melanogaster*. *Genetics* 94(2):383–423.

Beall EL, Rio DC. 1996. *Drosophila* IRBP/Ku p70 corresponds to the mutagen-sensitive *mus309* gene and is involved in P-element excision in vivo. *Genes Dev.* 10(8):921–933.

Belote JM, McKeown MB, Andrew DJ, Scott TN, Wolfner MF, *et al.* 1985. Control of sexual differentiation in *Drosophila melanogaster*. *Cold Spring Harb. Symp. Quant. Biol.* 50:605–614.

Bender A, Pringle JR. 1991. Use of a screen for synthetic lethal and multicopy suppressee mutants to identify two new genes involved in morphogenesis in *Saccharomyces cerevisiae*. *Mol. Cell. Biol.* 11:1295–1305.

Bickel SE, Wyman DW, Orr-Weaver TL. 1997. Mutational analysis of the *Drosophila* sister-chromatid cohesion protein ORD and its role in the maintenance of centromeric cohesion. *Genetics* 146: 1319–1331.

Böhm S, Riedmayr LM, Nguyen ONP, Gießl A, Liebscher T, *et al.* 2017. Peripherin-2 and Rom-1 have opposing effects on rod outer segment targeting of retinitis pigmentosa-linked *peripherin-2* mutants. *Sci. Rep.* 7:2321.

Bosco G. 2012. Chromosome pairing: a hidden treasure no more. Copenhaver GP, editor. *PLoS Genet.* 8:e1002737.

Burghes AH, Vaessin HE, de La Chapelle A 2001. Genetics. The land between Mendelian and multifactorial inheritance. *Science* 293:2213–2214.

Cairns BR, Henry NL, Kornberg RD. 1996. TFG/TAF30/ANC1, a component of the yeast SWI/SNF complex that is similar to the leukemogenic proteins ENL and AF-9. *Mol. Cell. Biol.* 16:3308–3316.

Carthew RW, Rubin GM. 1990. *Seven in absentia*, a gene required for specification of R7 cell fate in the *Drosophila* eye. *Cell* 63(3):561–577.

Chambeyron S, Popkova A, Payen-Groschêne G, Brun C, Laouini D, *et al.* 2008. piRNA-mediated nuclear accumulation of retrotransposon transcripts in the *Drosophila* female germline. *Proc. Natl. Acad. Sci.* 105:14964–14969.

Cline TW. 1978. Two closely linked mutations in *Drosophila melanogaster* that are lethal to opposite sexes and interact with *daughterless*. *Genetics* 90(4):683–698.

Cline TW. 1984. Autoregulatory functioning of a *Drosophila* gene product that establishes and maintains the sexually determined state. *Genetics* 107(2):231–277.

Denell RE. 1972. The nature of reversion of a dominant gene of *Drosophila melanogaster*. *Mutat. Res.* 15(2):221–223.

Duncan IW, Kaufman TC. 1975. Cytogenic analysis of chromosome 3 in *Drosophila melanogaster*: mapping of the proximal portion of the right arm. *Genetics* 80(4):733–752.

Elkins T, Zinn K, McAllister L, Hoffmann FM, Goodman CS. 1990. Genetic analysis of a *Drosophila* neural cell adhesion molecule: interaction of fasciclin I and Abelson tyrosine kinase mutations. *Cell* 60:565–575.

Fincham JRS. 1966. *Genetic Complementation*. New York: W.A. Benjamin.

Forsburg SL. 2001. The art and design of genetic screens: yeast. *Nat. Rev. Genet* 2:659–668.

Fuller MT. 1986. Genetic analysis of spermatogenesis in *Drosophila*: the role of the testis-specific beta-tubulin and interacting genes in cellular morphogenesis. In: Gall JG, editor. *Gametogenesis and the Early Embryo*. New York: Alan R. Liss. pp. 19–42; 24 p.

Gelbart WM. 1982. Synapsis-dependent allelic complementation at the *decapentaplegic* gene complex in *Drosophila melanogaster*. *Proc. Natl. Acad. Sci. U.S.A.* 79(8):2636–2640.

Gibson G, Hogness DS. 1996. Effect of polymorphism in the *Drosophila* regulatory gene Ultrabithorax on homeotic stability. *Science* 271:200–203.

Gilliland WD, Wayson SM, Hawley RS. 2005. The meiotic defects of mutants in the *Drosophila* mps1 gene reveal a critical role of Mps1 in the segregation of achiasmate homologs. *Curr. Biol.* 15:672–677.

Gilliland WD, Hughes SE, Cotitta JL, Takeo S, Xiang Y, *et al.* 2007. The multiple roles of *mps1* in *Drosophila* female meiosis. *PLoS Genet.* 3:e113.

Goldberg AF, Molday RS. 1996. Defective subunit assembly underlies a digenic form of retinitis pigmentosa linked to mutations in peripherin/rds and rom-1. *Proc. Natl. Acad. Sci. U.S.A.* 93:13726–13730.

Golic MM, Golic KG. 1996. A quantitative measure of the mitotic pairing of alleles in *Drosophila melanogaster* and the influence of structural heterozygosity. *Genetics* 143(1):385–400.

Green MM. 1977. X-ray induced reversions of the mutant *forked-3N* in *Drosophila melanogaster*, a reappraisal. *Mutat. Res.* 43(2):305–308.

Green LL, Wolf N, McDonald KL, Fuller MT. 1990. Two types of genetic interaction implicate the whirligig gene of *Drosophila melanogaster* in microtubule organization in the flagellar axoneme. *Genetics* 126:961–973.

Guarente L. 1993. Synthetic enhancement in gene interaction: a genetic tool come of age. *Trends Genet.* 9:362–366.

Halsell SR, Kiehart DP. 1998. Second-site noncomplementation identifies genomic regions required for *Drosophila* nonmuscle myosin function during morphogenesis. *Genetics* 148:1845–1863.

Halsell S, Chu B, Kiehart D. 2000a. Genetic analysis demonstrates a direct link between rho signaling and nonmuscle myosin function during *Drosophila* morphogenesis. *Genetics* 156:469.

Halsell SR, Chu BI, Kiehart DP. 2000b. Genetic analysis demonstrates a direct link between rho signaling and nonmuscle myosin function during *Drosophila* morphogenesis. *Genetics* 155:1253–1265.

Hartman JL, Garvik B, Hartwell L. 2001. Principles for the buffering of genetic variation. *Science* 291:1001–1004.

Hawley RS, Gilliland WD. 2006. Sometimes the result is not the answer: the truths and the lies that come from using the complementation test. *Genetics* 174:5–15.

Hays TS, Deuring R, Robertson B, Prout M, Fuller MT. 1989. Interacting proteins identified by genetic interactions: a missense mutation in alpha-tubulin fails to complement alleles of the testis-specific beta-tubulin gene of *Drosophila melanogaster*. *Mol. Cell. Biol.* 9:875–884.

Hazelrigg T, Kaufman TC. 1983. Revertants of dominant mutations associated with the *Antennapedia* gene complex of *Drosophila melanogaster*: cytology and genetics. *Genetics* 105(3):581–600.

Hirano M, Satake W, Moriyama N, *et al.* 2020. Bardet–Biedl syndrome and related disorders in Japan. *J. Hum. Genet.* 65:847–853.

Jack JW, Judd BH. 1979. Allelic pairing and gene regulation: a model for the *zeste-white* interaction in *Drosophila melanogaster*. *Proc. Natl. Acad. Sci. U.S.A.* 76(3):1368–1372.

Jackson SM, Berg CA. 1999. Soma-to-germline interactions during *Drosophila* oogenesis are influenced by dose-sensitive interactions between cut and the genes cappuccino, ovarian tumor and agnostic. *Genetics* 153:289–303.

Kajiwara K, Berson EL, Dryja TP. 1994. Digenic retinitis pigmentosa due to mutations at the unlinked peripherin/RDS and ROM1 loci. *Science* 264:1604–1608.

Katsanis N, Ansley SJ, Badano JL, Eichers ER, Lewis RA, *et al.* 2001. Triallelic inheritance in Bardet–Biedl syndrome, a Mendelian recessive disorder. *Science* 293:2256–2259.

Kidd T, Bland KS, Goodman CS. 1999. Slit is the midline repellent for the robo receptor in *Drosophila*. *Cell* 96:785–794.

Knowles BA, Hawley RS. 1991. Genetic analysis of microtubule motor proteins in *Drosophila*: a mutation at the ncd locus is a dominant enhancer of nod. *Proc. Natl. Acad. Sci. U.S.A.* 88:7165–7169.

Kusano A, Staber C, Ganetzky B. 2001a. Nuclear mislocalization of enzymatically active RanGAP causes segregation distortion in *Drosophila*. *Dev. Cell* 1(3):351–361.

Kusano K, Johnson-Schlitz DM, Engels WR. 2001b. Sterility of *Drosophila* with mutations in the Bloom syndrome gene--complementation by *Ku70*. *Science* 291(5513):2600–2602.

Lemmers RJLF, Tawil R, Petek LM, Balog J, Block GJ, *et al.* 2012. Digenic inheritance of an SMCHD1 mutation and an FSHD-permissive D4Z4 allele causes facioscapulohumeral muscular dystrophy type 2. *Nat. Genet.* 44:1370–1374.

Lewis EB. 1954. The theory and application of a new method of detecting chromosomal rearrangements in *Drosophila melanogaster*. *Am. Nat.* 88(841):225–239.

Lewis EB. 2003. C. B. Bridges' repeat hypothesis and the nature of the gene. *Genetics* 164:427–431.

Lewis TR, Makia MS, Kakakhel M, Al-Ubaidi MR, Arshavsky VY, *et al.* 2020. Photoreceptor disc enclosure occurs in the absence of normal Peripherin-2/rds oligomerization. *Front. Cell. Neurosci.* 14:92.

Matzke M, Mette MF, Jakowitsch J, Kanno T, Moscone EA, *et al.* 2001. A test for transvection in plants: DNA pairing may lead to *trans*-activation or silencing of complex heteroalleles in tobacco. *Genetics* 158(1):451–461.

McLaren A. 1999. Too late for the midwife toad: stress, variability and Hsp90. *Trends Genet.* 15:169–171.

Miller DE, Vaerenberghe KV, Li A, Grantham E, Cummings C, *et al.* 2020. Germline gene de-silencing by a transposon insertion is triggered by an altered landscape of local piRNA biogenesis. *bioRxiv* 2020(06):https://doi.org/10.1101/2020.06.26.173187.

Mounkes LC, Fuller MT. 1999. Molecular characterization of mutant alleles of the DNA repair/basal transcription factor *haywire*/ERCC3 in *Drosophila*. *Genetics* 152:291–297.

Mounkes LC, Jones RS, Liang BC, Gelbart W, Fuller MT. 1992. A *Drosophila* model for xeroderma pigmentosum and Cockayne's syndrome: *haywire* encodes the fly homolog of ERCC3, a human excision repair gene. *Cell* 71:925–937.

Nagel AC, Yu Y, Preiss A. 1999. *Enhancer of split* [*E(spl)(D)*] is a gro-independent, hypermorphic mutation in *Drosophila*. *Dev. Genet.* 25(2):168–179.

Niederlova V, Modrak M, Tsyklauri O, Huranova M, Stepanek O. 2019. Meta-analysis of genotype-phenotype associations in Bardet–Biedl syndrome uncovers differences among causative genes. *Hum. Mutat.* 40:2068–2087.

Patch K, Stewart SR, Welch A, Ward RE. 2009. A second-site noncomplementation screen for modifiers of Rho1 signaling during imaginal disc morphogenesis in *Drosophila*. *PLoS ONE* 4:e7574.

Perkins DD. 1997. 6 chromosome rearrangements in Neurospora and other filamentous fungi. *Adv. Genet.* 36:239–398.

Pontecorvo G. 1958. *Trends in Genetic Analysis*. Columbia University Press.

Rancourt DE, Tsuzuki T, Capecchi MR. 1995. Genetic interaction between *hoxb-5* and *hoxb-6* is revealed by nonallelic noncomplementation. *Genes Dev.* 9:108–122.

Rasooly RS, New CM, Zhang P, Hawley RS, Baker BS. 1991. The *lethal(1)TW-6cs* mutation of *Drosophila melanogaster* is a dominant antimorphic allele of *nod* and is associated with a single base change in the putative ATP-binding domain. *Genetics* 129(2):409–422.

Raz E, Schejter ED, Shilo BZ. 1991. Interallelic complementation among DER/flb alleles: implications for the mechanism of signal transduction by receptor-tyrosine kinases. *Genetics* 129:191–201.

Regan CL, Fuller MT. 1990. Interacting genes that affect microtubule function in *Drosophila melanogaster*: two classes of mutation revert the failure to complement between *haync2* and mutations in tubulin genes. *Genetics* 125:77–90.

Reinke R, Zipursky SL. 1988. Cell-cell interaction in the *Drosophila* retina: the *bride of sevenless* gene is required in photoreceptor cell R8 for R7 cell development. *Cell* 55(2):321–330.

Russell S. 2000. The *Drosophila* dominant wing mutation *Dichaete* results from ectopic expression of a Sox-domain gene. *Mol Gen Genet.* 263(4):690–701.

Rutherford SL, Lindquist S. 1998. Hsp90 as a capacitor for morphological evolution. *Nature* 396:336–342.

Schmidt A, Hall MN. 1998. Signaling to the actin cytoskeleton. *Annu. Rev. Cell Dev. Biol.* 14:305–338.

Simon MA, Bowtell DD, Dodson GS, Laverty TR, Rubin GM. 1991. Ras1 and a putative guanine nucleotide exchange factor perform crucial steps in signaling by the Sevenless protein tyrosine kinase. *Cell* 67:701–716.

Smolik-Utlaut SM, Gelbart WM. 1987. The effects of chromosomal rearrangements on the *zeste-white* interaction in *Drosophila melanogaster*. *Genetics* 116(2):285–298.

Stearns T, Botstein D. 1988. Unlinked noncomplementation: isolation of new conditional-lethal mutations in each of the tubulin genes of *Saccharomyces cerevisiae*. *Genetics* 119:249–260.

Tartof KD, Henikoff S. 1991. *Trans*-sensing effects from *Drosophila* to humans. *Cell* 65(2):201–203.

Vinh DB, Welch MD, Corsi AK, Wertman KF, Drubin DG. 1993. Genetic evidence for functional interactions between actin noncomplementing (*Anc*) gene products and actin cytoskeletal proteins in *Saccharomyces cerevisiae*. *Genetics* 135:275–286.

Waddington CH. 1942. Canalization of development and the inheritance of acquired characters. *Nature* 150:563–565.

Waddington CH. 1953. Genetic assimilation of an acquired character. *Evolution* 7:118.

Welch MD, Drubin DG. 1994. A nuclear protein with sequence similarity to proteins implicated in human acute leukemias is important for cellular morphogenesis and actin cytoskeletal function in *Saccharomyces cerevisiae*. *Mol. Biol. Cell* 5:617–632.

Welch MD, Vinh DB, Okamura HH, Drubin DG. 1993. Screens for extragenic mutations that fail to complement *act1* alleles identify genes that are important for actin function in *Saccharomyces cerevisiae*. *Genetics* 135:265–274.

Whyte WL, Irick H, Arbel T, Yasuda G, French RL, et al. 1993. The genetic analysis of achiasmate segregation in *Drosophila melanogaster*. III. The wild-type product of the *Axs* gene is required for the meiotic segregation of achiasmate homologs. *Genetics* 134(3):825–835.

Winnier G, Blessing M, Labosky PA, Hogan BL. 1995. Bone morphogenetic protein-4 is required for mesoderm formation and patterning in the mouse. *Genes Dev.* 9:2105–2116.

Wu CT, Morris JR. 1999. Transvection and other homology effects. *Curr Opin Genet Dev.* 9(2):237–246.

Yook KJ, Proulx SR, Jorgensen EM. 2001. Rules of nonallelic noncomplementation at the synapse in *Caenorhabditis elegans*. *Genetics* 158(1):209–220.

Box 4.1 The Molecular Biology of Synapsis

Geneticists and cytologists often use the words alignment, pairing, and synapsis when discussing the processes that bring homologous chromosomes together during meiotic prophase. Unfortunately, these words do not always mean the same things, but it still seems worthwhile to attempt to define them here. By **alignment**, most investigators refer to the process of bringing homologous chromosomes into rough apposition along their lengths. The term pairing refers to an intimate physical association of the homologs. Synapsis refers to the stage in which paired homologs are connected along their lengths by a railroad track-like structure referred to as the synaptonemal complex (SC). The SC is a meiosis-specific, tripartite structure comprising two lateral elements that flank the chromosomes and a central element. Lateral elements are derived from the axial cores of meiotic chromosomes and are initially assembled between the two sister chromatids of the leptotene chromosome and then move to lie to one side of both sister chromatids. Indeed, one can think of the lateral elements of the SC as a permutation of the cohesin-type complexes that mediate sister chromatid cohesion.

The interrelationship between DSB formation and SC assembly is complex and varies among species. Yeast mutants that completely suppress recombination also block the formation of a synaptonemal complex. In contrast, other yeast mutants can block synapsis while allowing nearly normal levels of gene conversion and only two- to threefold reductions in meiotic recombination. As noted by Roeder, in yeast, "... synapsis is not required for recombination; instead steps in the recombination pathway appear to be required for synapsis.... Mutants that do not sustain DSBs fail to make SC." Similar conclusions have been reached in *Arabidopsis* (Grelon et al. 2001) and in the mouse (Baudat et al. 2000; Mahadevaiah et al. 2001). Moreover, recombination is initiated quite normally in *S. pombe*, which does not assemble an SC, and is apparently fully sufficient to ensure segregation.

However, evidence that exchange was not necessary to support synapsis in *Drosophila* was obtained by McKim et al. (1998), who demonstrated that exchange is not required for synapsis in *Drosophila*. The *mei-W68* and *mei-P22* mutations in *Drosophila* both ablate DSB formation. However, the assembly of the synaptonemal complex appears to proceed normally in these mutants (McKim et al. 1998). This observation became all the more striking in light of the subsequent finding that the *mei-W68* gene encodes the *Drosophila* homolog of Spo11, the yeast protein required for DSB formation (McKim and Hayashi-Hagihara 1998). Similar observations were made in *C. elegans* by Dernburg et al. (1998).

How, then, are we to deal with this rather disturbing inconsistency of functional relationships? Zickler and Kleckner (1999) have suggested that all of these observations are compatible with a view "... in which commitment of a recombinational interaction to a crossover fate is tightly coupled with formation of an underlying SC patch which would then nucleate polymerization of SC ...." They explain the *Drosophila* and *C. elegans* exceptions by suggesting that strong secondary mechanisms may exist to facilitate SC formation in those organisms. We note in Box 4.2, that *cis*-acting chromosomal sites, possessing the genetic properties expected of pairing initiating or stabilization sites, have been found in both *Drosophila* and *C. elegans*. Perhaps these sites can facilitate SC formation in flies and worms, even in the absence of exchange initiation. As noted by Zickler and Kleckner (1999), the meiotic programs of various organisms may "differ only with respect to the potency of secondary SC nucleation mechanisms, which can promote SC formation in a relatively normal time frame even when the normal nucleation mechanisms are missing."

Box 4.2 Do Specific Chromosomal Sites Mediate Pairing?

As reviewed by Loidl (1990), a number of workers have proposed that various chromosomal organelles may play important roles either in initiating SC formation or in maintaining synapsis. Two types of specific pairing elements or putative pairing sites, one in *C. elegans* and one in *Drosophila*, have also been identified by genetic studies aimed at identifying sites critical for normal levels of recombination. The studies presented below suggest that both chromosomal elements and these putative pairing sites may well define elements required to establish or maintain synapsis in these organisms.

The Role of Telomeres in Early Pairing

Cytogenetic studies of the first meiotic prophase of many organisms have demonstrated the clustering of telomeres (usually in the vicinity of the centrosome or the spindle pole body) during the leptotene–zygotene transition (Chikashige et al. 1994; Dernburg et al. 1995; Chikashige et al. 1997; Scherthan 1997; Rockmill and Roeder 1998; Zickler and Kleckner 1998). Because this configuration, known as a **chromosomal bouquet** or **telomere bouquet**, precedes the initiation of synapsis, the possibility has been raised that this early localization of the telomeres to a small region of the nuclear envelope facilitates the alignment of homologous chromosomes. In addition, *S. pombe* mutants that disrupt telomere clustering reduce the frequency of recombination (Cooper et al. 1998; Nimmo et al. 1998), suggesting that telomere clustering acts to facilitate homolog alignment.

Although telomere clustering may provide a useful step in early chromosome alignments, the clustering of a large number of telomeres might also present problems in the creation of individualized bivalents. This possibility is suggested by the studies of Conrad et al. (1997) and Chua and Roeder (1997) of the *NDJ1* gene in *S. cerevisiae*. *NDJ1* encodes a protein that is required for the completion of homologous synapsis and that accumulates at the telomeres of chromosomes during meiotic prophase. Loss of the NDJ1 protein delays the formation of the axial elements of the synaptonemal complex and results in high levels of failed meiotic chromosome segregation. Moreover, there is no effect of the absence of the NDJ1 protein on the segregation of telomereless ring chromosomes, arguing that the NDJ1 protein is not required for meiotic chromosome separation per se, but rather that the NDJ1 protein is essential to separate segregational partners that have telomeres. The mechanism by which the NDJ1 protein facilitates synapsis remains unclear, but it's possible that the normal clustering of telomeres into a single bouquet might create three-dimensional chromosome arrangements, such as interlocked bivalents, that would impede proper synapsis.

The Role of Centric Heterochromatin in Chromosome Pairing

The importance of heterochromatic pairings in mediating pairing and segregation was first suggested by studies of the effects of homologous duplications on the segregation of the achiasmate chromosomes in *Drosophila* oocytes (Hawley et al. 1992). This hypothesis was verified by the work of Karpen et al. (1996) who showed that the frequency with which two deletion derivatives of a **minichromosome** (small chromosome with a centromere, telomeres, and replication origins) segregate from each other during female meiosis is directly proportional to the amount of centric heterochromatic homology shared by the pairing partners. Normal segregation was shown to require that the two minichromosomes share 800 kb of overlap in the centric heterochromatin; nearly random disjunction was observed when the two minichromosomes shared only 300 kb of heterochromatic homology. Dernburg et al. (1996) used three-dimensional fluorescence in situ hybridization to demonstrate that although euchromatic pairings not locked in by chiasmata quickly dissolve following pachytene, heterochromatic pairings are still preserved.

Specific Pairing Sites in *C. elegans*

Two types of genetic studies of chromosome rearrangements in *C. elegans* have suggested that a single region (known as the **homolog recognition region**, or HRR) exists at the end of each chromosome and that this region may play a primary role in the pairing of homologous chromosomes (McKim et al. 1988a). The central observation is that when a chromosome is split by a translocation or large deletion, only the pieces that maintain the HRR are capable of pairing in a sufficiently stable manner to allow them to recombine with their homologs. For example, only duplications that contain an HRR are capable of recombining with their intact homolog (Herman and Kari 1989; McKim et al. 1993). Similarly, in worms heterozygous for a reciprocal translocation, crossing over is limited to regions of each chromosome that are physically contiguous with the HRR elements (McKim et al. 1988b, 1993). Based on both the genetic and cytological data, HRRs have been proposed by several workers to function as major synapsis sites (Villeneuve 1994).

Specific Euchromatic Pairing Sites in *Drosophila*

Studies in *Drosophila* have also pointed to a role for internal sites or elements that facilitate stable pairing and synapsis. In her master's thesis in 1956, Iris Sandler examined X-chromosomal exchange in *D. melanogaster* females heterozygous for a normal sequence X chromosome and for one of two different translocations between the X and fourth chromosomes. She observed that the X chromosome could be divided into large intervals, such that heterozygosity for a breakpoint within one interval strongly suppressed recombination within that interval but did not suppress exchange in adjacent intervals. She proposed that so-called "pairing sites" bound these intervals and that both homologous chromosomes must be continuous between two sets of paired sites for normal levels of exchange to occur within the interval bounded by those sites.

Hawley (1980) refined these observations by analyzing 20 more rearrangements and precisely mapping the proposed sites. He also tested a series of duplications for these sites and found several genetic properties that might be expected for such pairing sites. As shown in Figure B4.1, such duplications suppress exchange only in intervals for which they carry a boundary or pairing site. Indeed, free duplications for a putative pairing site can suppress exchange throughout any interval bounded by that site, but not beyond the next pairing site(s) (Hawley 1980). In the absence of the completion of cytological studies of pairing in translocation heterozygotes in *Drosophila*, it is not possible to discern whether these sites function at the level of pairing or synapsis. However, given the *C. elegans* studies described above, we currently favor the possibility that these sites may also play a role in the maintenance of synapsis.

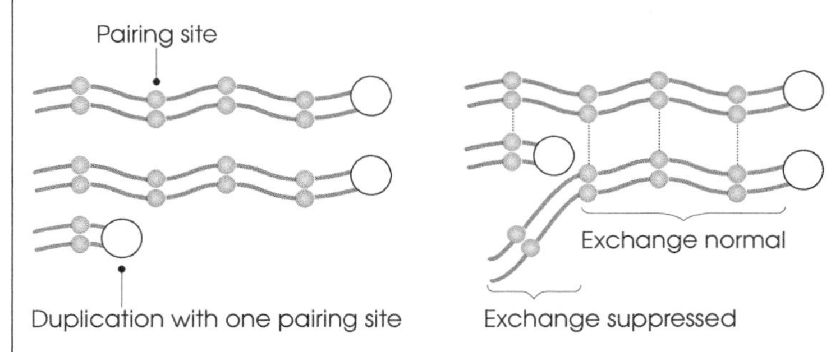

Figure B4.1 Euchromatic pairing sites in *Drosophila*. A free duplication that includes a pairing site suppresses exchange within the interval that encompasses that pairing site.

general – several excellent reviews of meiosis in a variety of organisms already exist for that purpose (Hawley et al. 1993; Roeder 1997; Zickler and Kleckner 1998, 1999; Davis and Smith 2001; Hassold and Hunt 2001; Hughes et al. 2018; Lake and Hawley 2012). However, for purposes of discussion, we do need to provide a brief summary of the meiotic process.

A Cytological Description of Meiosis

As shown in Figure 4.5, the two divisions of a generic meiosis may be divided into five steps. The term meiotic **prophase I** refers to the period after the last cycle of DNA replication. It is during this interval that homologous chromosomes pair and recombine. Pairing and recombination do not occur during any part of meiosis II. The breakdown of the nuclear envelope signals the end of meiotic prophase. The term **prometaphase** is used to describe the period during which the bivalents attach to or create the meiotic spindle and congress to the center of that spindle. **Metaphase I** is defined as the period before the first division during which the bivalents are lined up in the middle of the meiotic spindle, a position referred to as the **metaphase plate**. The chromosomes are primarily (but not exclusively) attached to the spindle at the **centromere**. The centromere of one homolog is attached to one pole, and the centromere of its partner is attached to the other pole. These bivalents are physically held together by **chiasmata**, which are the physical manifestation of meiotic recombination events. In most meiotic systems, meiosis will not continue until all of the homolog pairs are properly oriented on the metaphase plate.

Once the homologs are properly aligned, sister chromatid cohesion is released along the arms of the meiotic chromosomes but *not* in the region surrounding their centromeres. This event frees the two chromosomes in each bivalent from their chiasmate attachments, and thus allows each chromosome to proceed toward the closest spindle pole. This event is referred to as the **metaphase–anaphase transition** and initiates **anaphase I**. Because the release of sister chromatid cohesion at the metaphase–anaphase transition of meiosis I occurs *only* along the arms of the bivalent, and not at the centromeres, each homolog still comprises two sister chromatids held together by cohesion at the centromeres. In a very real sense, the key to understanding the two meiotic divisions is to realize that the release of sister chromatid cohesion is differentially regulated at the two meiotic divisions. At meiosis I, cohesion is released only along the arms, allowing chiasmata to be resolved,

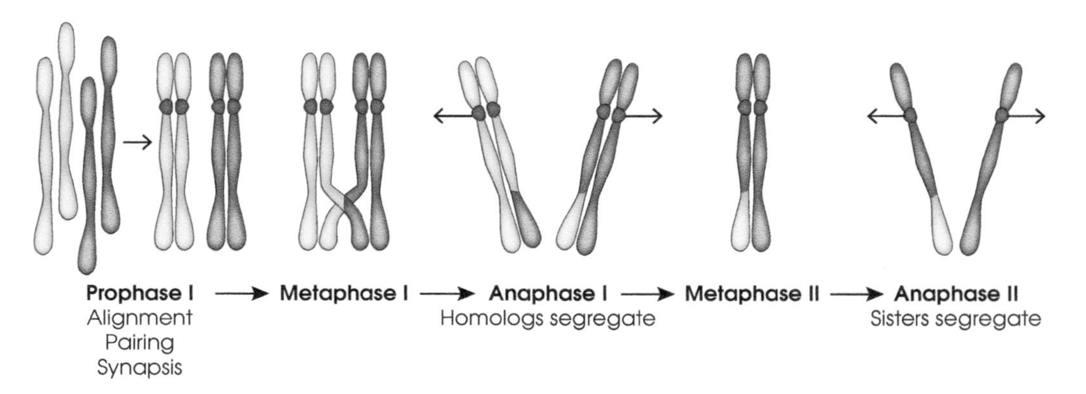

Figure 4.5 The meiotic cell cycle. During meiosis, chromosomes go through one round of genome duplication and two rounds of cell division. After replication, sister chromatids pair with their homologs and synapse during prophase I. Homologs are then locked together by recombination during metaphase I and segregate away from one another at anaphase I. Sisters segregate away from one another during the second meiotic division to form haploid gametes.

but sister centromeres remain connected in the vicinity of the centromeres. Sister chromatid cohesion around the centromeres is released during the mitosis-like meiosis II division.

Some organisms culminate the first meiotic division with a true **telophase I** (a time in which nuclei reform), while others simply proceed directly into meiosis II. In those organisms that do reform nuclei at the end of meiosis I, there may also be a brief **prophase II**. DNA replication does not occur during prophase II; each chromosome still consists of the two sister chromatids. There are no opportunities for pairing or recombination at this stage, due to the prior separation of homologs at anaphase I.

Following the completion of the first meiotic division, the chromosomes of each meiosis I product align themselves on a new pair of spindles with their sister chromatids oriented toward opposite poles. This stage is referred to as **metaphase II**. **Anaphase II** is signaled by the separation of sister centromeres, and the movement of sister chromatids to opposite poles. At **telophase II**, the sisters have reached opposite poles and nuclei begin to reform. Thus, at the end of the second meiotic division, there are four daughter nuclei, each with a single copy of each chromosome.

The thumbnail description of meiotic cytology presented above belies the enormous diversity that has been documented with respect to the meiotic process. For example, female meiosis in many (but not all) animals is acentriolar (at least for meiosis I); the chromosomes themselves organize an asterless meiotic spindle (Theurkauf and Hawley 1992). Female meiosis in *Drosophila* also has an unusual end to prophase I in which the chromosomes condense into a dense mass where they remain until the beginning of prometaphase. Finally, in yeast, the entire meiotic and mitotic processes take place without nuclear envelope breakdown. But more critically, this variation in common experimental systems is quite small given the large number of unusual and often bizarre meiotic systems documented by insect cytogeneticists in the last century (cf. Wolf 1994).

A More Detailed Description of Meiotic Prophase

Because pairing and recombination occur during the first meiotic prophase, much attention has been focused on this stage of the process (Zickler and Kleckner 1998; Woglar and Jantsch 2013). Prophase I is subdivided into five stages: leptotene, zygotene, pachytene, diplotene, and diakinesis. **Leptotene** describes the initial phase of chromosome individualization during which initial homolog alignments are made, and by **zygotene**, homologous chromosomes are associated at various points along their length. There are three critical events that occur during leptotene and zygotene. These are:

1) homolog recognition and alignment,
2) synaptonemal complex formation, and
3) the initiation of meiotic recombination.

The mature SC is a tripartite structure consisting of two lateral elements, which flank the chromatin, and a central element situated between the lateral elements. Lateral elements are derived from the axial cores of meiotic chromosomes. The function of the SC remains obscure (see Boxes 4.2 and 4.3). Although SC formation may be essential to initiate recombination in organisms such as *Drosophila* (Walker and Hawley 2000), meiotic recombination occurs quite happily in *Schizosaccharomyces pombe* in which no SC is formed (Davis and Smith 2001). In organisms that do possess SC, sites of recombination are marked along the meiotic chromosomes during pachytene by spherical structures known as **recombination nodules** that sit on top of the SC. Adelaide

Box 4.3 Crossing Over in Compound-X Chromosomes

Morgan recovered the first attached-X, or compound-X, chromosome in *Drosophila* on 12 February 1921. This chromosome, perhaps more properly designated an **isochromosome**, consisted of two full-length X chromosomes whose left arms appeared to be fused at a medial centromere. Because there are no essential genes on the very small right arm of the fly X chromosome, a female carrying only an attached-X chromosome is fully viable and fertile. Presumably, this chromosome arose by a translocation-like event with one break on the right arm of one chromosome and the other very proximal on the left arm of the other X chromosome. The rejoining then attached virtually the entire left arm of the X chromosome to the centromere of the other. We refer to this type of attached-X as a *C(1)RM* chromosome (for *compound-one-reversed metacentric*).

Other types of compound-X chromosomes exist and their recombinational properties have also been studied in detail (Lindsley and Sandler 1958), as have those of compound chromosomes carrying two copies of the four autosomal arms in *Drosophila* (Holm 1976). However, for our purposes here we need only consider the original type of compound-X chromosome. This chromosome segregates as a univalent at meiosis in females with no other sex chromosomes, and thus it segregates to a single pole at anaphase I. At meiosis II, the two sister chromatids comprising the compound chromosome divide normally, transmitting a full copy of the compound-X (as a single chromatid) to each daughter cell. Thus, the compound is transmitted to only half of her ova; the remaining ova receive no X-chromosomal material.

We are concerned here with the case in which the two arms used to construct the compound-X chromosome were heterozygous for multiple genetic markers (Figure B4.2). Note that a single crossover between the arms of the compound-X chromosome can result in the formation of a crossover product that is homozygous for all markers distal to that crossover event (Anderson 1925; Beadle and Emerson 1935). This fact alone demonstrates that exchange must occur at the four-strand stage. Exchange at the two-strand stage would not result in homozygosity.

One can thus easily detect those cases where a given mutant marker has become homozygous. But how does one distinguish homozygosity of the wildtype markers from retention of heterozygosity? The ability to passage the attached-X chromosome in question through meiosis in several subsequent generations of females allows us to fully determine the genotype of that chromosome. If a marker on the compound-X is still heterozygous, it will eventually be revealed as a homozygote by a crossover event that makes it homozygous. This process of genotyping an attached-X after one passage through meiosis allows us to determine the type of tetrad from which it arose. This then allows us to perform a reasonably complete tetrad analysis.

The frequency with which any given recessive marker can be made homozygous will increase with its distance from the centromere. (This observation allowed the first accurate mapping of the X centromere in *Drosophila*.) If there is no sister chromatid exchange, then the maximum frequency of homozygosity ought to be 25%. However, if sister chromatid exchange were to

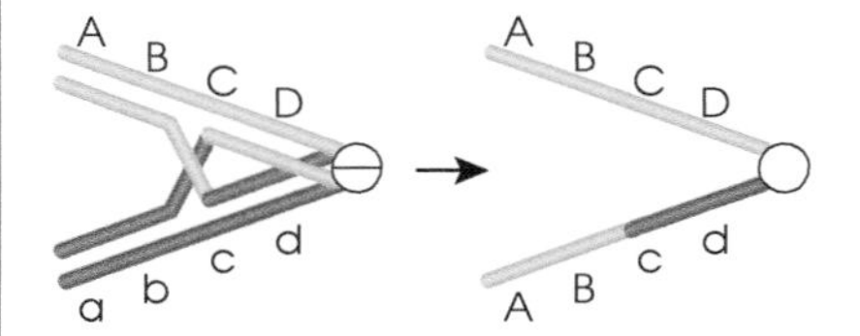

Figure B4.2 Exchange between the arms of an attached-X chromosome can result in homozygosity for distal markers.

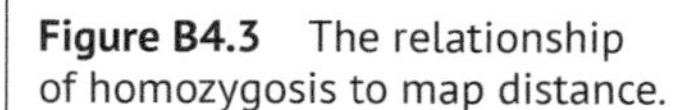

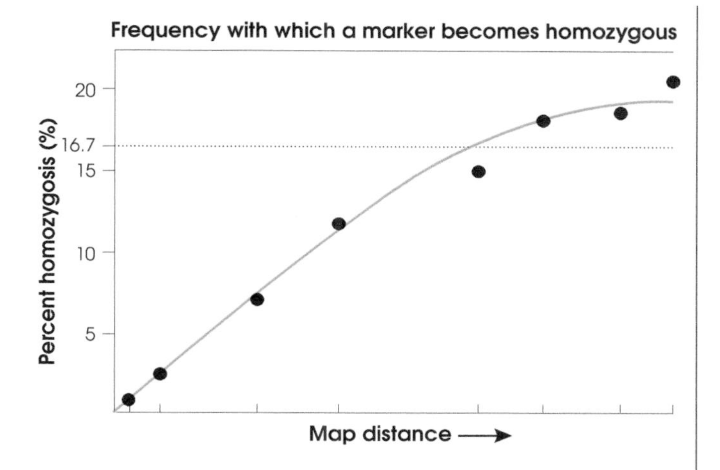

Figure B4.3 The relationship of homozygosis to map distance.

occur with the same frequency as interhomolog exchange, then the maximum frequency of homozygosity would be 16.7% (see Figure B4.3). As seen in the figure, the frequency of homozygosity for distal markers clearly and repeatedly exceeds 16.7% and approaches the maximum value of 25% expected if no sister chromatid exchange is allowed. Then, as expected from the accumulation of double and higher levels of exchanges, the frequency of homozygosity of recessive markers will decline toward a final value of 16.7%.

The logic here is identical to that presented in the text for the relationship of second division segregation frequencies to map length. Over long distances, the double (and higher level) exchanges will effectively randomize the order of the four alleles of each marker within the attached-X. So, if you have four alleles of a very distal gene, *AAaa*, and assign the *a* allele to the first of the four chromatids, the odds that the other arm of the same attached X chromatid would pick up the *a* allele are only 1/3. For this reason, the maximum possible frequency of marker homozygosity (both *AA* and *aa*) is going to be 33.3%, and recessive homozygotes will only account for half of these. The critical point here is that the medial values in this curve did exceed 16.7%. This would not be expected if sister chromatid exchanges were present along with interhomolog exchanges.

Carpenter discovered recombination nodules in the 1970s and demonstrated that their number and distribution parallel that of exchange events (1975, 1979a, b, 1981, 1984). Evidence that recombination nodules do indeed mature into chiasma is reviewed by Wettstein and Rasmussen (1984), Hawley (1988), and Carpenter (1987). Although most components of the recombination nodule remain elusive, a few have been identified, such as MLH1p in mouse (Moens et al. 2002) and Vilya and Narya in *Drosophila* (Lake et al. 2015, 2019).

The beginning of **pachytene** is signaled by the completion of a continuous SC running the full length of each bivalent, and its end is signaled by the dissolution of the SC. During **diplotene**, the attractive forces that mediated homologous pairing disappear, and the homologs begin to repel each other. At this stage in most organisms, homologs are held together only by their chiasmata. Indeed, in organisms without a backup system for recombination, those rare chromosome pairs that have failed to undergo recombination will prematurely fly apart from each other at this stage. The final stage in meiotic prophase is **diakinesis**, during which the homologs shorten and condense in preparation for nuclear division.

4.2 Crossing Over and Chiasmata

Recombination involves the physical interchange of genetic material and ensures homolog separation. Chiasmata can be visualized at diplotene–diakinesis as sites on the bivalent in which two nonsister chromatids appear to cross over from one homolog to the other. At the beginning of the twentieth century, there were two views of the origin of chiasmata. According to the **chiasmatype hypothesis** of Janssens (1909), chiasmata were the cytological manifestation of genetic exchanges (McClung 1927). The competing **classical hypothesis** of Sax (1930, 1936) said that chiasmata were simply sites at which chromosomes had traded sister chromatid regions without breakage and reunion. In these hypothetical structures, the chromatid-swapping events could be resolved by exchanges but were not the consequences of them. These two models are portrayed in Figure 4.6. There are many proofs that the chiasmatype hypothesis is correct, the most compelling of which involved the cytological analysis of bivalents that were in some way **heteromorphic** (Fu and Sears 1973; Whitehouse 1982). For example, the two heteromorphic homologs might differ by a heterochromatic knob or a block of heterochromatin at their tips. The two models make very different predictions for the structure of such chiasmate bivalents, and it is the structure consistent with the chiasmatype hypothesis that is always observed.

The chiasmatype model predicts that exchanges will result in the physical interchanges of homologous material. Thus, one predicts that the two chromosomes produced by this type of exchange will be heteromorphic after the bivalent has separated at anaphase I (Figure 4.6). Each chromosome that experiences a crossover proximal to the heteromorphic marker will then possess a chromosome of each type (i.e. knobbed or knobless). Exactly such heteromorphic dyads have been observed at anaphase by a number of researchers (see Maguire and Riess 1991). Finally, the inheritance of interchanges of distal knobs and heterochromatic blocks following exchange was documented by Creighton and McClintock (1931) in maize and by Stern (1936) in *Drosophila*

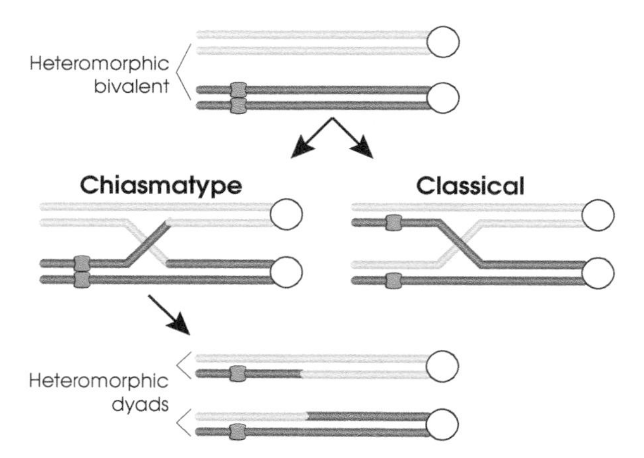

Figure 4.6 Classical versus chiasmatype hypotheses. In the chiasmatype hypothesis, chiasmata form as a consequence of crossing over between two nonsister chromatids. The classical hypothesis, on the other hand, assumes that chiasmata form without chromatid breakage and repair; furthermore, chromatids that have formed chiasmata may or may not be resolved as crossovers. Examining heteromorphic bivalents that either have or lack a knob of distal heterochromatin, for example, can be used to follow the inheritance of physical exchanges. The chiasmatype hypothesis corresponds with the observed 1:1 ratio of chiasmata to crossovers, while the classical hypothesis does not. A heteromorphic bivalent thus makes two heteromorphic dyads at anaphase I.

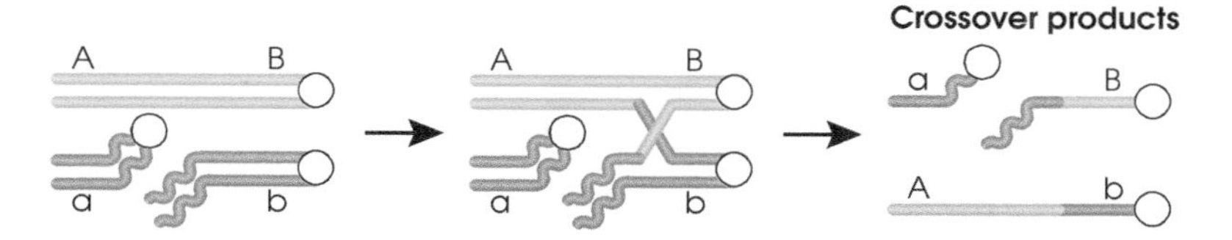

Figure 4.7 **Crossing over.** Crossing over involves the physical exchange of genetic material between the two homologs – not only the genetic markers (A and B), but the aberrations as well.

(Figure 4.7), which together demonstrated unambiguously that chiasmata were the result of exchange. But it was the work of Nicklas (1974) that demonstrated exchanges serving the vital function of ensuring homolog separation at the first meiotic division; these experiments are described in detail in Section 9.4.

4.3 The Classical Analysis of Recombination

Larry Sandler began his Chromosome Mechanics class at the University of Washington by handing out a document he referred to as *The Ten Commandments of Crossing Over*. A modified and annotated version of this list, presented below, provides an excellent summary of the formal "rules" of meiotic exchange.

1) **In each chromosome, the genes are arranged in a linear series, and corresponding groups of genes are exchanged in crossing over.** This is self-evident to the modern biologist. To the early geneticist, however, this conclusion derived from the fact that it was possible to map three genes in a linear order (three-factor mapping).
2) **Exchanges are complementary and involve the physical exchange of material.** The evidence for this assertion was presented in Section 4.2.
3) **Exchanges occur when each chromosome consists of exactly two chromatids.** This is to say that recombination occurs after DNA replication. There are proofs of this assertion in many organisms, most notably in Neurospora (Perkins 1962) and in *Drosophila* (Anderson 1925; Beadle and Emerson 1935). The evidence in *Drosophila* is based on a truly elegant analysis of recombination within compound-X chromosomes and is presented in Box 4.3.
4) **Each exchange event involves only two chromatids, one from each chromosome.** This is a critical rule because it reminds us that a bivalent with a single exchange will produce an equal number of crossover (two) and noncrossover (two) chromatids (see Figure 4.3).
5) **A given bivalent may undergo more than one exchange event, but for each such event; crossing over is limited to two chromatids, again, one from each chromosome.** This statement means that a single paired bivalent may be involved in more than one exchange event. Thus, single, double, and triple, etc. crossover products are possible. However, any given exchange can involve only two chromatids, one from each chromosome.
6) **In bivalents with two or more exchanges, the choice of the two chromatids that crossover in one exchange does not affect the choice of which two chromatids participate in the other exchange(s).** In other words, chromatid choice is random (except for sister exclusion) for each of the exchanges. This rule is often referred to as the "no chromatid interference" rule. The easiest way to understand this rule is to look at the three classes of double crossover bivalents shown in Figure 4.8. In the uppermost bivalent, referred to as a

Figure 4.8 The three classes of double crossovers. Only two chromatids are involved in a two-strand double, a three-strand double involves three chromatids, and a four-strand double involves all four chromatids.

two-strand double, both crossovers involve the same two chromatids. In the two middle bivalents, referred to as a **three-strand double**, one chromatid is involved in both crossover events. In the lower bivalent, a **four-strand double**, different pairs of chromatids are involved in each of the two crossover events. One might imagine that if a given chromatid was involved in the first exchange event, it might be more or less likely to participate in the second. In fact, this is not at all the case. There is no chromatid interference or preference in the second event, as shown by the fact that the ratio of two-, three-, and four-strand doubles is very close to the predicted ratio of 1 : 2 : 1. As described in Appendix C, this has been clearly and directly demonstrated in Neurospora (Perkins 1962) and flies (Beadle and Emerson 1935).

7) **Meiotic recombination does not occur between sister chromatids, or if it does, it does not interfere with recombination between homologs.** Because exchange between sister chromatids generally has no genetic consequences, it has been difficult to determine whether it occurs meiotically and, if so, how often. Studies designed to examine sister chromatid recombination during meiosis are presented in Box 4.4.

Box 4.4 Does Any Sister-Chromatid Exchange Occur During Meiosis?

Throughout this chapter we use multiple lines of evidence to demonstrate either that sister chromatid exchange does not occur or that if it does occur, it does not compete with interhomolog exchange. Still, this does not answer the question: does sister chromatid exchange occur at a measurable frequency during meiosis. During the last several decades, workers have approached this question in *Drosophila* and yeast by genetic means, and Kleckner and her collaborators have addressed the question by direct molecular approaches in yeast (Schwacha and Kleckner 1994). Both approaches suggest that while sister chromatid recombination events do indeed occur, they are substantially less common than interhomolog recombination events.

Genetic Studies in Yeast

As shown in Figure B4.4, sister chromatid exchange that occurs within a ring chromosome is detectable because it produces a dicentric chromosome that is virtually always lost at the second meiotic (or first mitotic) division. Exchange between a ring and a normal sequence linear homolog can also create dicentric products. These two classes of events are easily distinguishable in an organism like yeast where one can recover all four products of meiosis. Sister chromatid exchange involving the ring will produce two viable spores, both carrying the linear chromosome. Exchange between the ring and the rod results in one viable rod and one viable ring-bearing spore. Based on such studies, Haber et al. (1984) estimated that the ring chromosome used in their studies underwent sister chromatid exchange in approximately 15% of cells. This estimate of the frequency of sister chromatid exchange in a ring was confirmed

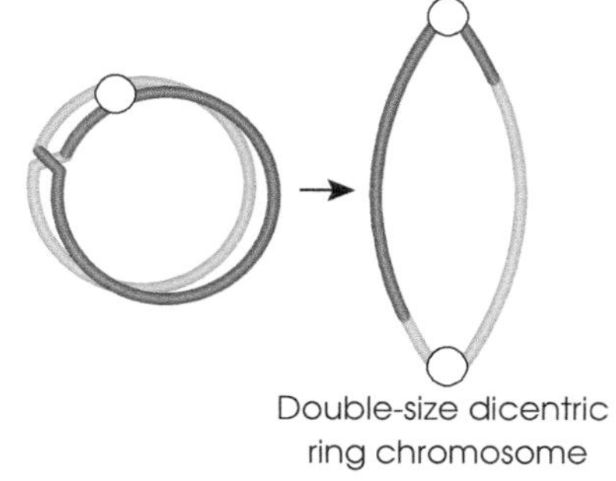

Figure B4.4 Sister chromatid exchange in a ring chromosome produces a dicentric ring.

by Game et al. (1989), who used pulsed-field DNA gels to detect the dicentric chromosomes produced by sister chromatid exchanges involving a ring chromosome in yeast. While this frequency of sister chromatid exchange may seem high, the per meiosis frequency of sister chromatid exchange for this chromosome (0.15) is substantially lower than the per meiosis frequency of interhomolog exchange (1.7).

Genetic Studies in *Drosophila*

Similar ring loss experiments have been reported in *Drosophila*. Using the null recombination mutant $c(3)G^{17}$ and the ring X chromosome [*R(1)2*], Jeff Hall observed a ring/rod ratio of 0.755 (*N* = 7,552) in controls and an elevated ratio of 0.894 (*N* = 5,355) in *c(3)G* females (Hall 1972). More recently, McKim et al. (1998) measured ring loss by crossing females carrying the $R(1)w^{vc}$ ring X chromosome and *FM7* (a multiply-inverted X chromosome balancer that strongly suppresses interhomolog exchange) to appropriate tester males. They assayed ring loss (presumably dicentric formation) by comparing the frequency *of* $R(1)w^{vc}$-bearing female progeny with that of the corresponding *FM7*-bearing sisters. In the case of otherwise genetically normal females, the ring/rod ratio was 0.854 (*N* = 2,731). However, when the investigators did the same experiment in females that lacked Mei-P22, a fly protein required for recombination, the ring/rod ratio increased to 1.079 (*N* = 659). Thus, blocking the initiation of recombination seemed to increase the frequency of ring recovery by approximately 15%. The background frequency of sister chromatid exchange in meiosis is seen by a reduced recovery of ring chromosomes in control experiments. Because both *mei-P22* and *c(3)G* inhibit sister chromatid exchange events, it is reasonable to assume they would increase the transmissibility of the ring X chromosome.

A Direct Molecular Assessment in Yeast

We have already noted that Schwacha and Kleckner (1995) succeeded in recovering double Holliday recombination intermediates, known as **joint molecules** (JMs), produced during meiosis in yeast. Subsequently, Schwacha and Kleckner (1997) also examined the relative frequency with which such molecules involved homologs versus sister chromatids. Their data suggested that interhomolog recombination events outnumbered sister chromatid events 2.4–1. Of perhaps equal interest is the observation that several yeast recombination mutants impaired this preference for interhomolog events. The authors concluded that "... most meiotic recombination occurs via an interhomolog-only pathway along which interhomolog bias is established early, prior to or during DSB formation, and then enforced, just at the time when DSBs initiate JM formation. A parallel, less differentiated pathway yields intersister and, probably, a few interhomolog events."

8) **There is crossover (or regional) interference.** In many meiotic systems, the occurrence of an exchange in one interval interferes with the occurrence of other exchanges in neighboring intervals (Copenhaver et al. 2002; Housworth and Stahl 2003; Hillers 2004; Stahl 2012). In some organisms or on some chromosomes, this interference can be strong enough to limit exchange to approximately one crossover per bivalent. Interference usually decreases with distance before finally vanishing (Berchowitz and Copenhaver 2010). The mechanism of regional or crossover interference remains a mystery, although many models have been proposed and there are many useful reviews (Foss et al. 1993; Foss and Stahl 1995; Zickler and Kleckner 2016; Otto and Payseur 2019; Wang et al. 2019; Smith and Nambiar 2020; von Diezmann and Rog 2021; Pazhayam et al. 2021). Both the mechanism(s) underlying crossover interference and its function remain unclear, but there is evidence in *Drosophila* that it does not occur at the time DSBs are made (Crown et al. 2018).
9) **The frequency of recombination is not simply proportional to unit length of DNA, but rather is under tight genetic control from both *cis*- and *trans*-acting elements.** As shown in Figure 4.4, the amount of exchange per chromosome arm is not uniform. Exchange is reduced in the centromere-proximal and distalmost euchromatin due to centromere and telomere effects and is absent in heterochromatic intervals. However, even within small euchromatic intervals, careful studies of sites of exchange initiation reveal obvious hotspots and cold areas for DSB formation in many organisms (Lichten and Goldman 1995; Kirkpatrick et al. 1999; Hey 2004; Manzano-Winkler et al. 2013).
10) **Exchanges are generally sufficient to ensure the segregation of two chromosomes, even if their centromeres are nonhomologous.** Although very distal exchanges do not always ensure segregation (Carpenter 1973; Rasooly et al. 1991; Koehler et al. 1996; Lamb et al. 1996; Ross et al. 1996; Thomas and Hassold 2003), in the vast majority of cases, one meiotic exchange will ensure that two chromosomes segregate from each other. This is true even in those cases (such as in a translocation heterozygote) where the homologous intervals that participated in the exchange are attached to nonhomologous centromeres.

4.4 Measuring the Frequency of Recombination

There are multiple ways to determine the frequency of meiotic recombination. Perhaps the most direct approach is to simply count chiasmata (Lawrie et al. 1995; Barlow and Hultén 1998). Unfortunately, such estimates are not possible in all organisms and they do not usually permit the mapping of specific markers along the length of chromosomes. Such genetic, or recombinational, mapping requires the recovery and assessment of crossover chromatids and the calculation of recombination distances (more commonly referred to as **map lengths**). This section discusses the tools required for using the frequency of crossover chromatids to do exactly that.[3]

3 A brief note on terminology: Unfortunately, most geneticists now use the terms recombination, exchange, and crossing over interchangeably (no pun intended) to describe the process of meiotic recombination. Because we are as guilty of this as anyone, we cannot and will not try to change that practice here. But we will reserve the term "crossover chromatid" to mean *only* a chromatid that carries the nonparental combination of flanking markers.

determine directly the number of exchange events per bivalent. This method is referred to as **tetrad analysis**. Oftentimes, however, only one chromatid is recoverable. In these cases, tetrad analysis must be done algebraically.

Algebraic tetrad analysis is used to infer the frequency of recombination on the chromatids you did not recover. For example, in human females, three of the four products of meiosis are discarded as polar bodies. The chromatids in those polar bodies may have undergone recombination, but because they are difficult to recover you can't directly measure the rate of recombination on those chromatids. Your only readout is from the one chromatid you did recover. Algebraic tetrad analysis is the back-calculation of the number of single, double, and triple crossovers that occur during each meiosis. A detailed discussion of tetrad analysis can be found in Appendix C.

Statistical Estimation of Recombination Frequencies

The human geneticist mapping human genetic loci (locations of genes, markers, phenotypes, etc.) is often faced with using statistical tools to estimate the map distance between a given set of markers or between a marker and a locus. In such cases, the available data may sometimes come from large families showing simple Mendelian inheritance, but frequently such studies pool information from multiple small families or even groups of affected relative pairs. In traditional **parametric linkage analysis**, assumptions about the mode of inheritance are included in the evaluation of the results. In **nonparametric** forms of analysis (e.g. analytical approaches used in studying pooled data on sibling pairs), the calculations are carried out in a way that allows for evaluation of the data even if your assumptions about the mode of inheritance are wrong or nonexistent.

Two-Point Linkage Analysis

Two-point linkage analysis evaluates apparent linkage between two points in the genome. Frequently, when attempting to map a phenotype being transmitted in a family, what is being evaluated is linkage of a single marker (which can be a DNA marker such as a SNP or a microsatellite, a blood type, or whatever other marker you can measure) and a locus (the region of the genome that is responsible for the phenotype that is being mapped). Conceptually, what takes place is a test of the hypothesis that two items are linked and will be transmitted together between generations more frequently than would happen if they were unlinked (this implies that the phenotype and the maker are on the same chromosome). We can consider such a test of linkage for two different traits (e.g. blond hair and blue eyes), for two different DNA-based markers (e.g. D1S210 and D1S452), for two non-DNA markers (e.g. a blood group and a tissue transplantation antigen), for two different disease phenotypes (e.g., aniridia [absence of the iris] and Wilms tumor [a cancer of the kidney that typically occurs in children]), or for any combination of two items from this list. Realize, however, that if the geneticist is after the location of a disease gene, then what will usually be tested is the hypothesis of linkage between a particular allele of a genetic marker (again, these days that would be a DNA-based marker such as a SNP or a microsatellite repeat marker) and the presence of the disease.

One of the key pieces of information that tells us whether the two items are linked is the apparent distance between the two items. An allele of a marker that is located very close to the locus being mapped will nearly always co-segregate with the locus, giving an apparent marker–locus distance of zero. A marker and a locus that are located on different chromosomes (or far apart on the same chromosome) will segregate independently as shown by 50% recombination between the marker and the locus. The distance between the marker allele and the locus is called the recombination fraction and is typically denoted by r or theta (θ). Note that we now may know what the real distance is as long as we know what the marker and the gene are. But, even today we may not know exactly where our gene of interest is, so we can only estimate what the distance appears to be based on the small slice of information we can see in this particular family or data set.

We need to clearly state here that the value of θ that we arrive at by the time we are done is a theoretical rather than a real value, so, as in statistics, we use the symbol $\hat{\theta}$ (theta hat) to indicate this as our best estimate of the real value. In carrying out linkage analysis, we will test many possible values of θ in an effort to determine which value of $\hat{\theta}$ is closest to the real (but unknown) value of θ.

To obtain the best estimate of $\hat{\theta}$, we need to know which values of θ are most consistent with the data that we have in hand. This will require a set of statistical tests. These tests are designed to "model" the inheritance of two markers for any given value of theta. Using that model, we can ask how likely it would be to obtain a given dataset for that value of theta. This model can then be compared to other possible models generated using different values of theta. Such a comparison allows us to determine the best model (the one that gives the best fit to the data) and to determine the statistical likelihood that any given model is correct.

The key piece of information we need if we are going to assess the significance of any particular value of θ, or the likelihood that it is the best fit to the observed data, is a measure called the **LOD score** (indicated by the symbol Z). As we test each value of θ across the range from 0.00 (complete linkage) to 0.50 (unlinked), we obtain a LOD score for each value of θ. For any particular marker, we end up considering the value of θ that gives us the highest LOD score (also sometimes called maxLOD) to be the most statistically significant indicator of the real value of θ. Thus, what we actually do is test a range of values across the entire interval of possible values (from completely linked to unlinked) and select the value of θ that is associated with the highest LOD score for that particular marker.

The term LOD comes from the phrase "logarithm of the odds ratio" or "log odds." The odds ratio effectively compares the probability that the two items are linked to the probability that they are unlinked. Taking the $\log_{10}$ of this ratio then provides values that are easier to talk about and compare than the values resulting from the actual odds ratio calculation, which can sometimes get to be very large and cumbersome to deal with. One of the big advantages of dealing with logarithmic values instead of the odds ratio itself is that when data on several different families are obtained, the LOD scores can simply be added together to give a combined LOD score.

$$\mathrm{LOD} = \log_{10}\left(\frac{\text{the probability of obtaining the observed data if the two genes are linked with are combination frequency of } \theta}{\text{the probability of obtaining the observed data if the two genes are unlinked } (\theta = 0.5)}\right)$$

The odds ratio in this case compares the odds (or probability) that the marker and locus are linked (close together on the same chromosome) to the odds that the marker and locus are unlinked (on separate chromosomes or far apart on the same chromosome). In doing this comparison, we have to test specific distances between the items, such as testing for whether they are so close together that only 1% of meiotic events show recombination ($\theta = 0.01$) or linked a bit less closely so that 20% of meiotic events show recombination ($\theta = 0.20$).

If there are no recombination events observed, then the observed value for θ is taken to be zero. If $\theta = 0.00$, then the LOD score can be calculated from the simple formula:

$$Z = n\log(2)$$

where n is the number of informative meioses in that family. Interestingly, one of the things we can see from this equation is that in the case where there are no recombination events, each individual contributes about 0.301 (that is to say, $\log_{10} 2$) to the LOD score.

If the recombination fraction is a value other than zero, then the LOD score is calculated using the formula:

$$Z = n\log(2) + k\log(\theta) + (n-k)\log(1-\theta)$$

where n is the number of informative meioses and k is the number of recombinant individuals. When we look at these two equations, we see that the LOD scores will potentially be quite different if we test 2 individuals rather than 20 or 200. The LOD scores will be potentially higher when more of the tested meioses are informative. The LOD scores will be lower if there are more recombinants. In fact, when two families of the same size are compared, the one with more recombinants will have a lower LOD score.

These equations really apply only to the simplest situation of autosomal dominant inheritance in a two-generation family (parent plus children) in which we have complete information available for both markers and phenotypes for every single person screened. As soon as any of those conditions change, including more complex family structure, missing information, or a different mode of inheritance, the computations become much more complex. Most linkage calculations these days are carried out using computer programs. Those who are interested in learning more about LOD score calculations in more complex circumstances could consult Ott (2001).

But what about this phrase **informative meioses**? The number of people we test is not actually the critical determinant of the value of n. What actually determines n is how many of the people tested gave us information that told us whether they are recombinant or nonrecombinant, which requires that we be able to tell which of two copies of the marker got passed along from the affected parent. Two different markers can give very different LOD scores for the same recombination fraction if one of the markers is less informative.

An example of an uninformative situation would be a family in which both parents and all children were homozygous for the same allele of a marker. (Note that most of the markers that actually get used have been selected to have properties that should avoid this eventuality, but it still happens from time to time. Also, it is now easier to analyze millions of SNPs using next-generation sequencing.) Another situation in which reduced information is obtained happens when both parents are heterozygous for the same alleles, in which case they can produce heterozygous (uninformative) children in addition to homozygous children. To see how this works, look at Figure 4.11 to see that child A received allele 2 from her father and allele 2 from her affected mother, which makes her informative since we can tell which allele the affected parent gave her. Child B received allele 7 from his affected mother and allele 7 from his father, once again informative. Child C received one copy of allele 2 and one copy of allele 7, but we cannot tell whether allele 2 came from the affected mother or from the father. The same situation holds for allele 7. The result is that we cannot tell which allele the affected mother passed to child C. So even though we tested three children in this family, only two of them are informative. Thus, this marker will produce a lower LOD score than a different marker that is fully informative in all three children.

Unfortunately, unlike researchers lucky enough to be working with flies or yeast, the human geneticist cannot always simply decide to generate larger numbers. Geneticists studying human beings also face other limitations to their experimental design that are not present in most of the

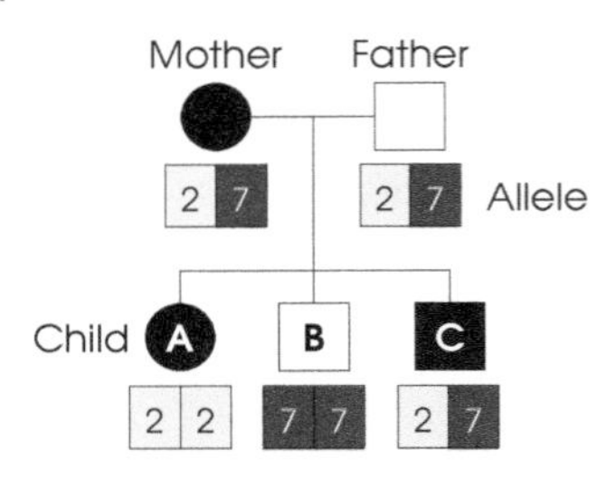

Figure 4.11 Informative meioses. In this partially informative family, filled symbols represent affected individuals and open symbols represent unaffected individuals. Individuals A and B are both informative because we can tell which allele they received from each parent, and specifically, we can tell which allele they received from the affected mother. Individual C is uninformative because we cannot tell whether his mother passed him allele 2 or allele 7.

model systems being used in genetics. One of the most common problems in human pedigrees is missing information. In some cases, this is because members of the pedigree are deceased, some of which may be critical relatives that connect some members to the rest of the pedigree, and in other cases, a family member may decide that they do not want to participate in the research project. In these instances, the computations become more complex. Among other things, in cases where someone is not available for screening with genetic markers but clearly is still a part of the pedigree structure, part of the computational process attempts to recreate what the genotype of that person might have been. Sometimes, more than one alternative genotype is possible, and the computation must be retried with each of the possible genotypes considered. One of the key pieces of information that are often needed on such occasions will be the frequencies of the different alleles in the population. This matters because an allele that is very common in the population will be much more likely to be part of the missing person's genotype than an allele that is rare. Another key issue here is that there are many genetic markers for which the allele frequencies are different for different populations, so it is important to know whether the allele frequencies being considered have been derived from a relevant control population.

In still other cases, the missing piece of data may be medical information needed to confirm the disease phenotype, such as in cases where a relative has not had some special test done that would provide the conclusive diagnosis or when someone died before they were old enough to show signs of the trait. Often, pieces of medical information are missing for spouses who marry into a family but are not in the line of descent. Can we simply consider them to be unaffected and not at risk? No, they actually also have a level of risk that can be evaluated from our knowledge about the frequency of the disorder in the population. Thus, one of the additional complications to the LOD score computation is the need to know something about how common the disease is in the population, and once again it is important that the disease frequencies being used come from a population of similar origins to the person being "reconstructed."

What Constitutes Statistically Significant Evidence for Linkage?

The standard in the field is that a LOD score of 3.0 or greater is taken as providing highly significant evidence in favor of the linkage of two items at a distance indicated by the value of θ associated with that LOD score. When a very large number of tests for linkage are carried out, issues of multiple comparisons arise (this refers to the probability of obtaining a chance result because one is testing so many markers), and in such cases, a LOD score higher than 3.0 might be required. In certain cases, a LOD score lower than 3.0 may also be allowed as indicating significant evidence of linkage. For instance, for a disease that is already known to be X-linked prior to carrying out any marker testing, a LOD score lower than 3.0 could sometimes be taken as an indication of significant evidence of linkage to a marker on the X chromosome.

Because the LOD score is $\log_{10}$ of the odds ratio, a LOD score of 3.0 indicates 1000-to-1 odds in favor of linkage. The standard in the field for determining that two things are not linked is that a

LOD score of −2.0 or lower indicates statistically significant evidence against linkage. A LOD score of −2.0 translates to odds of 100-to-1 against linkage.

An Example of LOD Score Analysis

Let's go through a simple example of calculating the LOD score for a family with a predisposition to cancer, but no gene has been identified as the cause. To simplify things, let's assume that the only other piece of information we have is the blood type of each of the family members and that the tumor phenotype is 100% penetrant before the age of 30. We realize this is a somewhat contrived example, especially in the era of simplified genotyping via microarray or genome sequencing, but teaching the fundamentals is what we are concerned with here.

Let's construct a family tree in which a parent with blood type AB had a tumor in their 20s while the other parent is type O (meaning they are OO) with no history of cancer. They have 10 children, and six have blood type A (they are AO) and four are blood type B (they are BO). Five of the six with blood type A had a tumor before they were 30, so they have the phenotype of interest, meaning one individual with blood type A did not have the phenotype. Furthermore, all four children with blood type B do not have any evidence of tumor. This suggests that the tumor phenotype is linked with blood type A, but the question is, how closely is it linked?

If there were no linkage between blood type and tumor phenotype then half of the offspring with blood type A would have the tumor phenotype and half with blood type B would have the tumor phenotype. This would make the probability of inheriting both genes 50%, or 0.5. We also need to determine the frequency of recombination between blood type and our unknown tumor gene. We can assume there is one recombination event between blood type and the tumor-causing gene in the family, which means our recombination frequency between these genes is 1/10, or 0.1 – a genetic distance of 10 cM. This makes the probability of a nonrecombinant event 0.9, or 1–0.1.

Finally, we have to divide the recombination frequency, nonrecombination frequency, and frequency of independent segregation by half, because any recombination event affects two of the four chromatids during meiosis. We can now determine the LOD score for this trait of 1.6.

$$\log_{10}\left(\frac{0.45^{9} * 0.05^{1}}{0.25^{10}}\right) = 1.6$$

Multipoint Linkage Analysis

A **multipoint linkage analysis** is calculated in a way that takes information from several adjoining markers into account when evaluating the theoretical separation between two items being mapped. This analysis will also produce a LOD score, but the analysis may come out with its maximized LOD score at a recombination fraction near to but not exactly the value of θ for which the two-point LOD score is maximized. Results of such a multipoint analysis may sometimes be displayed graphically, which makes the probable location of the locus easy to visualize. As shown in Figure 4.12, the bottom of the graph marks the positions of the various markers along the chromosome relative to the estimated most likely position of the locus, which is placed in the center at the position marked with zero. As you can see, the highest LOD scores on the graph occur near the center. Marker 3 provides the highest multipoint LOD score, but it is only slightly higher than that found for several markers on either side. This kind of multipoint graph is usually easy to interpret – look for the mountain peak and you have the location of the marker that is providing the strongest evidence for linkage.

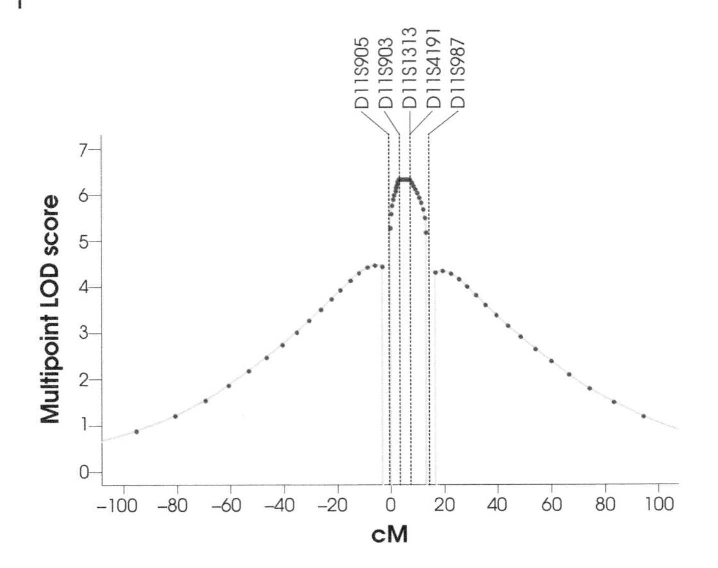

Figure 4.12 Multipoint linkage analysis. In this example after Othman et al. (1998), dots represent markers for which multipoint LOD scores were calculated to determine linkage between *nanophthalmos* (*NNO1*) and markers surrounding this locus. The marker map (D11S905 – 3.8 cM – D11S903 – 3.8 cM – D11S1313 – 0.1 cM – D11S4191 – 7.0 cM – D11S987) indicates the markers with the highest LOD scores, which show the strongest evidence for linkage. *Source:* Othman et al. (1998)/with permission of Elsevier.

Local Mapping via Haplotype Analysis

Once a gene has been mapped to a particular region of a chromosome, studies of additional families and more markers in the immediate vicinity of the locus can assist with narrowing the region thought to contain the gene that is being sought. Another multipoint approach to evaluating co-transmission of human marker information for markers located close together on the same chromosome is through **haplotype analysis**. By displaying markers geographically arrayed as they are located on the chromosomes, we can evaluate in more detail which alleles are located on the same strand of DNA together and likely to be transmitted together to the next generation.[6]

We are conveniently able to represent the alleles at multiple markers along a chromosome by placing the alleles in a column in the order in which the markers occur along the chromosome, as in Figure 4.13. In this family, we want to identify which alleles are transmitted together on a strand of DNA that contains the disease gene. As we can see in Figure 4.13, affected individuals always have allele 9 at marker 3, allele 6 at marker 4 and allele 2 at marker 5, but never have allele 4 at marker 3, allele 1 at marker 4 or allele 1 at marker 5. This suggests there is a region of contiguous DNA containing alleles 9, 6, and 2 on one chromosome and alleles 4, 1, and 1 on the other. The affected father has passed along an **affected haplotype** (a linked set of alleles characteristic of chromosomes passed to affected individuals) consisting of alleles 9–6–2 at markers 3, 4, and 5 to his affected children and an unaffected haplotype consisting of alleles 4–1–1 at those same markers to his unaffected children.

6 More information on haplotypes and efforts to catalog commonly transmitted haplotypes can be found at the HapMap Project (International HapMap Consortium 2005).

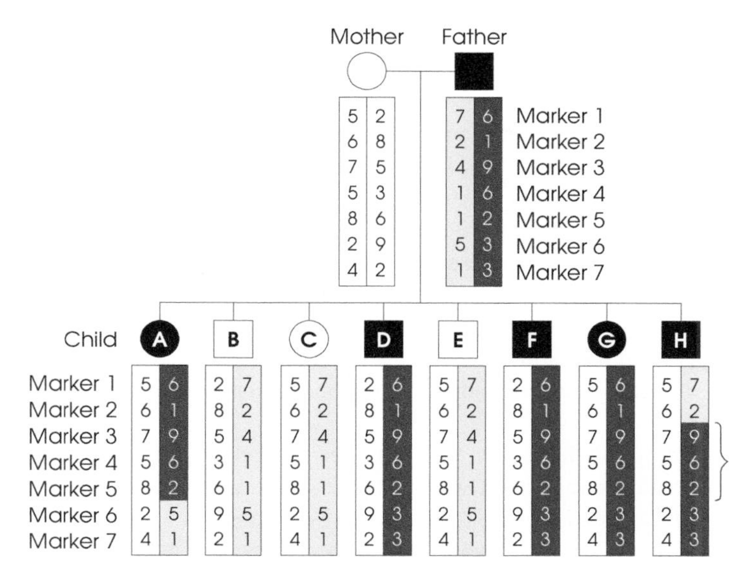

Figure 4.13 Haplotype analysis. Circles and squares denote individuals in the family; filled symbols represent individuals affected with the disease, which the family history indicates displays autosomal dominant inheritance. Beneath each circle or square are two columns diagramming the two copies of the chromosome present in that individual, with the seven marker alleles arranged along the haplotype box in the same order in which they sit on the chromosome. Dark-shaded boxes with white numbers show the alleles that co-segregate with the disease phenotype. White boxes show alleles inherited from the unaffected parent. The bracket indicates the region most likely to contain the disease gene. The affected haplotype is defined as a region of adjoining markers that are present on the same piece of DNA and transmitted together to all the affected members of the next generation – in this case, alleles 9–6–2 at markers 3, 4, and 5.

In addition to letting us identify an affected haplotype, this kind of analysis allows us to see where recombination events have taken place. In Figure 4.13, the chromosome on the right in child A was produced by a recombination event during paternal meiosis between marker 5 and marker 6, which leaves child A carrying unaffected alleles at markers 6 and 7. The chromosome on the right in child H was produced by recombination between marker 2 and marker 3 during paternal meiosis, which leaves child H carrying unaffected alleles at markers 1 and 2. This suggests to us that the disease gene should be located between marker 2 and marker 6, in the region that contains the affected haplotype.

When comparing information on multiple different families with the same disease, it is possible to compare the allele sizes found in the different families. In some cases, it is possible to identify an affected haplotype from one family that turns up in another family with the same disease. This constitutes evidence that the families might share a common ancestor, although it is not proof. It is important to be careful when deciding that a shared haplotype can be interpreted as meaning shared ancestry. Consider that if a very small number of markers define the affected haplotype and the alleles in that haplotype are very common alleles, there should be a concern that perhaps this is simply a fairly common haplotype present in many individuals in the population. This can be resolved by doing two things: looking at haplotypes in the general population to see whether the haplotype is present in individuals and families that lack the disease phenotype and looking at additional markers in the same region in the two families being studied to see whether additional markers spaced between the original markers continue to show allele-sharing between the two families.

The Endgame

Once the chromosomal region containing the gene has been identified (sometimes called the **genetic inclusion interval** or the **critical region**), you will want to identify the specific gene affected. Because the genetic markers are pieces of sequence whose positions are known within the human genome, it is possible first to map a gene, identify the flanking markers whose recombination events define the boundaries of the genetic inclusion interval, and then go look up the genes within that interval. You can then evaluate all of the genes listed within that interval as possible candidate disease genes. To prioritize candidate genes to decide which ones to screen first, examine information on which tissues or cell types express the gene, information on what stages in development express the gene, and information on gene product function, biochemical pathway or gene family. This process of prioritizing the candidate genes is important since there can often be hundreds or even thousands of genes located within a genetic inclusion interval under consideration.

Once the candidate genes have been prioritized, then it is a matter of screening to determine which genes in the interval contain sequence variants. The expectation is that more than one gene in the interval will have sequence variants present, but likely only one of them will be the culprit causing the disease. How can we tell when we find the real disease gene? Today, this would involve sequencing the individual and other family members. Ideally, you would sequence both affected and unaffected family members, but that may not always be possible. You might also be limited in the type of sequencing you can do. Exome sequencing will target the protein-coding regions of the genome, but the mutation may lie in an intronic or regulatory region, which exome sequencing may miss. Alternatively, the phenotype may be caused by a change in copy number, which can usually be accurately assayed using whole-genome sequencing but may be more difficult to assay using exome sequencing. However you go about collecting your variants, you'll have to consider the trade-off, which really comes down to cost versus the amount of information collected. Data are beginning to show that if you can get it, you (and the families you are working with) would be best served by performing either whole-genome sequencing or RNA sequencing of the affected individuals and controls (Gilissen et al. 2014; Hong et al. 2022).

We can also look at the variant frequency in a population of individuals who do not have the disease or in the population as a whole. A variant that is rare (seen in fewer than 1% of individuals sequenced) is much more likely to be disease-causing than one commonly seen (present in >1% of individuals). Another thing we can do is to look for variants in that particular gene in other families or individuals with that same disease. The formal proof of principle normally calls for doing something like showing a change in a measurable biochemical function associated with that mutation and gene or constructing a transgenic animal (this can be done in flies or yeast – not everything requires a mouse) that has the mutation and showing that the animal then develops the disease characteristics. This final step can be complicated on many levels, only one of which centers around the terribly obvious observation that laboratory animals are not humans and thus will not always develop the same phenotype in response to mutations in a particular gene. However, as a general rule so far, this approach has worked well:

- Map the gene;
- define boundaries to the genetic inclusion interval;
- reduce the size of the interval;
- identify the genes within the interval;
- prioritize them based on expression, function, and sequence homology information;
- test for mutations;

- evaluate other populations for the presence of those mutations;
- look for co-segregation of mutation and disease phenotype in other families;
- and get confirmation in an animal model or through biochemical testing.

Resources that offer additional information on linkage analysis are available and may be of use to the interested reader (Lander and Botstein 1989; Taylor et al. 1997; Ott 2001; Rice et al. 2001). Additional information on other methods of mutant mapping in human genetics can be found by looking at works on linkage disequilibrium (Reich et al. 2001), sibling-pair analysis (Farrall 1997), and population genetics (Weiss and Clark 2002). Some large-scale population studies include gnomAD (Karczewski et al. 2020), the 1000 Genomes Project (Consortium et al. 2010), and those from the Netherlands (Genome of the Netherlands Consortium 2014) and Iceland (Gudbjartsson et al. 2015).

The Actual Distribution of Exchange Events

The actual distribution of exchanges is not random among the bivalents that comprise a karyotype; the distribution of the number of chiasma per bivalent is much narrower than that predicted by a Poisson distribution (for a review see Jones 1984) so that there are almost no achiasmate bivalents and few with high numbers of chiasmata. For example, in rye (*Secale cereale*) only 1.4% of the bivalents are achiasmate. The remaining 98% of the bivalents possess 1 (26.2%), 2 (70.5%), or 3 (1.7%) exchanges. These observations differ markedly from those predicted by a Poisson distribution of chiasmata per bivalent, in that there are large excesses of bivalents with one exchange and corresponding deficiencies of bivalents with zero or two exchanges. Similarly, for the acrocentric X chromosomes in a genetically normal *Drosophila* female, the frequency of achiasmate bivalents is approximately 10%, the frequency of bivalents with a single exchange is approximately 60%, and the frequency of bivalents with two exchanges is approximately 30% (Page et al. 2007). (Bivalents with three or more exchanges are observed only very rarely.) Each arm of the metacentric second and third chromosomes displays a similar pattern of nonexchange frequency. Thus, the probability of a major autosome being achiasmate is the product of the nonexchange frequencies for each of the two arms or approximately 2%.

The position of recombination events is also tightly controlled. Exchange occurs only in the euchromatin, and the amount of exchange is not proportional to physical distance (Lindsley and Sandler 1977; Jones 1984; Berger et al. 2001; Miller et al. 2016b). Rather, for each of the five major chromosome arms in *Drosophila*, the frequency of exchange is extremely low near the base and tip of the euchromatin and reaches its highest levels in an interval beginning approximately 30% of the distance from the tip to the base of the euchromatin and ending some 50–60% of that distance (see Figure 4.4). This pattern is not unique to *Drosophila* females, but rather is a general feature of chiasma distribution in many organisms (Jones 1984).

The distribution of meiotic recombination results from the combined action of three types of genetic control:

1) *trans*-acting regulators of exchange position, which appear to act at the level of entire chromosome arms (Baker and Hall 1976);
2) local *cis*-acting regulators of exchange (Szauter 1984); and
3) chromosomal elements such as centromeres and telomeres that suppress exchange in a polar fashion over long chromosomal distances.

All three of these levels of regulation appear to act to keep exchanges, and hence chiasmata, a substantial distance away from the centromeres and the telomeres. Evidence in several organisms suggests that exchanges too close to the centromere may interfere with homolog separation at meiosis I. Exchanges too close to the telomeres fail to properly orient homologous centromeres, perhaps as a consequence of insufficient sister chromatid cohesion distal to the chiasma (Ross, Maxfield, et al. 1996; Koehler et al. 1996; Lamb et al. 1996). The mechanisms by which these centromere and telomere effects are mediated remain obscure.

The Centromere Effect

As noted in Section 4.1, the frequency of crossing over along a euchromatic arm of a given pair of chromosomes diminishes as one approaches the centromere (Hughes et al. 2018; Pazhayam et al. 2021). However, a clear experimental demonstration of the centromere effect, which requires moving a genetically defined interval away from or closer to the centromeres and observing the expected increase or decrease in its map length, has been accomplished in only two species, *Drosophila melanogaster* and maize. In *Drosophila*, this was accomplished by studying homozygous paracentric inversions (Beadle and Sturtevant 1935; Szauter 1984), while the evidence for a true effect of proximity to the centromeres and the result of being next to large blocks of heterochromatin comes from the study of the recombinational effects of deletions of the pericentric heterochromatin (Yamamoto and Miklos 1978). Similarly, studies of homozygous translocations in maize have shown that moving a small interval away from the centromere increases the frequency of exchange in that interval (Yu and Peterson 1973).

We can say very little about the mechanism of the centromere effect other than to note that, at least in flies, neither DSB position nor gene conversions are affected by the centromere effect (Miller et al. 2016b; Crown et al. 2018). Thus, it seems likely the centromere effect exerts its effect by altering the fate of DSBs so that in the vicinity of a centromere, a DSB is less likely to be resolved as a crossover (and is rather repaired by other mechanisms). In the long run, this mechanism can only be resolved by the recovery of mutants that specifically ablate this reduction in exchange. Finding such mutants may be a bit of a unicorn hunt, but it will be well worth the effort when it succeeds.

The Effects of Heterozygosity for Aberration Breakpoints on Recombination

When heterozygous, the breakpoints of both translocations and inversions create polar exchange suppressions emanating in both directions from the breakpoint (Sturtevant and Beadle 1936; Novitski and Braver 1954; Roberts 1972; Hawley 1980). The role of the exchange suppressions generated by inversion breakpoints in the functioning of balancers has been discussed (Miller et al. 2016a, 2019). In addition, similar studies have been performed in maize (Burnham et al. 1972; Maguire 1977, 1986). Unfortunately, despite extensive work, very little is known about the mechanism underlying this phenomenon.

Heterozygous breakpoints can paradoxically increase the frequency of recombination in distant intervals and on other chromosomes. This phenomenon is referred to as the **interchromosomal effect** (Sturtevant 1919, 1921; Sturtevant and Beadle 1936; Crown et al. 2018; Miller 2020). Recently, whole-genome sequencing was used to quantify the impact of the interchromosomal effect (Crown et al. 2018) and revealed that the distribution of DSBs was not changed in the presence of multiple inversions, suggesting the mechanism acts when DSB fate is determined.

Practicalities of Mapping

The effectiveness of any mapping method you use will be only as good as your markers and your ability to score them. We can recall a failed, but instructive, attempt to map a mutant that affects meiotic chromosome segregation by using two flanking mutants that had identical phenotypic effects. In one case an attempt was made to precisely map a meiotic mutant (*m*), which by itself produced no visible phenotype, by collecting recombinants between the genes *singed* (*sn*) and *forked* (*f*). Because the meiotic mutant was known to lie between these two genes, it was hoped that recovering a large number of recombinants between *singed* and *forked* from $+ \ m \ +/sn \ + f$ mothers and testing whether each recombinant carried the mutant would position the meiotic mutant within this interval. Unfortunately, although the cross looked fine on paper, in reality, it was impossible to distinguish the *singed–forked* recombinants from one of the parental chromosomes ($sn + f$) because all three mutants have a similar effect on bristle morphology. It was then necessary to restart the experiment, replacing the *sn* mutant with a closely linked mutant known as *cut* (*ct*) that affects wing morphology. The $+ \ m \ +/ct \ + f$ experiment worked nicely. But had the investigator thought beforehand about just how this cross was to be scored, they would have been spared much lost time and a fair amount of teasing. We tell this story to urge you to stick to clean, easily observable markers.

It's also good to use common sense in setting up the size of your experiment. Although methods exist for the statistical evaluation of recombination frequencies (Szauter 1984), it will be very hard to convince anyone that two map lengths of, say, 18 and 20 cM are meaningfully different, no matter how large your numbers are. Thus, 1000 progeny are generally fully sufficient. (Obviously, those looking for very rare exchange events will need larger numbers.) If you need to measure the length of long distances, use a larger number of markers that lie within the larger interval.

4.5 The Mechanism of Recombination

We will now turn our attention from using recombination as a tool (i.e. for mapping) to the mechanism of recombination itself. There are several recent and not-so-recent reviews on the evolution of our current models of recombination and the interested reader is encouraged to read them (Kauppi et al. 2004; Keeney and Neale 2006; LaFave and Sekelsky 2009; Kohl and Sekelsky 2013; Keeney et al. 2014). These are wonderful reviews that dive further into the details of recombination that we can only hint at here.

Gene Conversion

The story of recombination models begins with the description of a phenomenon called **gene conversion**. The simplest way to think of gene conversion is as a repair event that results in the 3 : 1 segregation of an allele after the repair of a double-strand break. There are two types of gene conversion that must be mentioned: **crossover-associated gene conversion** and **noncrossover gene conversion**. The difference is what it sounds like. Crossover-associated gene conversions occur in conjunction with a crossover event while noncrossover gene conversions are stand-alone events.

The theory of gene conversion was first proposed by Winkler (1930) and described by Lindegren (1953) in *Saccharomyces*. Gene conversion explained early observations in Neurospora (Lindegren 1932a, b) in which sporulating an *Aa* diploid produced one of three types of unusual asci (Figure 4.14). In the first case, known as a 6+ : 2– ascus, there are six *A* spores and only two *a*-bearing spores (the reciprocal 6– : 2+ class is observed as well). The easiest way to understand this result is to imagine that one of the *A* alleles had reached over and "converted" an *a* allele into an *A* allele – hence the term gene conversion. That gene conversion was indeed precise was demonstrated in yeast (where the 6+ : 2– or 2+ : 6– segregations seen in Neurospora could be visualized

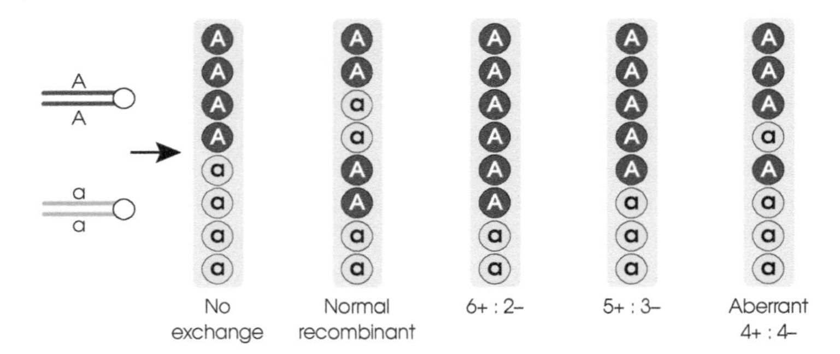

Figure 4.14 Gene conversion in Neurospora. In Neurospora, meiosis produces a linear array of four meiotic products and is immediately followed by one round of mitosis. This all occurs in a sac, or ascus, that holds the eight spores in a specific order such that pairs of adjacent ascospores should be identical, as shown in the "No exchange" or "Normal recombinant" asci. Gene conversion explains the recovery of 6+ : 2−, 6− : 2+, 5+ : 3−, 5− : 3+, and aberrant 4+: 4− asci.

as 3+: 1− or 1+: 3− asci) by Fogel and Mortimer (1969, 1971) using nonsense mutants. They observed that the newly converted mutant spore was suppressible by the same tRNA suppressor that restored function to the original (converting) mutant, suggesting that conversion was an accurate process in which information from one allele was faithfully replaced by the allele on its homolog.

As odd as the 6+ : 2− (or 6− : 2+) segregations in Neurospora might be, the 5+ : 3− (or 5− : 3+) and the aberrant 4+ : 4− segregations were even more curious. The only way to understand these observations was to suggest that at the end of meiosis, at least some chromatids carried *both* the *A* and *a* alleles. This suggests that these chromatids carried *A* information on one strand of the DNA duplex and *a* information on the other, such that replication produced both *A* and *a* haploid daughter cells. This is to say that the DNA of these chromosomes must include a region of base-pair mismatch corresponding to the base pair(s) that differs between the *A* and *a* alleles. This structure is referred to as a **heteroduplex**, the concept of which has a long history in phage genetics (Wildenberg and Meselson 1975; Wagner and Meselson 1976).

It was clear from the beginning that gene conversion was associated with the process of DSB repair. Most notably, gene conversion was accompanied by the exchange of flanking markers, or crossovers, in approximately 50% of the cases in which it occurred in the yeast *Saccharomyces cerevisiae*.[7] Better yet, a similar fraction of crossovers were associated with gene conversion events (Hurst et al. 1972; Borts and Haber 1987). In addition, those cases of gene conversion that were associated with flanking marker exchange displayed interference, while those not associated with flanking marker exchange did not display interference. These observations led to the development of the first set of recombination models, which are described in the following section.

Early Models of Recombination

The Holliday Model

The first, and in many ways most important, model of recombination was proposed by Robin Holliday (1964). This **Holliday model** is diagrammed in Figure 4.15. The initiating step is an identical single-strand nick on both homologs. The two nicked single strands then swap DNA over a

7 We might note, however, that far too much emphasis was placed on the observation that gene conversion was accompanied by exchange of flanking markers in approximately 50% of the cases in which it occurred – it was by no means true for all loci.

Figure 4.19 Synthesis-dependent strand annealing (SDSA). Gene conversions not associated with crossover events are known as noncrossover gene conversions. They are created by a different pathway than crossover-associated gene conversions. (Note: line pairs in this diagram represent DNA duplexes, not sister chromatids.)

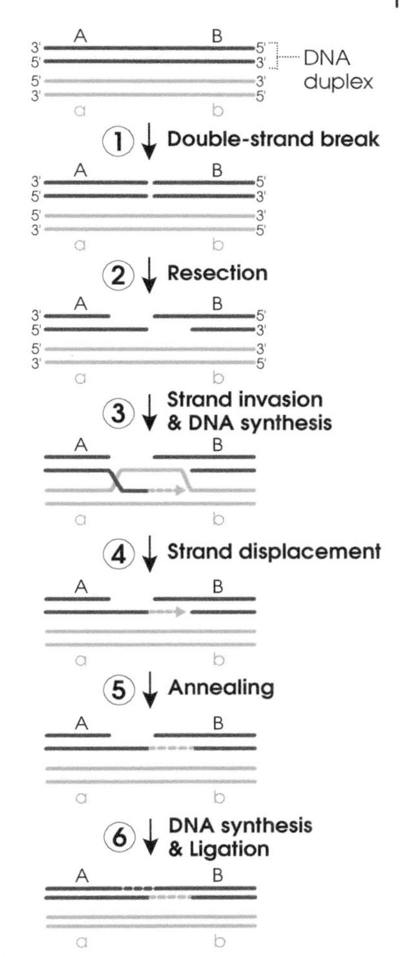

crossovers, which are known as Class I events. Class I events are defined by the properties listed earlier in this chapter, they create and respond to interference and their formation is associated with (perhaps requires) recombination nodules. There is, however, a less commonly used set of pathways for the generation of crossovers, known as Class II crossovers (de los Santos et al. 2003; Hughes et al. 2018). Class II events do not create or respond to interference and are not associated with recombination. They also appear to be generated by different molecular mechanisms. Organisms differ in the extent to which the Class II pathway is used. All that said, the crucial fact is that both types of crossover events can form functional chiasmata.

4.6 Summary

This chapter has focused primarily on the use of meiotic recombination to generate maps and on the analysis of the recombination process itself. We introduced meiosis and discussed how double-strand breaks are repaired by both crossing over and noncrossover gene conversion. Studying where and how frequently recombination occurs allows us to understand how mutants affect the distribution of recombination and allows us to map genes. Understanding the distribution and frequency of exchange also allows us to infer recombination frequency on chromatids that we did not recover, which is discussed in more detail in Appendix C. However, the biological function of exchange is not to help us map genes, but to ensure chromosome segregation – a topic covered in Chapter 9.

References

1000 Genomes Project Consortium, Abecasis GR, Altshuler D, Auton A, Brooks LD, *et al*. 2010. A map of human genome variation from population-scale sequencing. *Nature* 467:1061–1073.

Allers T, Lichten M. 2001. Differential timing and control of noncrossover and crossover recombination during meiosis. *Cell* 106:47–57.

Anderson EG. 1925. Crossing over in a case of attached X chromosomes in *Drosophila melanogaster*. *Genetics* 10:403–417.

Baker BS, Hall JC. 1976. Meiotic mutants: genetic control of meiotic recombination and chromosome segregation. In: Ashburner M and Novitski E, editors. *The Genetics and Biology of Drosophila*. Vol. 1a. London: Academic Press. pp. 352–434.

Barlow AL, Hultén MA. 1998. Crossing over analysis at pachytene in man. *Eur. J. Hum. Genet.* 6:350–358.

Barton NH, Charlesworth B.1998. Why sex and recombination? *Science* 281:1986–1990.

Baudat F, Manova K, Yuen JP, Jasin M, Keeney S. 2000. Chromosome synapsis defects and sexually dimorphic meiotic progression in mice lacking Spo11. *Mol. Cell* 6:989–998.

Beadle GW, Emerson S. 1935. Further studies of crossing over in attached-X chromosomes of *Drosophila melanogaster*. *Genetics* 20:192–206.

Beadle GW, Sturtevant AH. 1935. X chromosome inversions and meiosis in *Drosophila* Melanogaster. *Proc. Natl. Acad. Sci.* 21:384–390.

Berchowitz LE, Copenhaver GP. 2010. Genetic interference: don't stand so close to me. *Curr. Genomics* 11:91–102.

Berger J, Suzuki T, Senti KA, Stubbs J, Schaffner G, *et al*. 2001. Genetic mapping with SNP markers in *Drosophila*. *Nat. Genet.* 29(4):475–481.

Borts RH, Haber JE. 1987. Meiotic recombination in yeast: alteration by multiple heterozygosities. *Science* 237:1459–1465.

Buonomo SB, Clyne RK, Fuchs J, Loidl J, Uhlmann F, *et al*. 2000. Disjunction of homologous chromosomes in meiosis I depends on proteolytic cleavage of the meiotic cohesin Rec8 by Separin. *Cell* 103:387–398.

Burnham CR, Stout JT, Weinheimer WH, Kowles RV, Phillips RL. 1972. Chromosome pairing in maize. *Genetics* 71:111–26.

Cao L, Alani E, Kleckner N. 1990. A pathway for generation and processing of double-strand breaks during meiotic recombination in *S. cerevisiae*. *Cell* 61:1089–1101.

Carpenter AT. 1973. A meiotic mutant defective in distributive disjunction in *Drosophila melanogaster*. *Genetics* 73:393–428.

Carpenter AT. 1975. Electron microscopy of meiosis in *Drosophila melanogaster* females: II. The recombination nodule – a recombination-associated structure at pachytene? *Proc. Natl. Acad. Sci. U.S.A.* 72:3186–3189.

Carpenter AT. 1979a. Recombination nodules and synaptonemal complex in recombination-defective females of *Drosophila melanogaster*. *Chromosoma* 75:259–292.

Carpenter AT. 1979b. Synaptonemal complex and recombination nodules in wild-type *Drosophila melanogaster* females. *Genetics* 92:511–541.

Carpenter AT. 1981. EM autoradiographic evidence that DNA synthesis occurs at recombination nodules during meiosis in *Drosophila melanogaster* females. *Chromosoma* 83:59–80.

Carpenter AT. 1984. Recombination nodules and the mechanism of crossing-over in *Drosophila*. *Symp. Soc. Exp. Biol.* 38:233–243.

Carpenter AT. 1987. Gene conversion, recombination nodules, and the initiation of meiotic synapsis. *Bioessays* 6:232–236.

Case ME, Giles NH. 1958. Recombination mechanisms at the *pan-2* locus in *Neurospora crassa*. *Cold Spring Harb. Symp. Quant. Biol.* 23:119–135.

Case ME, Giles NH. 1964. Allelic recombination in *Neurospora*: tetrad analysis of a three-point cross within the *pan-2* locus. *Genetics* 49:529–540.

Chikashige Y, Ding DQ, Funabiki H, Haraguchi T, Mashiko S, *et al.* 1994. Telomere-led premeiotic chromosome movement in fission yeast. *Science* 264:270–273.

Chikashige Y, Ding DQ, Imai Y, Yamamoto M, Haraguchi T, *et al.* 1997. Meiotic nuclear reorganization: switching the position of centromeres and telomeres in the fission yeast *Schizosaccharomyces pombe*. *EMBO J.* 16:193–202.

Chua PR, Roeder GS. 1997. Tam1, a telomere-associated meiotic protein, functions in chromosome synapsis and crossover interference. *Genes Dev.* 11:1786–1800.

Conrad MN, Dominguez AM, Dresser ME. 1997. Ndj1p, a meiotic telomere protein required for normal chromosome synapsis and segregation in yeast. *Science* 276:1252–1255.

Cooper JP, Watanabe Y, Nurse P. 1998. Fission yeast Taz1 protein is required for meiotic telomere clustering and recombination. *Nature* 392:828–831.

Copenhaver GP, Housworth EA, Stahl FW. 2002. Crossover interference in *Arabidopsis*. *Genetics* 160:1631–1639.

Creighton HB, McClintock B. 1931. A correlation of cytological and genetical crossing-over in *Zea mays*. *Proc. Natl. Acad. Sci. U.S.A.* 17:492–497.

Cromie GA, Leach DR. 2000. Control of crossing over. *Mol. Cell* 6:815–826.

Crown KN, Miller DE, Sekelsky J, Hawley RS. 2018. Local inversion heterozygosity alters recombination throughout the genome. *Curr. Biol.* 28:1–19.

Davis L, Smith GR. 2001. Meiotic recombination and chromosome segregation in *Schizosaccharomyces pombe*. *Proc. Natl. Acad. Sci. U.S.A.* 98:8395–8402.

Dawson DS, Murray AW, Szostak JW. 1986. An alternative pathway for meiotic chromosome segregation in yeast. *Science* 234:713–717.

Dernburg AF, Sedat JW, Cande WZ, Bass HW. 1995. Cytology of telomeres. In: Blackburn EH, Greider CW, editor. *Telomeres*. Cold Spring Harbor Monograph Archive. pp. 295–338.

Dernburg AF, Sedat JW, Hawley RS. 1996. Direct evidence of a role for heterochromatin in meiotic chromosome segregation. *Cell* 86:135–146.

Dernburg AF, McDonald K, Moulder G, Barstead R, Dresser M, *et al.* 1998. Meiotic recombination in *C. elegans* initiates by a conserved mechanism and is dispensable for homologous chromosome synapsis. *Cell* 94:387–398.

von Diezmann L, Rog O. 2021. Let's get physical – mechanisms of crossover interference. *J. Cell Sci.* 134:jcs255745.

Farrall M. 1997. Affected sibpair linkage tests for multiple linked susceptibility genes. *Genet. Epidemiol.* 14:103–115.

Fogel S, Mortimer RK. 1969. Informational transfer in meiotic gene conversion. *Proc. Natl. Acad. Sci. U.S.A.* 62:96–103.

Fogel S, Mortimer RK. 1971. Recombination in yeast. *Annu. Rev. Genet.* 5:219–236.

Fogel S, Choi T, Kilgore D, Lusnak K, Williamson M. 1982. The molecular genetics of non-tandem duplications at *ADE8* in yeast. In: *Recent Advances in Yeast Molecular Biology: Recombinant DNA* 1:269–288.

Foss EJ, Stahl FW. 1995. A test of a counting model for chiasma interference. *Genetics* 139:1201–1209.

Foss E, Lande R, Stahl FW, Steinberg CM. 1993. Chiasma interference as a function of genetic distance. *Genetics* 133:681–691.

Fu TK, Sears ER. 1973. The relationship between chiasmata and crossing over in *Triticum aestivum*. *Genetics* 75:231–246.

Game JC, Sitney KC, Cook VE, Mortimer RK 1989. Use of a ring chromosome and pulsed-field gels to study interhomolog recombination, double-strand DNA breaks and sister-chromatid exchange in yeast. *Genetics* 123:695–713.

Genome of the Netherlands Consortium. 2014. Whole-genome sequence variation, population structure and demographic history of the Dutch population. *Nat. Genet.* 6(8):818–825.

Gilissen C, Hehir-Kwa JY, Thung DT, van de Vorst M, van Bon BWM, *et al.* 2014. Genome sequencing identifies major causes of severe intellectual disability. *Nature* 511 344–347.

Grelon M, Vezon D, Gendrot G, Pelletier G. 2001. AtSPO11-1 is necessary for efficient meiotic recombination in plants. *EMBO J.* 20:589–600.

Guacci V, Kaback DB. 1991. Distributive disjunction of authentic chromosomes in *Saccharomyces cerevisiae*. *Genetics* 127:475–488.

Gudbjartsson DF, Helgason H, Gudjonsson SA, Zink F, Oddson A. 2015. Large-scale whole-genome sequencing of the Icelandic population. *Nat. Genet.* 47(5):435–444.

Haber JE, Thorburn PC, Rogers D. 1984. Meiotic and mitotic behavior of dicentric chromosomes in *Saccharomyces cerevisiae*. *Genetics* 106:185–205.

Haldane JBS. 1919. The combination of linkage values and the calculation of distances between the loci of linked factors. *J. Genet.* 8:299–309.

Hall JC. 1972. Chromosome segregation influenced by two alleles of the meiotic mutant *c(3)G* in *Drosophila* melanogaster. *Genetics* 71:367–400.

Hassold T, Hunt P. 2001. To err (meiotically) is human: the genesis of human aneuploidy. *Nat. Rev. Genet.* 2:280–291.

Hawley RS. 1980. Chromosomal sites necessary for normal levels of meiotic recombination in *Drosophila melanogaster*. I. Evidence for and mapping of the sites. *Genetics* 94:625–646.

Hawley RS. 1988. Exchange and chromosome segregation in eukaryotes. In: Kucherlapati R, (editor). *Genetic Recombination*. Washington, DC: ASM Press. pp.497–527.

Hawley RS, Irick H, Zitron AE, Haddox DA, Lohe A, *et al.* 1992. There are two mechanisms of achiasmate segregation in *Drosophila* females, one of which requires heterochromatic homology. *Dev. Genet.* 13:440–467.

Hawley RS, McKim KS, Arbel T. 1993. Meiotic segregation in *Drosophila melanogaster* females: molecules, mechanisms, and myths. *Annu. Rev. Genet.* 27:281–317.

van Heemst D, Heyting C. 2000. Sister chromatid cohesion and recombination in meiosis. *Chromosoma* 109:10–26.

Herman RK, Kari CK. 1989. Recombination between small X chromosome duplications and the X chromosome in *Caenorhabditis elegans*. *Genetics* 121:723–737.

Hey J. 2004. What's so hot about recombination hotspots? *PLoS Biol.* 2:e190.

Hillers KJ. 2004. Crossover interference. *Curr. Biol.* 14:R1036–R1037.

Holliday R. 1964. A mechanism for gene conversion in fungi. *Genet. Res.* 5:282–304.

Holm GD. 1976. Compound autosomes. In: M. Ashburner and E. Novitski (eds.) *The Genetics and Biology of Drosophila*. Academic Press, London. pp. 529–561.

Hong SE, Kneissl J, Cho A, Kim MJ, Park S, *et al.* 2022. Transcriptome-based variant calling and aberrant mRNA discovery enhance diagnostic efficiency for neuromuscular diseases. *J. Med. Genet.*: https://doi.org/10.1136/jmedgenet-2021-108307.

Housworth EA, Stahl FW. 2003. Crossover interference in humans. *Am. J. Hum. Genet.* 73:188–197.

Hughes SE, Miller DE, Miller AL, Hawley RS. 2018. Female meiosis: synapsis, recombination, and segregation in *Drosophila melanogaster*. *Genetics* 208:875–908.

Hurst DD, Fogel S, Mortimer RK. 1972. Conversion-associated recombination in yeast (hybrids-meiosis-tetrads-marker loci-models). *Proc. Natl. Acad. Sci. U.S.A.* 69:101–105.

International HapMap Consortium. 2005. A haplotype map of the human genome. *Nature* 437:1299–1320.

Janssens FA. 1909. La Théorie de la Chiasmatypie. *Genetics* 191:319–346.

Jones GH. 1984. The control of chiasma distribution. *Symp. Soc. Exp. Biol.* 38:293–320.

Karczewski KJ, Francioli LC, Tiao G, Cummings BB, Alföldi J, *et al.* 2020. The mutational constraint spectrum quantified from variation in 141,456 humans. *Nature* 581:434–443.

Karpen GH, Le MH, Le H. 1996. Centric heterochromatin and the efficiency of achiasmate disjunction in *Drosophila* female meiosis. *Science* 273:118–122.

Kauppi L, Jeffreys AJ, Keeney S.2004. Where the crossovers are: recombination distributions in mammals. *Nat. Publ. Group* 5:413–424.

Keeney S, Neale MJ. 2006. Initiation of meiotic recombination by formation of DNA double-strand breaks: mechanism and regulation. *Biochem. Soc. Trans.* 34:523–525.

Keeney S, Lange J, Mohibullah N. 2014. Self-organization of meiotic recombination initiation: general principles and molecular pathways. *Annu. Rev. Genet.* 48:187–214.

Kirkpatrick DT, Fan Q, Petes TD. 1999. Maximal stimulation of meiotic recombination by a yeast transcription factor requires the transcription activation domain and a DNA-binding domain. *Genetics* 152:101–115.

Klar AJ, Strathern JN, Abraham JA. 1984. Involvement of double-strand chromosomal breaks for mating-type switching in *Saccharomyces cerevisiae*. *Cold Spring Harb. Symp. Quant. Biol.* 49:77–88.

Koehler KE, Boulton CL, Collins HE, French RL, Herman KC, *et al.* 1996. Spontaneous X chromosome MI and MII nondisjunction events in *Drosophila melanogaster* oocytes have different recombinational histories. *Nat. Genet.* 14:406–414.

Kohl KP, Sekelsky J. 2013. Meiotic and mitotic recombination in meiosis. *Genetics* 194:327–334.

LaFave MC, Sekelsky J. 2009. Mitotic recombination: why? when? how? where? Copenhaver GP, editor. *PLoS Genet.* 5:e1000411.

Lake CM, Hawley RS. 2012. The molecular control of meiotic chromosomal behavior: events in early meiotic prophase in *Drosophila* oocytes. *Annu. Rev. Physiol.* 74:425–451.

Lake CM, Nielsen RJ, Guo F, Unruh, JR, Slaughter BD, *et al.* 2015. Vilya, a component of the recombination nodule, is required for meiotic double-strand break formation in *Drosophila*. *Elife* 4:e08287

Lake CM, Nielsen RJ, Bonner AM, Eche S, White-Brown S, *et al.* 2019. Narya, a RING finger domain-containing protein, is required for meiotic DNA double-strand break formation and crossover maturation in *Drosophila melanogaster*. *PLoS Genet.* 15:e1007886.

Lamb NE, Freeman SB, Savage-Austin A, Pettay D, Taft L, *et al.* 1996. Susceptible chiasmate configurations of chromosome 21 predispose to non-disjunction in both maternal meiosis I and meiosis II. *Nat. Genet.* 14:400–405.

Lander ES, Botstein D. 1989. Mapping mendelian factors underlying quantitative traits using RFLP linkage maps. *Genetics* 121(1):185–199.

Lawrie NM, Tease C, Hultén MA. 1995. Chiasma frequency, distribution and interference maps of mouse autosomes. *Chromosoma* 104:308–314.

Lee JY, Orr-Weaver TL. 2001. The molecular basis of sister-chromatid cohesion. *Annu. Rev. Cell Dev. Biol.* 17:753–777.

Lichten M. 2001. Meiotic recombination: breaking the genome to save it. *Curr. Biol.* 11:R253–R256.

Lichten M, Goldman A. 1995. Meiotic recombination hotspots. *Annu. Rev. Genet.* 29:423–445.

Lindegren CC. 1932a. The genetics of Neurospora. I. The inheritance of response to heat-treatment. *Bull. Torrey Bot. Club* 59:85.

Lindegren CC. 1932b. The genetics of Neurospora-II. segregation of the sex factors in asci of *N. crassa*, Sitophila, N, Tetrasperma, N, editor. *Bull Torrey Bot. Club* 59:119.

Lindegren CC. 1953. Gene conversion in Saccharomyces. *J. Genet.* 51:625–637.

Lindsley DL, Sandler L. 1958. The meiotic behavior of grossly deleted X chromosomes in *Drosophila melanogaster*. *Genetics* 43:547–563.

Lindsley DL, Sandler L. 1977. The genetic analysis of meiosis in female *Drosophila melanogaster*. *Philos. Trans. R Soc. Lond. B Biol. Sci.* 277(955):295–312.

Loidl J. 1990. The initiation of meiotic chromosome pairing: the cytological view. *Genome* 33:759–778.

de los Santos T, Hunter N, Lee C, Larkin B, Loidl J, *et al.* 2003. The Mus81/Mms4 endonuclease acts independently of double-Holliday junction resolution to promote a distinct subset of crossovers during meiosis in budding yeast. *Genetics* 164:81–94.

Maguire MP. 1977. Homologous chromosome pairing. *Philos. Trans. R. Soc. Lond. Ser. B Biol. Sci.* 277:245–258.

Maguire MP. 1986. The pattern of pairing that is effective for crossing over in complex B-A chromosome rearrangements in maize: III. Possible evidence for pairing centers. *Chromosoma* 94:71–85.

Maguire MP, Riess RW. 1991. Synaptonemal complex behavior in asynaptic maize. *Genome* 34:163–168.

Mahadevaiah SK, Turner JM, Baudat F, Rogakou EP, de Boer P. *et al.* 2001. Recombinational DNA double-strand breaks in mice precede synapsis. *Nat. Genet.* 27:271–276.

Manzano-Winkler B, McGaugh SE, Noor MAF. 2013. How hot are *Drosophila* hotspots? Examining recombination rate variation and associations with nucleotide diversity, divergence, and maternal age in *Drosophila pseudoobscura*. Palsson A, editor *PLoS One* 8:e71582.

McClung CE. 1927. The chiasmatype theory of Janssens. *Q. Rev. Biol.* 2:344–366.

McKim KS, Hayashi-Hagihara A. 1998. *mei-W68* in *Drosophila melanogaster* encodes a Spo11 homolog: evidence that the mechanism for initiating meiotic recombination is conserved. *Genes Dev.* 12:2932–2942.

McKim KS, Heschl MF, Rosenbluth RE, Baillie DL. 1988a. Genetic organization of the *unc-60* region in *Caenorhabditis elegans*. *Genetics* 118:49–59.

McKim KS, Howell AM, Rose AM. 1988b. The effects of translocations on recombination frequency in *Caenorhabditis elegans*. *Genetics* 120:987–1001.

McKim KS, Peters K, Rose AM. 1993. Two types of sites required for meiotic chromosome pairing in *Caenorhabditis elegans*. *Genetics* 134:749–768.

McKim KS, Green-Marroquin BL, Sekelsky JJ, Chin G, Steinberg C, *et al.* 1998. Meiotic synapsis in the absence of recombination. *Science* 279:876–878.

McVey M, Radut D, Sekelsky JJ. 2004. End-joining repair of double-strand breaks in *Drosophila melanogaster* is largely DNA ligase IV independent. *Genetics* 168:2067–2076.

Meselson MS, Radding CM. 1975. A general model for genetic recombination. *Proc. Natl. Acad. Sci. U.S.A.* 72:358–361.

Miller DE. 2020. The interchromosomal effect: different meanings for different organisms. *Genetics* 216:621–631.

Miller DE, Cook KR, Kazemi NY, Smith CB, Cockrell AJ, *et al.* 2016a. Rare recombination events generate sequence diversity among balancer chromosomes in *Drosophila melanogaster*. *Proc. Natl. Acad. Sci.* 113:E1352–E1361.

Miller DE, Smith CB, Kazemi NY, Cockrell AJ, Arvanitakis AV, *et al.* 2016b. Whole-genome analysis of individual meiotic events in *Drosophila melanogaster* reveals that noncrossover gene conversions are insensitive to interference and the centromere effect. *Genetics* 203:159–171.

Miller DE, Cook KR, Hawley RS. 2019. The joy of balancers. *PLoS Genet.* 15:e1008421.

Moens PB, Kolas NK, Tarsounas M, Marcon E, Cohen PE, *et al.* 2002. The time course and chromosomal localization of recombination-related proteins at meiosis in the mouse are compatible with models that can resolve the early DNA-DNA interactions without reciprocal recombination. *J. Cell Sci.* 115:1611–1622.

Molnar M, Bähler J, Kohli J, Hiraoka Y. 2001. Live observation of fission yeast meiosis in recombination-deficient mutants: a study on achiasmate chromosome segregation. *J. Cell Sci.* 114:2843–2853.

Nicklas RB. 1974. Chromosome segregation mechanisms. *Genetics* 78:205–213.

Nicolas A, Petes TD. 1994. Polarity of meiotic gene conversion in fungi: contrasting views. *Experientia* 50:242–252.

Nimmo ER, Pidoux AL, Perry PE, Allshire RC. 1998. Defective meiosis in telomere-silencing mutants of *Schizosaccharomyces pombe*. *Nature* 392:825–828.

Novitski E, Braver G. 1954. An analysis of crossing over within a heterozygous inversion in *Drosophila melanogaster*. *Genetics* 39:197–209.

Othman MI, Sullivan SA, Skuta GL, Cockrell DA, Stringham HM, *et al.* 1998. Autosomal dominant nanophthalmos (NNO1) with high hyperopia and angle-closure glaucoma maps to chromosome. *Am. J. Hum. Genet.* 63(5):1411–1418.

Ott J. 2001. Major strengths and weaknesses of the lod score method. *Adv. Genet.* 42:125–132.

Otto SP, Payseur BA. 2019. Crossover interference: shedding light on the evolution of recombination. *Annu. Rev. Genet.* 53:19–44.

Page SL, Hawley RS. 2001. *c(3)G* encodes a *Drosophila* synaptonemal complex protein. *Genes Dev.* 15:3130–3143.

Page SL, Nielsen RJ, Teeter K, Lake CM, Ong S, *et al.* 2007. A germline clone screen for meiotic mutants in *Drosophila melanogaster*. *Fly (Austin)* 1(3):172–181.

Pazhayam NM, Turcotte CA, Sekelsky J. 2021. Meiotic crossover patterning. *Front. Cell. Dev. Biol* 9:681123.

Perkins DD. 1962. Crossing-over and interference in a multiply marked chromosome arm of Neurospora. *Genetics* 47:1253–1274.

Rasooly RS, New CM, Zhang P, Hawley RS, Baker BS. 1991. The *lethal(1)TW-6cs* mutation of *Drosophila melanogaster* is a dominant antimorphic allele of *nod* and is associated with a single base change in the putative ATP-binding domain. *Genetics* 129:409–422.

Reich DE, Cargill M, Bolk S, Sabeti PC, Richter DJ, *et al.* 2001. Linkage disequilibrium in the human genome. *Nature* 411(6834):199–204.

Rice JP, Saccone NL, Corbett J. 2001. The lod score method. *Adv. Genet.* 42:99–113.

Riddle DL, Blumenthal T, Meyer BJ, Priess JR (eds.). 1997 *C. elegans II*. New York: Cold Spring Harbor Laboratory Press.

Roberts PA. 1972. Differences in synaptic affinity of chromosome arms of *Drosophila melanogaster* revealed by differential sensitivity to translocation heterozygosity. *Genetics* 71:401–415.

Rockmill B, Roeder GS. 1998. Telomere-mediated chromosome pairing during meiosis in budding yeast. *Genes Dev.* 12:2574–2586.

Roeder GS. 1997. Meiotic chromosomes: it takes two to tango. *Genes Dev.* 11:2600–2621.

Romanienko PJ, Camerini-Otero RD. 2000. The mouse *Spo11* gene is required for meiotic chromosome synapsis. *Mol. Cell* 6:975–987.

Ross LO, Rankin S, Shuster MF, Dawson DS. 1996. Effects of homology, size and exchange of the meiotic segregation of model chromosomes in *Saccharomyces cerevisiae. Genetics* 142:79–89.

Rossignol JL, Paquette N, Nicolas A. 1979. Aberrant 4:4 asci, disparity in the direction of conversion, and frequencies of conversion in *Ascobolus immersus. Cold Spring Harb. Symp. Quant. Biol.* 43(Pt 2):1343–1352.

Rossignol JL, Nicolas A, Hamza H, Langin T. 1984. Origins of gene conversion and reciprocal exchange in *Ascobolus. Cold Spring Harb. Symp. Quant. Biol.* 49:13–21.

Sax K. 1930. Chromosome structure and the mechanism of crossing over. ***J. Arnold. Arbor.*** 11:193–220.

Sax K. 1936. Chromosome coiling in relation to meiosis and crossing over. *Genetics* 21:324–338.

Scherthan H. 1997. Chromosome behaviour in earliest meiotic prophase. In: Henriques-Gil N, Parker JS, Puertas MJ (editors) *Chromosomes Today*. Netherlands, Dordrecht: Springer. PP. 217–248.

Schwacha A, Kleckner N. 1994. Identification of joint molecules that form frequently between homologs but rarely between sister chromatids during yeast meiosis. *Cell* 76:51–63.

Schwacha A, Kleckner N. 1995. Identification of double Holliday junctions as intermediates in meiotic recombination. *Cell* 83:783–791.

Schwacha A, Kleckner N. 1997. Interhomolog bias during meiotic recombination: meiotic functions promote a highly differentiated interhomolog-only pathway. *Cell* 90:1123–1135.

Smith GR, Nambiar M. 2020. New solutions to old problems: molecular mechanisms of meiotic crossover control. *Trends Genet.* 36:337–346.

Stadler DR, Towe AM. 1963. Recombination of allelic cysteine mutants in Neurospora. *Genetics* 48:1323–1344.

Stahl FW. 1979. *Genetic Recombination: Thinking About It in Phage and Fungi*. San Francisco: W. H. Freeman and Company.

Stahl F. 2012. Defining and detecting crossover-interference mutants in yeast. Lichten M, editor *PLoS One* 7:e38476.

Stern C. 1936. Somatic crossing over and segregation in *Drosophila melanogaster. Genetics* 21:625–730.

Strathern JN, Klar AJ, Hicks JB, Abraham JA, Ivy JM, *et al.* 1982. Homothallic switching of yeast mating type cassettes is initiated by a double-stranded cut in the MAT locus. *Cell* 31:183–192.

Sturtevant AH. 1919. Inherited linkage variations in the second chromosome. In: *Contributions to the Genetics of Drosophila melanogaster*. Carnegie Institution of Washignton, pp. 305–341

Sturtevant AH. 1921. A case of rearrangement of genes in *Drosophila*. *Proc. Natl. Acad. Sci.* 7:235–237.

Sturtevant AH, Beadle GW. 1936. The relations of inversions in the X chromosome of *Drosophila melanogaster* to crossing over and disjunction. *Genetics* 21:554–604.

Sun H, Treco D, Schultes NP, Szostak JW. 1989. Double-strand breaks at an initiation site for meiotic gene conversion. *Nature* 338:87–90.

Szauter P. 1984. An analysis of regional constraints on exchange in *Drosophila melanogaster* using recombination-defective meiotic mutants. *Genetics* 106:45–71.

Szostak JW, Orr-Weaver TL, Rothstein RJ, Stahl FW. 1983. The double-strand-break repair model for recombination. *Cell* 33:25–35.

Taylor EW, Xu J, Jabs EW, Meyers DA. 1997. Linkage analysis of genetic disorders. *Methods Mol. Biol.* 68:11–25.

Theurkauf WE, Hawley RS. 1992. Meiotic spindle assembly in *Drosophila* females: behavior of nonexchange chromosomes and the effects of mutations in the *nod* kinesin-like protein. *J. Cell Biol.* 116:1167–1180.

Thomas NS, Hassold TJ. 2003. Aberrant recombination and the origin of Klinefelter syndrome. *Hum. Reprod. Update* 9:309–317.

Villeneuve AM. 1994. A cis-acting locus that promotes crossing over between X chromosomes in *Caenorhabditis elegans*. *Genetics* 136:887–902.

Wagner R, Meselson M. 1976. Repair tracts in mismatched DNA heteroduplexes. *Proc. Natl. Acad. Sci. U.S.A.* 73:4135–4139.

Walker MY, Hawley RS. 2000. Hanging on to your homolog: the roles of pairing, synapsis and recombination in the maintenance of homolog adhesion. *Chromosoma* 109:3–9.

Wang S, Liu Y, Shang Y, Zhai B, Yang X, *et al.* 2019. Crossover interference, crossover maturation, and human aneuploidy. *Bioessays* 41:1800221.

Weiss KM, Clark AG. 2002. Linkage disequilibrium and the mapping of complex human traits. *Trends Genet.* 18(1):19–24.

Wettstein D, Rasmussen SW. 1984. The synaptonemal complex in genetic segregation. *Annu. Rev. Genet.* 38:195–231.

White CI, Haber JE. 1990. Intermediates of recombination during mating type switching in *Saccharomyces cerevisiae*. *EMBO J.* 9:663–673.

Whitehouse HLK. 1982. *Genetic Recombination*. John Wiley & Sons.

Wildenberg J, Meselson M. 1975. Mismatch repair in heteroduplex DNA. *Proc. Natl. Acad. Sci. U.S.A.* 72:2202–2206.

Winkler H. 1930. *Die Konversion der Gene*. Jena: Verlag von Gustav Fischer.

Woglar A, Jantsch V. 2013. Chromosome movement in meiosis I prophase of *Caenorhabditis elegans*. *Chromosoma* 123:15–24.

Wolf KW. 1994. How meiotic cells deal with non-exchange chromosomes. *Bioessays* 16:107–114.

Yamamoto M, Miklos GLG. 1978. Genetic studies on heterochromatin in *Drosophila melanogaster* and their implications for the functions of satellite DNA. *Chromosoma* 66:71–98.

Yu M-H, Peterson PA. 1973. Influence of chromosomal gene position on intragenic recombination in maize. *Theor. Appl. Genet.* 43:121–133.

Zickler D, Kleckner N. 1998. The leptotene–zygotene transition of meiosis. *Annu. Rev. Genet.* 32:619–697.

Zickler D, Kleckner N. 1999. Meiotic chromosomes: integrating structure and function. *Annu. Rev. Genet.* 33:603–754.

Zickler D, Kleckner N. 2016. A few of our favorite things: pairing, the bouquet, crossover interference and evolution of meiosis. *Semin. Cell Dev. Biol.* 54:135–148.

5

Identifying Homologous Genes

Now that you have your mutant and you've mapped it to a specific gene, your next question might be something like, "What does this gene do?" If your mutant is in an uncharacterized gene in your organism, then one way to ascertain the gene's function is to see if a similar gene sequence has been described in another organism. Alternatively, you may wish to investigate whether your model organism contains a gene homologous to a known gene in a different organism. The identification of homologous genes requires that you have the nucleotide or protein sequence of a gene of interest and access to some basic bioinformatic tools. Keep in mind that this can be a tricky process, as evolution has a way of changing the function of a gene to best suit the needs of the organism. Sometimes, parts of genes are reused and rearranged in a unique way that may confuse your search. We will begin our discussion with a consideration of what a homolog is and then move into methods for identifying similar genes among different organisms.

5.1 Homology

Homologous genes, or **homologs**, are related to one another by common descent, either through speciation or gene duplication.[1] Homologs may be further classified as orthologs, paralogs, or, less frequently, xenologs (Figure 5.1) (Fitch 1970, 2000). Orthologs are genes in different species that evolved from a common ancestor, while paralogs are two genes related by an ancestral gene duplication event. Xenologs are somewhat distinct in that they are homologous genes derived from horizontal transfer.

You may also encounter two genes in different organisms that serve similar functions but that evolved independently from one another. These are known as **analogs**, and they are often mistakenly described as homologs when, in fact, they are not. Although two analogs may share a similar function, their gene or protein sequences are usually quite different. Rather than evolving from a common ancestor, as homologs do, analogs arise by **convergent evolution**, whereby two unrelated species develop a similar characteristic in response to some similar environmental condition, such as echolocation in dolphins and bats (Parker et al. 2013). It is not uncommon to confuse these terms. Thus, before we discuss the process of finding homologous genes, we must first distinguish between the different types of similarity.

1 Remember that throughout this chapter, the term homolog refers to homologous *genes* or *proteins*, not homologous chromosomes.

Genetic Theory and Analysis: Finding Meaning in a Genome, Second Edition.
Danny E. Miller, Angela L. Miller, and R. Scott Hawley.

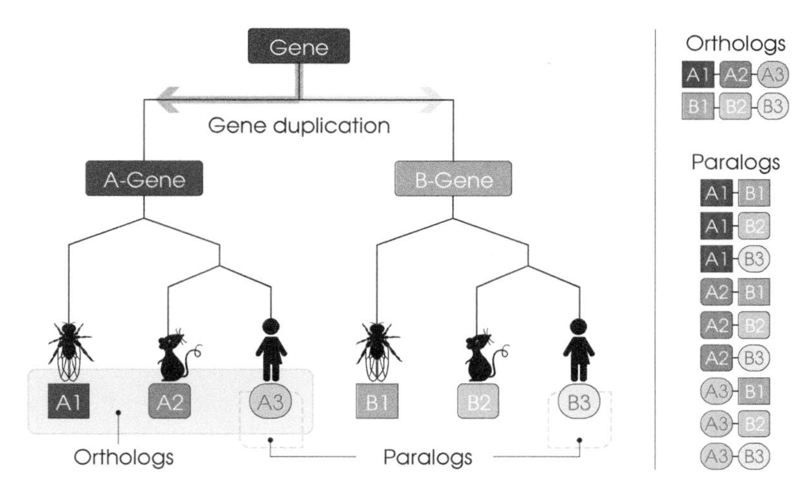

Figure 5.1 Homologs: orthologs and paralogs. Orthologs are genes in different species that evolved from a common ancestor, while paralogs are two genes related by an ancestral gene duplication event.

Orthologs

Genes that share a common ancestor because of a speciation event are known as orthologs. Because our most common ancestor existed approximately 10 million years ago, it is easy to grasp that humans and chimpanzees have thousands of orthologous genes (Arnason et al. 1998). It may be a bit more difficult, however, to imagine that humans and the budding yeast *Saccharomyces cerevisiae* also have thousands of orthologous genes – genes that existed in our last common ancestor nearly a billion years ago (Kachroo et al. 2015). This is because we share fundamental similarities at the cellular level. One way to think about genes that are highly similar in yeast and humans is that a gene may be so essential in function that changing it in any way has disastrous consequences for the organism. We see this in genes that code for the tubulin that cells use to transport material, histones that are used to control the compaction and expression of DNA, and enzymes that are used for cell metabolism.

Paralogs

Similar genes that were created by an ancestral duplication event within an organism are classified as **paralogs**. Duplications, also known as copy number variants, occur frequently during meiosis and can have a variety of effects. While a deleterious duplication may increase the quantity of a gene that then gives the organism a relative disadvantage, an advantageous duplication event may give an individual an extra copy of a gene (and thus its protein product) that confers a competitive advantage. The duplication could also be essentially neutral, having little to no impact on the organism. Over time, through random mutation, the two copies of the gene diverge from one another until one copy gains a new function and is selected for in the population. Duplications leading to paralogs can be on the scale of whole-chromosome duplications or even whole-genome duplications. Indeed, it has been shown that there were at least two rounds of whole-genome duplications in the vertebral lineage from which humans descended (Dehal and Boore 2005). Interestingly, paralogs can also be formed through transposable element movement. When a transposon moves, it may inadvertently copy a nearby coding segment and move it to a new location in the genome.

Xenologs

Homologs that arise through **horizontal gene transfer** are known as **xenologs**. Horizontal gene transfer, also known as **lateral gene transfer**, is the transmission of genetic information between organisms via transformation, conjugation, or transduction. It has been shown to be important in prokaryotic evolution. Indeed, xenologs play a larger role in prokaryotic genome expansion than duplication does (Treangen and Rocha 2011). Perhaps the most practical – and worrisome – example of xenologs in medicine or biology today is the transfer of antibiotic resistance genes among bacteria. The transfer of these genes gives the recipient bacteria an obvious competitive advantage, allowing for their selection and expansion.

5.2 Identifying Sequence Homology

So how exactly do you find the homolog of a gene of interest? The short answer is that you use a search program to compare your sequence against a database of annotated genes looking for sequence that is more similar to your gene than would be expected by chance. The most widely used search tool today is **BLAST**, which stands for basic local alignment search tool. Although there are other search tools, BLAST is usually the best place to start a homology search. BLAST may be run either as a stand-alone program on your computer or from one of several online locations. If you're just starting out, the National Center for Biotechnology Information (NCBI; https://blast.ncbi.nlm.nih.gov/Blast.cgi) is a great place to begin. However, some organism databases, such as FlyBase (*Drosophila*; http://flybase.org/blast/), SGD (*Saccharomyces*; http://www.yeastgenome.org/blast-sgd), or TAIR BLAST (*Arabidopsis*; https://www.arabidopsis.org/Blast/), also offer their own BLAST instances. Organism-specific databases such as these allow you to easily narrow your search to a specific organism, a specific set of genes, or some other specified database related to your organism, which may not be offered by NCBI BLAST.

BLAST uses an algorithm to compare your nucleotide or amino acid sequence (the query) to the BLAST database (the subject). BLAST assigns scores (bits) to sequences in its database that are similar to the query sequence – the more similar the sequence, the higher the score.[2] This score is used to calculate an **E-value**, or an **Expect value**, that is then used to classify two sequences as either homologous or nonhomologous. The E-value is the number of hits you would expect to find by chance when searching a selected database. It is a function of several things, but the size of the database and the length of your query sequence are the two most important items. A match with an E-value of 1 means that with the query sequence you provided, you would expect to find that match once in the database you are searching. An E-value of 1 is not significant. A very small E-value such as 1×10^{-15} suggests that your match is very unlikely to have occurred by chance. An E-value of 0.001 – meaning there is a 1 in 1000 possibility the match occurred by chance – is a typical cutoff used to decide whether a match is significant. There are several different types of BLAST searches you can run. Each has its own advantages and disadvantages, and understanding the differences is important. Here, we will discuss the five main search types: blastn, blastx, blastp, tblastn, and tblastx (Table 5.1).

2 For a more detailed explanation of the similarity scores that BLAST provides see The Statistics of Sequence Similarity Scores on the NCBI website.

Table 5.4 Protein BLAST (blastp) of *PYK2* from *S. cerevisiae* against *D. melanogaster* using the nr database.

Description	Max score	Total score	Query cover (%)	E-value	% Identity	Accession length
Pyruvate kinase, isoform B (*Drosophila melanogaster*)	471	471	98	2.00E−162	47.74	512
Pyruvate kinase, isoform A (*Drosophila melanogaster*)	471	471	98	4.00E−162	47.74	533
Pyruvate kinase (*Drosophila melanogaster*)	468	468	98	7.00E−161	47.35	533
GH09258p (*Drosophila melanogaster*)	357	715	79	6.00E−116	44.6	679
Uncharacterized protein Dmel_CG7069 (*Drosophila melanogaster*)	358	358	79	1.00E−115	44.6	744
AT19392p (*Drosophila melanogaster*)	338	338	94	6.00E−110	36.29	554
Uncharacterized protein Dmel_CG2964 (*Drosophila melanogaster*)	337	337	94	1.00E−109	36.29	554
Uncharacterized protein Dmel_CG7362 (*Drosophila melanogaster*)	218	278	68	2.00E−62	43.94	793
MIP06618p1 (*Drosophila melanogaster*)	218	278	69	3.00E−62	43.01	841
AT01479p (*Drosophila melanogaster*)	112	112	28	2.00E−26	40.27	440

Translated BLASTx (tblastx) and Translated BLASTn (tblastn)

The final two methods to discuss are tblastx and tblastn. Translated blastx (**tblastx**) translates a nucleotide sequence as the query into six-frame amino acid sequences (three forward and three reverse) and then compares them to a nucleotide database that has been translated on all six reading frames. This is useful if you are looking for more distant matches and you have a sequence that does not have any strong matches when searching using other methods, or it can be used to find novel genes in a sequence. This is a computationally expensive process, so keep that in mind, especially when searching large databases.

Alternatively, **tblastn** takes a protein sequence and searches a nucleotide database that was translated from protein using all six frames. This can be useful for searching a recently assembled but unannotated genome. For example, you could take the PYK sequence above and use tblastn to find the homolog in a genome you just assembled but have not yet annotated.

5.3 How Similar is Similar?

At what point can we say that two genes are homologous? For example, would we consider two genes that are 21% identical to be homologs and two genes that are 19% identical not to be homologs? The answer is, don't focus too much on identity as a proxy for similarity. Some proteins

Table 5.5 Protein BLAST (blastp) of *PYK2* from *S. cerevisiae* against *D. melanogaster* using the curated database Swis-Prot.

Description	Max score	Total score	Query cover (%)	E-value	% Identity	Accession length
RecName: full = pyruvate kinase; short = PK (*Drosophila melanogaster*)	471	471	98	3E−163	47.74	533

may be 30% identical but not homologous, while other proteins may be considered homologous with only 15% identity. Just make sure you search the proper database using the proper query sequence and then use your E-values as your guides. Also keep in mind that only a portion of your gene of interest may be similar to another gene in another species, and you can use that small segment of similarity to identify homology in distant or rapidly evolving genes.

For example, Xiang and colleagues were interested in identifying protein components of a meiotic structure called the synaptonemal complex in the freshwater flatworm *Schmidtea mediterranea* (Xiang et al. 2014). To do this, the authors first used RNA-seq to build a BLAST database of genes expressed in *S. mediterranea*. They then ran BLAST searches against this database to identify proteins of interest. But they quickly ran into a problem – synaptonemal complex proteins are notoriously difficult to identify by BLAST searching because they are coded for by very rapidly evolving genes (Fraune et al. 2012a, b). To circumvent this problem, the authors searched a second time using smaller fragments of these amino acid sequences from other organisms. For example, using a small 92-amino acid segment of the mouse synaptonemal complex protein SYCP-1 (out of a total of 1205 amino acids), the authors were able to find the homologous protein in *S. mediterranea*. While this fragment was 35% identical and 52% similar between the two species, a comparison of the entire amino acid sequences from the two organisms yielded less than 10% identity and similarity. RNAi knockdown experiments and immunofluorescence confirmed that they had indeed correctly identified the *S. mediterranea* homolog of SYCP-1.

5.4 Summary

This chapter was a brief introduction to using a common computational resource, BLAST, to identify genes homologous to a gene of interest. After identifying homologs of your gene of interest, the next steps often depend on how well characterized those homologs are and whether the gene exists within a well-established model organism. Regardless of whether they are orthologs, paralogs, or xenologs, you will likely want to find genes that share not only sequence homology but functional homology as well. For example, it is possible that two proteins may share a great deal of sequence homology, but that one of those proteins lacks a particular domain that allows it to function in a specified manner. You could choose to further explore the function of your gene in a commonly used laboratory organism, or you could work through those questions in your organism of interest.

References

Arnason U, Gullberg A, Janke A. 1998. Molecular timing of primate divergences as estimated by two nonprimate calibration points. *J. Mol. Evol.* 47:718–727.

Dehal P, Boore JL. 2005. Two rounds of whole genome duplication in the ancestral vertebrate. Holland P, editor. *PLoS Biol.* 3:e314–e319.

Fitch WM. 1970. Distinguishing homologous from analogous proteins. *Syst. Zool.* 19:99–113.

Fitch WM. 2000. Homology: a personal view on some of the problems. *Trends Genet.* 16:227–231.

Fraune J, Schramm S, Alsheimer M, Benavente R. 2012a. The mammalian synaptonemal complex: protein components, assembly and role in meiotic recombination. *Exp. Cell Res.* 318:1340–1346.

Fraune J, Alsheimer M, Volff J-N, Busch K, Fraune S, *et al.* 2012b. Hydra meiosis reveals unexpected conservation of structural synaptonemal complex proteins across metazoans. *Proc. Natl. Acad. Sci.* 109:16588–16593.

Kachroo AH, Laurent JM, Yellman CM, Meyer AG, Wilke CO, *et al.* 2015. Systematic humanization of yeast genes reveals conserved functions and genetic modularity. *Science* 348:921–925.

Nei M, Xu P, Glazko G. 2001. Estimation of divergence times from multiprotein sequences for a few mammalian species and several distantly related organisms. *Proc. Natl. Acad. Sci.* 98:2497–2502.

Parker J, Tsagkogeorga G, Cotton JA, Liu Y, Provero P, *et al.* 2013. Genome-wide signatures of convergent evolution in echolocating mammals. *Nature* 502:1–9.

Pearson WR. 2013. An introduction to sequence similarity ("homology") searching. *Curr. Protoc. Bioinform.* 42:3–1.

Treangen TJ, Rocha EPC. 2011. Horizontal transfer, not duplication, drives the expansion of protein families in prokaryotes. Moran NA, editor. *PLoS Genet.* 7:e1001284. 12.

Xiang Y, Miller DE, Ross EJ, Alvarado AS, Hawley RS. 2014. Synaptonemal complex extension from clustered telomeres mediates full-length chromosome pairing in *Schmidtea mediterranea*. *Proc. Natl. Acad. Sci.* 111:E5159–E5168.

6

Suppression

This chapter describes cases in which two mutations interact to produce a normal, or near normal, phenotype.[1] These cases are examples of **suppression**. We define suppression as follows: Mutant *a1* produces some measurable phenotype in an otherwise wildtype background. However, that phenotype is masked in individuals who also carry the *b1* mutation, such that a normal phenotype is produced. In this instance, the *b1* mutation is said to suppress the phenotypic effects of the *a1* mutant. Thus, we call *b1* a **suppressor mutant**.[2]

Now, we left a fair amount unsaid in this definition of suppression. For instance, we did not tell you whether the *b1* mutant creates a phenotype on its own and, if so, whether that phenotype is related to the phenotype exhibited by the *a1* mutant. Nor did we stipulate whether *a1* and *b1* were dominant or recessive or whether they define the same or different genes. We left these issues unspecified because suppression can encompass virtually all of these possible cases. Both dominant and recessive mutations can be suppressed, and suppressor mutations themselves can be either dominant or recessive as well. In some cases, suppressor mutants are defined only by their ability to suppress some other mutations, while, in other cases, the suppressor mutants may exert phenotypes of their own. Finally, there are examples in which *a1* and *b1* lie within the same gene (intragenic suppression) as well as cases where they lie in separate genes (extragenic or intergenic suppression).

We care about suppression because it can tell us a great deal about the functions of the genes involved, both in terms of the relative functions of the gene products and in terms of the mechanisms of gene expression. For example, suppression can result from the physical interaction of the two mutant proteins in a fashion that corrects the defect(s) caused by the initial mutation. For this reason, screens for suppressor mutations are a commonly used technique to identify genes whose products interact with the product of a given gene of interest.

---

1 If your mother ever told you that two wrongs never make a right, you can call her right now and tell her she was wrong.

2 Why don't we say that *b1* rescues *a1*? Rescue is typically used when we are talking about transgenes, where we put in a gene that compensates for the function of either the same gene or another similarly functioning gene. We would say, "I made a rescue construct for the *muctinase* gene." Rescue constructs are typically considered proof that the gene you say is causing the phenotype is, in fact, causing the phenotype. This construct would either rescue or fail to rescue. In our example here, *b1* does not rescue *a1* because it is not doing the exact same thing that *a1* is doing.

Genetic Theory and Analysis: Finding Meaning in a Genome, Second Edition.
Danny E. Miller, Angela L. Miller, and R. Scott Hawley.

Unfortunately, not all suppressor mutants define genes whose protein products physically interact with the product of the gene being suppressed. Many suppressor mutations occur in genes whose protein products are involved in various aspects of gene expression. These mutants alter the transcription or translation of the suppressed mutant gene in a way that at least partially restores function. Other suppressor mutants act by altering one or another metabolic or biosynthetic pathway in the cell in a fashion that compensates for the defect caused by the original mutation.

So, then you ask: "Can't you just tell me the answer? Like everyone else, I am *only* interested in using suppression as a tool for finding genes whose protein products interact with that of the gene I am currently studying. Must I read these many pages just to be told how to find the suppressors I want?" Much as we discourage such impatience, we can give you an answer. You will have your best luck when you can identify extragenic suppressors that on their own convey a phenotype similar to that of the mutation you are trying to suppress. If the rationale for that assertion is patently obvious to you, you can skip this chapter. If not, we hope that it will become obvious by the end of the chapter. We begin our discussion of suppression with a consideration of intragenic suppression and then focus our attention on extragenic suppression by a variety of mechanisms, including tRNA modification in *Escherichia coli*, the physical interaction of mutant proteins in bacteriophage P22, and multiple cases of suppression in higher eukaryotes.[3]

6.1 Intragenic Suppression

The term **intragenic suppression**, also known as **pseudoreversion**, refers to the case where the two interacting mutations both lie within the same gene. Suppose you had just done a large screen for mutants that suppressed the phenotypic effects of mutant *a1* and have recovered 1,000 animals that carry *a1* but fail to express the expected phenotype. The first thing you should now do is to ask whether you can easily recover the original *a1* mutant from these animals by doing one or two crosses. In the case where your suppressor mutant defines different genes than does the *a1* mutant, it should be straightforward to separate the *a1* mutant and its suppressor. But if the original *a1* mutation cannot be easily separated from the newly isolated suppressor mutation, then the possibility of intragenic suppression must be considered. In this case, we concede that the fastest determination of this issue can be accomplished by sequencing the mutant and suppressor alleles.[4] If your gene is small enough, you could PCR and Sanger sequence your suppressor alleles (as long as you have a manageable number of them), but keep in mind that you may not fully capture all the nearby regulatory regions associated with your gene that could be altering your phenotype. If you have a large gene, PCR may be cumbersome and whole-genome sequencing of a selected number of suppressors may be appropriate.

Intragenic Suppression of Loss-of-Function Mutations

The most extreme example of an intragenic suppressor mutation is a true reversion of the original mutation. As shown in Figure 6.1, it's possible that mutation *a1* is a single base-pair change that creates a nonsense codon midway through the coding region. If mutation *b1* is a second change

3 There are numerous excellent and highly detailed reviews of suppression in both eukaryotic and prokaryotic systems (e.g. Hartman and Roth 1973; Hawthorne and Leupold 1974; Guarente 1993; Prelich 1999; Manson 2000). Our intent here is to be instructive rather than exhaustive, so readers wishing a broader array of examples are encouraged to peruse those articles.

4 You have a copy of the original stock you mutagenized, right?

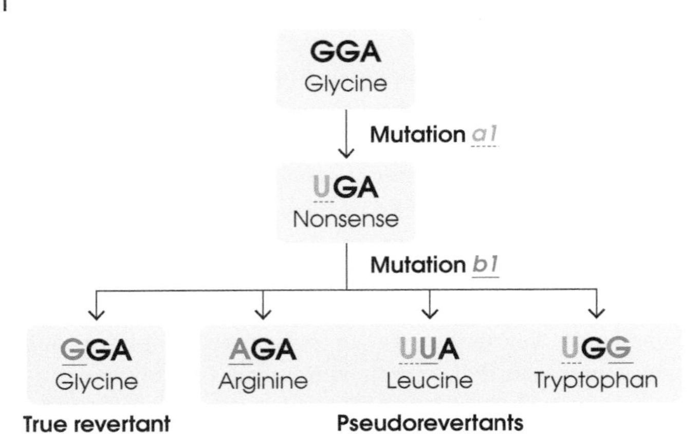

Figure 6.1 Intragenic reversion of a nonsense mutation. Both true revertants and pseudorevertants exhibit intragenic suppression by restoring some or all wildtype protein function.

that precisely reverses, or reverts, the original mutant, wildtype function will be reestablished. In this case, *b1* would be a **true revertant**. Suppose instead that mutation *b1* is not a precise reversal of the original *a1* mutation, but rather converts mutation *a1* into a different sense codon than the wildtype one. Or perhaps *b1* is a change in another base pair within that same codon that allows the doubly mutant gene to specify some amino acid (even the wrong one) at that codon. The mRNA produced by these mutant genes can now be translated. If the newly created codon specifies the same amino acid as the original wildtype codon, a normal protein will be produced. Indeed, even if the doubly mutant codon specifies a different amino acid, the resulting protein might still be capable of function depending on where the substitution occurs in the protein and the nature of the amino acid substitution. The critical distinction between these cases is that in the latter example, the *a1* and *b1* mutations occurred at different sites within the gene. The *b1* mutant is thus referred to as a **second-site revertant** or a **pseudorevertant**.

Second-site intragenic revertants can mediate suppression in several different fashions. We will consider only three possible cases. First, we will discuss the suppression of a frameshift mutation that ablates gene function, focusing on the ability of a second compensatory frameshift mutation to suppress the effect of the first frameshift mutation by restoring the reading frame downstream of the first mutation. Second, we will consider cases in which the initial mutation is a missense mutation that alters the folding of the protein product of this gene. In this case, the ability of a second missense mutation to suppress that mutation results from the ability of the second amino acid substitution to undo the damage to protein structure caused by an original missense mutation. Third, we will consider the suppression of a dominant antimorphic (poisonous) or neomorphic mutation by a second mutation that prevents the expression of the poisonous gene product.

Intragenic Suppression of a Frameshift Mutation by the Addition of a Second, Compensatory Frameshift Mutation

The first evidence that the genetic code was a triplet came from studies of intragenic suppressors of mutants within the *rII* gene in phage T4 (Crick et al. 1961). The studies began with a single mutation in the *rII* gene known to have been caused by a mutagen (proflavin) that causes base insertions or deletions. This mutation was denoted *FC0*. Putative revertants of this mutation were easily obtained. However, most of these revertants turned out to be due to the occurrence of second mutations in the *rII* gene rather than precise reversals of the original mutation. These additional mutations within the *rII* gene acted as suppressors of the original mutation.

As noted here, a suppressor mutation can be distinguished from a true revertant because it should be possible to re-isolate the original mutation. (This is especially true for bacteriophage genetics, where it is possible to score for recombinants that occur at very low frequencies. It is not practical for most higher eukaryotes.) Consider the case drawn here, where *FC0* is the original mutation and *SF* is the suppressor mutation. A crossover with a wildtype phage should allow one to re-isolate the original *FC0* mutation.

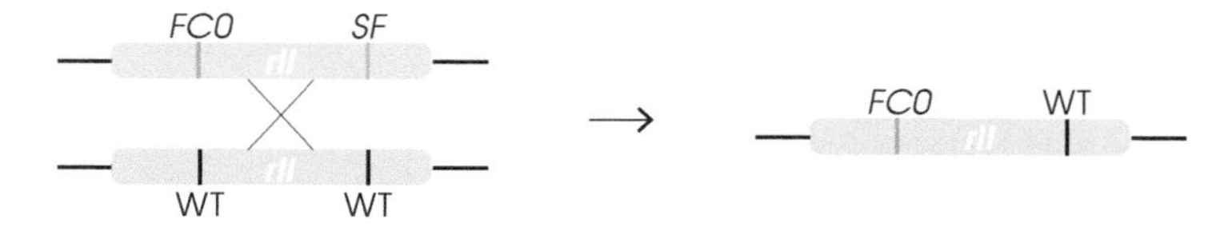

Using this technique, it was straightforward to demonstrate that these apparent revertants were the result of intragenic suppressor mutations that were themselves frameshift mutants in the *rIIb* gene. One can push this trick a bit further, and Crick et al. (1961) did just that by isolating suppressors of these new frameshift mutants. These suppressors also turned out to be *rIIb* mutants. Let's refine our nomenclature here. Let's call the *FC0* mutation the archetypal Class I mutant. All suppressors of Class I mutants will be called Class II mutants. All suppressors of Class II mutants will be thought of as Class I mutants, for reasons that will be clear in a moment. Using these terms, we can state some rules:

- Every observed case of suppression resulted from combining a Class I mutant and a Class II mutant.
- No pairwise combination of Class I mutants displayed suppression.
- No pairwise combination of Class II mutants displayed suppression.
- But not every combination of Class I and Class II mutants showed suppression.

To explain these results, Crick et al. (1961) supposed that the gene was read from a defined starting point using a consistent reading frame such that each set of *n* base pairs encoded one amino acid. If one then adds a single base, the entire frame is thrown off. Consider the model drawn below in which each underlined grouping of three bases represents the unit in which the code is read (*i.e.* the codon):

C-A-U-A-A-U-G-A-U-U-G-G-C-G-G-C-A-G-C-A-U-G-G-G-C-A-G-C-A-U-

Now insert a C after the second U:

C-A-U-A-A-U-C-G-A-U-U-G-G-C-G-G-C-A-G-C-A-U-G-G-G-C-A-G-C-A-U-

Because different codons have different meanings, the whole "sense" of the message will be disrupted after the insertion.

Now imagine a mutation that deletes the A at position 5 (so AAU becomes AUG). Again, from the deletion forward, the entire message will be changed:

C-A-U-A-U-G-A-U-U-G-G-C-G-G-C-A-G-C-A-U-G-G-G-C-A-G-C-A-U-

But now combine the insertion and the deletion:

C-A-U-A-U-C-G-A-U-U-G-G-C-G-G-C-A-G-C-A-U-G-G-G-C-A-G-C-A-U-

Note that the two mutations compensate for each other and restore the reading frame after the second mutation. Those codons including or separating the two mutations will be changed, but on either side of the two mutants, sense is restored. That the codon was, in fact, read in sets of three bases was demonstrated by the observation that three insertions or three deletions could, in some cases, restore functionality.[5]

It is critical to realize that this experiment made a strong case for the code being **degenerate** (i.e. that one amino acid can be specified by more than one codon). The fact that one could get rescue at all meant that the out-of-frame sequence between the insertion and deletion had to be readable. This is especially true when considering that the suppressor mutations were often well separated from the original mutant within the *rII* gene. If 44 of the 64 possible codons were gibberish, suppression would seem to be highly improbable. Clearly, most possible codons are readable. By definition, since there are only 20 amino acids, there must be more than one codon for some, and, in fact, virtually all, amino acids.

So, why doesn't combining a single-base insertion and a single-base deletion always create suppression? Crick et al. (1961) point out that, in some cases, "the shift of the reading frame produces some triplets the reading of which is 'unacceptable'; for example, they may be 'nonsense', or stand for 'end the chain', or be unacceptable in some other way due to the complications of protein structure." Indeed, they were right on both counts. Insertion/deletion combinations that create one of the three stop codons will result in premature chain termination. Similarly, combinations that turn a critical region of the protein into the gibberish that separates the two mutant sites will also fail to display suppression.

It is worth noting that decades later, many of the mutants studied by Crick and his colleagues would be sequenced (Shinedling et al. 1987). *FC0* turned out to be a 1-base pair insertion that could indeed be suppressed by 1-base pair deletion mutants. The triple insertion mutants that restored function were just that when analyzed by DNA sequencing – three separate insertions. The curious thing is that all these mutants fell within an approximately 120-base pair region of this gene, a region that was apparently malleable enough to withstand missense changes without seriously disrupting function. We confess to some degree of awe – well, actually, a lot of awe – with respect to the ability of Crick and his colleagues to get it all correct so long ago.

Intragenic Suppression of Missense Mutations by the Addition of a Second and Compensatory Missense Mutation

A fascinating example of this type of "corrective" suppression is provided by the analysis of intragenic suppressors of mutants in the human *p53* gene (Brachmann et al. 1998). The p53 protein is a critical regulator of cell division in humans, and the ablation of p53 activity is a characteristic of most human tumors. In most of these tumors, the inactivation of p53 is caused by an inherited missense mutant in one of the two *p53* genes (known as **Li–Fraumeni syndrome**) and the subsequent somatic loss of the other wildtype copy, a phenomenon known as **loss of heterozygosity**. These missense mutants can be divided into two classes. Mutants in the first class, functional mutations, create amino acid substitutions in a critical DNA-binding domain of the p53 protein. The second class, structural mutations, cause amino acid changes that alter the overall structural integrity and folding of the DNA-binding domain.

Jef Boeke's lab proposed that the creation of a second mutation that increased the stability of the folded state of p53 might restore functional activity to at least some of the so-called structural

5 In fact, Crick et al. (1961) only really proved that the codon was read in sets of three bases or in multiples of three bases. After all, the insertions and deletions *could* have involved deletions of sets of 2 or more base pairs. Such statements are, of course, correct. But these mutants have now been sequenced. The original *FC0* mutant was indeed a single base insertion and its suppressors are 1-base-pair deletions. The code is read in sets of three bases.

mutants (Brachmann et al. 1998). Using a clever screen for assaying the function of a human *p53* gene construct in yeast, Boeke and his collaborators were able to select for second-site intragenic revertants for three tumor-derived *p53* missense mutants. In several cases, the effects of these mutations could be understood as the consequence of compensatory structural alteration in the folding of the DNA-binding domain. Other examples of compensatory intragenic suppressors whose action can be understood in terms of known aspects of protein structure can be found in Hong and Spreitzer's work on the ribulose-bisphosphate carboxylase/oxygenase gene in *Chlamydomonas* (Hong and Spreitzer 1997), the work of di Rago et al. (1995) on the yeast mitochondrial cytochrome *b* gene, and the work of Jung and Spudich (1998) on bacterial rhodopsins. The extension of this technique to the suppression of antimorphic mutants is discussed next.

Intragenic Suppression of Antimorphic Mutations that Produce a Poisonous Protein

Suppose the original mutation that you are trying to suppress is a dominant antimorphic or neomorphic mutation. The easiest way to suppress this mutation is to simply knock out the function of the mutant gene.[6] Suppressing a poisonous mutation is as simple as preventing the expression of the mutant gene or rendering the protein product utterly nonfunctional.

The technique of reverting a dominant mutation was discussed in detail in Box 3.4. But as an example, let us once again consider a neomorphic mutant in *Drosophila* that causes a leg-determining gene to be expressed in the part of an embryo that should become the eye, resulting in an eye-to-leg transformation. The easiest way to suppress this dominant mutation is by a second mutation which renders the errant leg-determining gene inactive. A small deletion in the coding region, or even a frameshift mutation early in the coding region, would precisely serve this function. There are legions of examples of such phenomena. Indeed, reverting a dominant mutant is often a powerful method for creating true null alleles of a given gene.

6.2 Extragenic Suppression

Perhaps the more interesting circumstance is that of **intergenic** or **extragenic suppression**, where the suppressor mutant lies in a different gene than the original mutant. How might a mutant in one gene suppress the phenotypic effects of another gene? Unfortunately, there are too many excellent answers to this question to fully describe here. In some cases, suppression can act by "correcting" the mutational defect prior to or during translation, thus resulting in the production of a functional protein product. In other instances, suppression may alleviate a mutant phenotype by increasing or restoring the mutant protein's affinity for its binding partner or by permitting the use of an alternate pathway. The following sections present several different types of extragenic suppression.

6.3 Transcriptional Suppression

There are various kinds of mutations that affect the ability of a given gene to be transcribed or the ability of the resulting message to be properly processed and translated. The simplest case might be the effect of an i^s mutant in *E. coli*, which functions as a constitutive repressor of the lactose

---

6 Think of science fiction movies. No one tries to revert the process that created Godzilla and turn it back into a small lizard. They just try to kill the mutant lizard before it destroys the entire city.

operon. Such a mutant could be easily suppressed by a deletion of the regulatory DNA sequence to which this repressor binds, the lac operator. If the repressor can't bind to this gene cluster, it can't shut it off! Other such examples in eukaryotes are described next.

Suppression at the Level of Gene Expression

In *Drosophila*, there is a transposable element named *gypsy* that has the capacity to inactivate or partially inactivate genes by inserting itself between the promoter of that gene and one or more enhancer elements (Figure 6.2). Many of these insertion mutants are suppressible by mutants in a gene called *su(Hw)* (*suppressor of Hairy wing*). The *su(Hw)* gene encodes a protein (SUHW) that binds to *gypsy* elements and blocks communication between the flanking enhancers and promoters (Dorsett 1993; Kim et al. 1996; Tsai et al. 1997; Scott et al. 1999). Inactivating the *su(Hw)* gene removes this interloper protein and restores communication between the enhancers and promoters. Thus, the inability to produce functional SUHW protein allows a *su(Hw)* mutant to suppress the effect of a *gypsy* element on a gene into which it has inserted. A *gypsy* insertion that actually damaged an enhancer element or inserted into a coding region would not likely be suppressible by *su(Hw)*. (A detailed discussion of the effects of *gypsy* insertions on gene expression may be found in the discussion of the genetics of the *Drosophila cut* and *yellow* genes in Appendix B.4, and Appendix B.5.)

A CRISPR Screen for Suppression of Inhibitor Resistance in Melanoma

Suppression of transcription can also occur at enhancers, which arguably may be more difficult to detect than other types of suppressors. Sanjana and colleagues (2016) used the CRISPR/Cas9 system to screen a large region surrounding specific genes involved in BRAF inhibitor resistance in melanoma. Using a cell line carrying the BRAF V600E mutation, the investigators were able to select for cells in which guide RNAs targeted noncoding regions that suppressed inhibitor resistance. Because you can treat these cells with a drug that selects against cells with inhibitor resistance, the readout in this case is really nice: cells that live or are enriched relative to others are ones which carried a guide RNA that targeted a suppressor of inhibitor resistance. Using these data, the team was able to dissect the regions surrounding the three target genes and identify functional elements involved in gene regulation and cancer drug resistance.

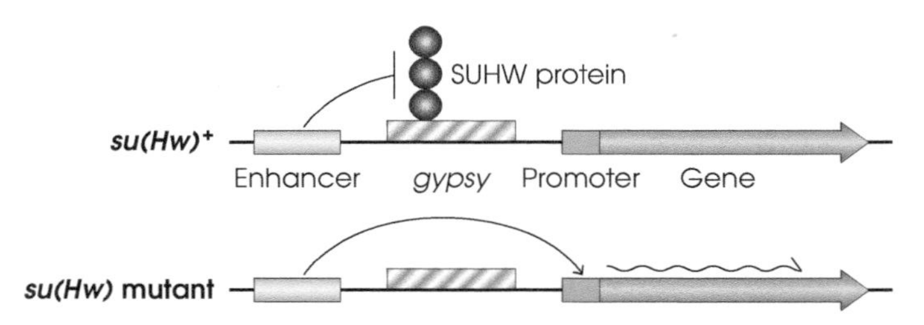

Figure 6.2 Transcriptional suppression by su(Hw) mutants. In this example, the initial mutation is the insertion of a *gypsy* transposable element. SUHW protein binds to the *gypsy* transposon, blocking communication between the flanking gene's enhancers and promoters and thereby inhibiting transcription. A second mutation in *su(Hw)* can prevent this interaction, suppressing the effect of the *gypsy* element's presence and restoring the normal transcription of the affected gene.

Suppression of Transposon-Insertion Mutants by Altering the Control of mRNA Processing

Let us suppose the transposon insertion you wish to suppress is not in some upstream regulatory region, but rather is sitting squarely within the transcribed region of the gene of interest. Now you are out of luck, right? In fact, no. Cases of exactly this type of insertion being suppressible are well known in *Drosophila*, and the mechanisms by which they are suppressed are instructive.

Some alleles of the *Drosophila vermilion* (*v*) gene are suppressed by recessive mutations at the *suppressor of sable* [*su(s)*] gene.[7] The *v* gene encodes a protein required for the normal brick-red pigment of the fly eye. In the absence of this protein, the flies have a brighter red phenotype called vermillion.

Previous work has shown that all of the *v* alleles that are suppressible by the *su(s)* mutations have identical insertions of a large transposon called *412* into the 5′ untranslated region of the *v* gene. Despite the transposon insertion into a nontranslated exon, *v* mutant flies do accumulate trace amounts of apparently wildtype-sized *v* gene transcripts in a $su(s)^{+}$ background, and the level of *v* transcript accumulation is increased by *su(s)* mutations (Fridell et al. 1990; Fridell and Searles 1994). This is to say, at some low frequency, *Drosophila* cells are capable of splicing the transposon sequences out of the *v* mRNA in the absence of the *su(s)* mutation, and the frequency of such splicing is increased in cells that are homozygous for the *su(s)* mutation. It turns out that these splicing events are apparently often imprecise, but that is probably tolerable for cells because the insertion occurs in a nontranslated exon. Examples of *su(s)* suppression are also known for alleles of other genes. In each of these cases, the suppressible mutations are also due to *412* insertions. Thus, a similar mechanism of suppression is likely (Searles et al. 1986).

Suppression of Nonsense Mutants by Messenger Stabilization

Some years ago, Phillip Anderson and Jonathan Hodgkins identified a class of mutants in *Caenorhabditis elegans* that were able to suppress specific alleles of many genes. The suppressible mutations were either nonsense mutations or mutations that resulted in aberrations in the 3′ noncoding region of the transcript. Because these suppressor mutations also caused an alteration in the male genitalia, they were named *smg* (*suppressor with a morphogenetic effect on genitalia*) mutations. Surprisingly, these *smg* genes do not encode tRNAs. Rather, they encode proteins required for a process called **mRNA surveillance**, which leads to the rapid decay of mRNAs that contain premature termination codons or altered 3′ noncoding regions (Page et al. 1999).

Also referred to as **nonsense-mediated decay**, mRNA surveillance has been demonstrated in many organisms. In *C. elegans*, the *smg* mutations block the destruction of messages with premature termination codons. So, why does that help? Isn't that nonsense codon still going to block translation? How can the cell tell a premature nonsense codon from a real stop codon, anyway? The answer to these questions lies in a process known as **translational read-through**, whereby ribosomes can read through "out-of-place" nonsense codons with a low probability, usually by inserting an improperly matched tRNA. As long as the resulting protein can function with this misincorporated amino acid, some degree of function will be restored. The *smg* mutants simply allow mutant messages more time in which read-through can occur.

---

7 The *su(s)* gene got its name because mutants in this gene also suppress the phenotypes of another mutation called *sable*.

6.4 Translational Suppression

As just mentioned, ribosomes can read past nonsense codons at a low frequency. For example, Hoja et al. (1998) showed that a UAG nonsense mutation in the yeast *BPL1* gene does not completely block the production of the normal-length protein. A low rate of read-through is accomplished by allowing the ribosome to use the glutamine (Gln) tRNA, which normally recognizes a CAG codon, to read the UAG nonsense codon. Glutamine is inserted at the site of the nonsense and translation is allowed to continue. Increasing the copy number of this normal Gln tRNA gene in these yeast cells fully suppresses the effects of the original nonsense mutant on cell growth. Samson et al. (1995) and Erdman et al. (1996) have documented other examples of this phenomenon in *Drosophila.*

Translational read-through of premature nonsense codons is a bit of an enigma. The frequency with which read-through occurs for any given nonsense mutant is dependent on the sequence, or context, in which the mutant stop codon resides (e.g. Kopczynski et al. 1992). Read-through is also usually extremely inefficient and thus unlikely to be a problem with respect to proper stop codons, which are often repeated several times in a proper context. Nonetheless, as long as read-through occurs at all, its existence suggests a mechanism by which nonsense mutants might be suppressed. If one could just keep those mutant messages around for a while and give read-through a chance to make even *some* protein, the cell might be able to squeak by on a small amount. Let's look further into how this may happen.

tRNA-Mediated Nonsense Suppression

Consider the simple case of a nonsense mutant in the *lacZ* gene of *E. coli*. Such a mutant converts a standard "sense" codon that specifies one of twenty-one amino acids into a stop codon. The presence of this stop codon causes the premature termination of the translation of the *lacZ* message and prevents the formation of its protein product. This protein, ß-galactosidase, is required for cells to grow if lactose is their only carbon source. If you mutagenize a population of *E. coli* cells that carry this stop mutant and then plate these cells on media in which lactose is the only available carbon source, the only cells that will form colonies are those that have regained the ability to make ß-galactosidase.

Some of these survivors will be true or pseudorevertants that change the stop codon into a codon that can specify the incorporation of an amino acid, and thus allow the completion of translation. In rare cases, this reversion event will create either the original codon or a synonymous codon that specifies the same amino acid as in wildtype. In other cases, the new mutation will create a codon that specifies the insertion of a different amino acid whose presence does not impair protein function.

But some of these rare survivors will still carry the original *lacZ* mutant gene. They survive because they also carry a second suppressor mutant in some other gene. Surprisingly, this second suppressor mutant also turns out to be able to suppress nonsense mutants in several other genes as well. How can a mutation in one gene suppress a nonsense mutant in other genes? More curiously, not all nonsense mutants are suppressible by the same suppressor mutation. Rather, there appear to be three classes of suppressor mutants (known as amber, ochre, and opal suppressors), each of which can suppress the corresponding set of nonsense mutants.

The answer to this curious puzzle lies in the fact that these suppressor mutations define tRNA genes. The mutation occurs in the portion of the tRNA gene that specifies the anticodon (Figure 6.3). This mutation allows the tRNA to recognize one of the three stop codons and insert an amino acid.

The finding that *sac6* mutants could suppress *act1* mutants was soon followed by the discovery that null alleles of *sac6* displayed allele-specific SSNC with the *act1-4* allele (Welch et al. 1993). Moreover, the Sac6 protein had already been identified as an actin-binding protein by biochemical studies (Drubin et al. 1988a, b). All lines of evidence suggested that the ability of *sac6* mutants to suppress the *act1* mutants reflected the physical interaction between actin and fimbrin, and thus possibly might be an example of conformational suppression.

This suggestion of conformational interaction was strengthened by the results of a screen for suppressors of the *sac6* temperature-sensitive mutations that yielded new temperature-sensitive mutations in the *ACT1* gene. As noted by Prelich (1999), this "mutual suppression" in which mutants in either of the two genes can suppress mutations in the other is a "genetic phenomenon frequently associated with interacting proteins." The *SAC6* gene was later shown to encode fimbrin, an actin filament-binding protein. Honts et al. (1994) would go on to show that eight of those *act1* mutants that could be suppressed by *sac6* mutants identified a small region on the actin protein (perhaps the fimbrin binding site?). Moreover, these *act1* mutants showed a weakened interaction with Sac6 protein in vitro.

Adams and her collaborators (Sandrock et al. 1997) further demonstrated that overexpression of *SAC6* was lethal to the yeast cell and that this effect could be specifically suppressed by mutations in the *ACT1* gene. Indeed, the specificity here was truly impressive. Out of 1,326 suppressors recovered, 1,324 simply reduced the expression of the *SAC6* gene. The remaining two suppressors were both missense mutants in the *ACT1* gene. Moreover, these two mutants altered amino acids within the same small region of actin that was identified by sequencing the *ACT1* alleles that are suppressible by *sac6* mutants.

Finally, Adams and her collaborators (Sandrock et al. 1999) would physically demonstrate that those mutant Sac6 proteins that suppressed the mutant Act1 protein bound more tightly to mutant Act1 than did wildtype Sac6 protein. Thus, one could think of the actin–fibrin interaction as the metaphorical lock-and-key mechanism in which a mutant key (fibrin) is able to interact only with a mutant lock (actin). Unfortunately, however – at least in this case – that explanation is likely to be wrong. It turns out that those mutant Sac6 proteins that suppressed the *act1* mutants also bound more tightly to wildtype Act1 than did wildtype Sac6 protein.[10]

Clearly, the ability of the altered Sac6 protein to bind more tightly to the mutant actin is not specific to the alteration in actin. Were this true lock-and-key suppression, one would not have expected the suppressing Sac6 protein to bind more tightly to wildtype actin than does normal Sac6. Quite the reverse, the suppressing Sac6 protein should bind mutant actin far better than it binds normal actin. Moreover, the sites altered in the suppressing *sac6* mutants are not normally in contact with actin (Hanein et al. 1998).

So, we are left with a simple explanation for all of this that does not really involve a modified lock opened by a reconfigured key. Simply put, mutants that reduce the ability of actin to bind fimbrin can be suppressed by any one of a number of mutants in fimbrin that increase the affinity of fimbrin for its partner. These mutants do not specifically increase the affinity of fimbrin for the mutant actin, but rather increase the affinity of fimbrin for actin in general. In this sense, these suppressor studies did not identify both the lock and the key. However, Adams and her colleagues did most likely identify the fimbrin binding site on actin (the lock), and that surely was a major accomplishment. We can leave the rest to the crystallographers.

---

10 This experiment, which eliminated the mutant-lock-meets-mutant-key model of suppression, is the ultimate case of a very careful scientist doing just one too many control experiments and killing a really cool hypothesis in the process.

Mediator Proteins and RNA Polymerase II in Yeast

The control of transcription in eukaryotic cells occurs at multiple levels. First, appropriate transcription factors must bind to the gene that is to be regulated. For simplicity, we can think of such transcriptional regulators as consisting of two functional domains. The first domain is a DNA-binding or -targeting domain that localizes the transcriptional regulator to the gene of interest. The second domain is an activation domain that either facilitates the formation of a pre-initiation complex, thus allowing transcriptional elongation, or results in the modification of chromatin structure. In the late 1980s, Rick Young and his collaborators identified multiple protein cofactors that directly influenced the activity of RNA polymerase II (RNAP II) by screening for suppressors of a unique class of internal-deletion mutants in RNAP II (Nonet and Young 1989).

In *S. cerevisiae*, the largest subunit of RNAP II is encoded by the *RPB1* gene. The carboxy-terminal domain (CTD) of this protein contains 26 or 27 repeats of a seven-amino-acid sequence Tyr-Ser-Pro-Thr-Ser-Pro-Ser. The CTD appears to play critical roles in both the initiation and elongation phase of transcription – reversible phosphorylation of the CTD plays an important role in regulating the activity of RNAP II in transcription. Although the unphosphorylated form appears to be essential for the recruitment of the pre-initiation complex, phosphorylation of the CTD is required for the transition from initiation to elongation and for the co-transcriptional processing of the primary transcript (Kang and Dahmus 1995).

Deletions within the *RPB1* gene have been recovered that reduce the number of these CTD repeats (Nonet et al. 1987). Of these *rpb1* mutants, deletions that retain at least 13 repeat units are fully viable, while deletion mutants that retain less than 10 repeats are lethal. However, those that retain 10–12 repeats are cold sensitive and unable to grow on media lacking inositol as a carbon source. Both the cold sensitivity and the inositol **auxotrophy** result from the inability of the mutant CTD to interact with the activator portions of various transcription factors, and thus to properly modulate gene expression.

To identify proteins involved in CTD function, Nonet and Young (1989) selected suppressors of the cold sensitivity of three CTD truncation mutants: *rpb1Δ101* and *rpb1Δ104*, both of which retain 11 copies of the repeat, and *rpb1Δ103*, which retains 10 copies of the repeat. Make sure you realize what they were assuming – and what they were asking for – in this selection. The first assumption was that there would be proteins that modulate the interaction between the CTD of RNAP II and the transcriptional activator proteins. The second assumption was that by reducing the repeat number in the CTD to 10 or 11 copies, they created a threshold where these accessory proteins could bind the altered CTD effectively at normal temperatures but not in the cold. At this threshold, these proteins are presumed to be capable of executing some transcriptional activation events but not others (for example, the induction of those genes required for growth on inositol). The third assumption was that one could create mutants in these accessory proteins that could bind effectively to the truncated CTD and, in doing so, reverse both the cold sensitivity and the inositol auxotrophy.

After selecting for suppressors of the cold sensitivity of the three *rpb1* mutants, the second step was to weed out intragenic revertants. The third step was testing for the restoration of the ability to grow in the absence of inositol. It was also expected that these mutants be at least semidominant, so they needed to be tested as heterozygotes. (The expectation of semidominance here is important. If the altered accessory protein can effectively mediate the interaction between the transcriptional activator and the mutant RNAP II protein, it ought to be able to do so even in the presence of a normal copy of the protein.) One might also predict that the mutants would display an interesting allele-specificity with respect to their ability to suppress other *rpb1* mutants; that is, these mutant accessory proteins ought to be able to suppress other partial CTD repeat deletions but should not be able to suppress other types of mutants in the *RPB1* gene. Finally, at least in the best of worlds, one would also predict that these suppressor mutants should have a phenotype similar to that exhibited by the CTD partial-repeat deletion mutants.

Young and his collaborators did this screen in a rather interesting fashion. They isolated suppressors in a strain that carried the deleted RNAP II gene on a plasmid and a nonfunctional large deletion on the chromosome. This trick allowed them to easily reextract the CTD-deletion-bearing plasmid from the strain after mutagenesis and thus easily distinguish intragenic from extragenic suppressors. In the case of intragenic suppressors, the CTD-deletion-bearing plasmid would no longer be able to produce the cold-sensitive mutant when transferred back into the unmutagenized parental line (it would now be viable at the restrictive temperature). But for extragenic suppressor mutations, the CTD-deletion-bearing plasmid would be unaltered.

Nonet and Young recovered 52 new suppressor mutants, of which 46 were fully characterized. The largest class of the suppressor mutants (31 out of 46) were **petite mutants**. These mutations are usually the result of gross deletions in the mitochondrial DNA and presumably function as suppressors because they slow down the growth of the yeast cells at the restrictive temperature enough to allow the residual function of the mutant CTD domain to be sufficient. But remember, they did not screen for mutants in genes that interact with RNAP II, they screened for *anything* that would suppress the cold sensitivity of the CTD-deletion mutant.

Unfortunately, 13 of the 15 remaining suppressor mutants were shown to be intragenic suppressors. Perhaps not surprisingly, most (10 out of 13) of these intragenic mutations simply enlarged the repeat domain by duplicating various portions of the repeat coding sequence, thus increasing the repeat copy number by two to five repeats. Such mutants are more properly considered to be partial revertants than suppressors. The other three intragenic suppressing mutations were independent, but identical, point mutations in a codon within a very highly conserved segment of RNAP II. All three were G- > T transversions at base pair 4,953 that changed amino acid 1,428 from valine to phenylalanine. It is not clear, but obviously of some interest, why such a mutation could suppress the repeat deletions of both *rpb1-101* and *rpb1-104*. Discarding these mutants left Nonet and Young with but two remaining candidates for extragenic suppressor mutations. One of these (denoted *s3*) was too weak to study further.

Fortunately, the remaining mutant (*s45*) was exactly what they were looking for: it suppressed both the cold sensitivity and the inositol auxotrophy. Moreover, this mutant was also semidominant. That is, it was capable of suppressing the CTD-deletion defect in both haploids and as a heterozygote. This mutation was said to define a gene known as *SRB2* (for *suppressor of RNA polymerase B*) and the mutant was named *srb2-1*. The *srb2-1* mutation was allele-specific in that it showed specificity for suppression of CTD internal deletions but did not suppress point mutations in other regions of the RNAP II gene. Moreover, deletion of the *SRB2* gene caused phenotypes similar to those of the original partial CTD-deletion mutants. These findings suggested that the Srb2 protein might functionally interact with the CTD of the RNAP II protein.

Pushing this successful approach to its limit, Young and coworkers isolated and analyzed additional suppressors (for an excellent review, see Carlson 1997). Nearly 30 dominant suppressors were isolated, all of which turned out to be alleles of four genes: *SRB2*, *SRB4*, *SRB5*, and *SRB6* (Thompson et al. 1993). All showed a pattern of suppression similar to that observed for the original *srb2-1* mutation. Biochemical analysis would eventually demonstrate that Srb2, Srb4, Srb5, and Srb6 proteins do indeed co-purify with a multisubunit complex termed "the mediator." The physical interaction of TATA-binding protein with Srb proteins was also reported. So, in this case, the screen clearly worked and was enormously informative.

Why did this screen work so incredibly well? The answer lies in the fact that the selection was well designed from the beginning and that appropriate secondary screens were able to quickly weed out the less interesting intragenic or petite mutants. But realize again that this suppression almost certainly does not reflect an "altered lock/altered key" type of suppression. Rather, all the investigators were seeking were proteins that had a higher affinity for binding the remaining repeat units.

"Lock-and-key" Conformational Suppression

> This rarest of rare circumstances constitutes the signal example of a modified lock opened by a reconfigured key, a mechanism that is prevalent in genetics textbooks but exceptional in the laboratory.
>
> – Manson (2000)

In both cases discussed here, the allele-specific suppression reflected protein–protein interactions without requiring lock-and-key conformational suppression. Rather, suppression was the result of compensatory mutations that increased the affinity of the suppressing protein for its mutant partner. All of this raises the question, does true lock-and-key suppression actually occur? Or more properly, has it been documented in any organism? Indeed, a small number of examples of this type of conformational suppression have been described in a variety of systems. A few of these, which involve the flagellar motor of *E. coli*, an acetylcholine transporter in *C. elegans* (Sandoval et al. 2006; Mathews et al. 2012), and the telomerase and shelterin proteins in humans (Armbruster et al. 2003; Schmidt et al. 2014), deserve mentioning (for review see Manson 2000).

Suppression of a Flagellar Motor Mutant in *E. coli*

The flagellar motor of *E. coli* contains two primary parts: the motor element, which is encoded by the *fliG* gene, and the stator element, which is encoded by the *motA* gene. Force generation requires the interaction of charged residues on each of these two proteins, specifically Arg90 and Glu98 in MotA and Arg281, Asp288, and Asp289 in FliG. In a truly elegant piece of work, Zhou et al. (1998) demonstrated that mutants that reversed one of these charges on MotA could be suppressed by a corresponding and compensatory change on the FliG protein. In this case, the suppressor protein is incapable of functioning when opposite a wildtype partner. Thus, the requirements for an altered key that only fits an altered lock are satisfied (Figure 6.4).

Suppression of a Mutant Transporter Gene in *C. elegans*

Acetylcholine is an excitatory neurotransmitter that functions in many animals. In *C. elegans*, acetylcholine is the major controller of motor function, and the acetylcholine transporter gene *unc-17* is essential for moving the neurotransmitter into synaptic vesicles. Mutant phenotypes in *unc-17* range from slow growth and lack of coordination to paralysis and lethality, depending on the severity of the allele (Alfonso et al. 1993; Mathews et al. 2012). A particular missense mutation in *unc-17*, *e245*, introduces a positive charge in a transmembrane domain of the UNC-17 protein, which results in impaired growth, movement, and behavior. This mutation can be suppressed by introducing a corresponding negative charge in a transmembrane domain in at least two other proteins, synaptobrevin (an integral membrane protein in synaptic vesicles) and SUP-1 (a small transmembrane protein of unknown function) (Sandoval et al. 2006; Mathews et al. 2012). Because each mutant protein independently suppresses the *unc-17* mutant, this suggests that these proteins directly interact with UNC-17.

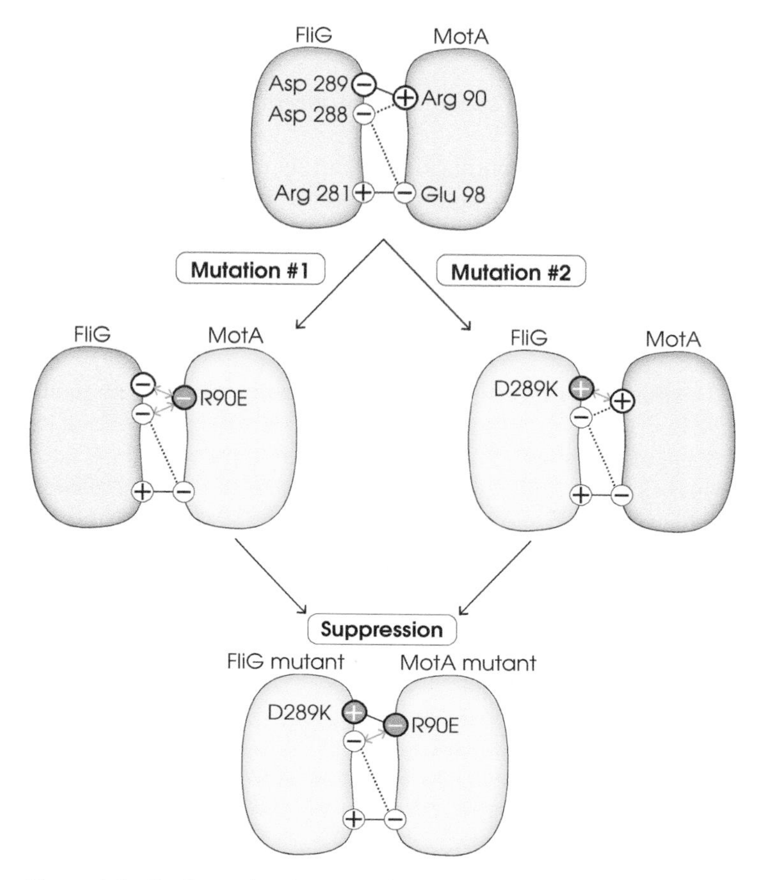

Figure 6.4 Conformational suppression. The *E. coli* proteins FliG and MotA normally interact through the charged residues indicated. A mutation in one of these residues in either protein results in a nonfunctional flagellar motor. Compensatory changes in corresponding residues on both proteins at once permits the association of the two proteins, suppressing the defect caused by a mutation in only one protein.

Suppression of a Telomerase Mutant in Humans

An excellent example of lock-and-key conformational suppression in humans is that of telomerase and shelterin mutants. Shelterin is a protein complex that binds to telomeres and shields them from being recognized as DNA damage sites during cell division. It also functions to recruit telomerase to the telomere, which in turn works to replace telomere sequence that has been lost due to the end-replication problem. The TEN-domain of human telomerase was suspected to be responsible for its recruitment to telomeres, and indeed, a mutation in the TEN-domain inhibits telomerase's ability to lengthen telomeres in vivo after chromosome duplication (Armbruster et al. 2003, 2004; Schmidt et al. 2014). Similarly, the oligosaccharide-binding-fold domain of the shelterin component TPP1 – and specifically, its TEL-patch region – was thought to mediate the recruitment of telomerase to the telomere (Nandakumar et al. 2012; Sexton et al. 2012; Zhong et al. 2012). Schmidt et al. (2014) exquisitely demonstrated that a specific two-amino-acid deletion in the TEN-domain of telomerase both prevented its interaction with shelterin and led to chromosomes with shortened telomeres. They found, however, that a compensatory change in the TEL-patch region of TPP1 suppressed the defect of the telomerase mutant, thus demonstrating a direct interaction between telomerase and TPP1.

6.7 Bypass Suppression: Suppression Without Physical Interaction

The examples presented above have dealt with the suppression of recessive mutations in the homozygous or hemizygous state. Obviously, the suppression of poisonous antimorphic or neomorphic mutations provides a powerful tool for the identification of partner proteins or of proteins functioning in the same pathway. But a more common type of suppression in higher organisms is a phenomenon referred to as **bypass suppression**. In this case, the second mutation simply allows the cell or organism to bypass a defect caused by the first mutation. In its simplest sense, one could imagine that the pathway shown in Figure 6.5 is blocked by some loss-of-function mutant that prevents the formation of an intermediate (4C), leading to the accumulation of an earlier intermediate (4B) and the absence of the final product (4D). A mutant in another gene might allow the accumulating intermediate 4B to be shunted into some related but alternative biochemical pathway that produces the necessary product. This type of bypass suppression is well characterized in prokaryotes and played a major role in identifying the various recombination pathways in *E. coli*. One of the nicest examples of bypass suppression in eukaryotes occurs in the yeast *S. pombe* (Box 6.1), in which a mutant that blocked proper telomere function was suppressed by the circularization of all three chromosomes.

There is a bit of a tendency among geneticists to view bypass suppression as less interesting than conformational suppression. However, it strikes us that the kinds of mutants recovered in bypass suppression, and thus the new genes defined, can be as interesting as those identified by screens for protein–protein interactions. Conformational suppression should only detect proteins that

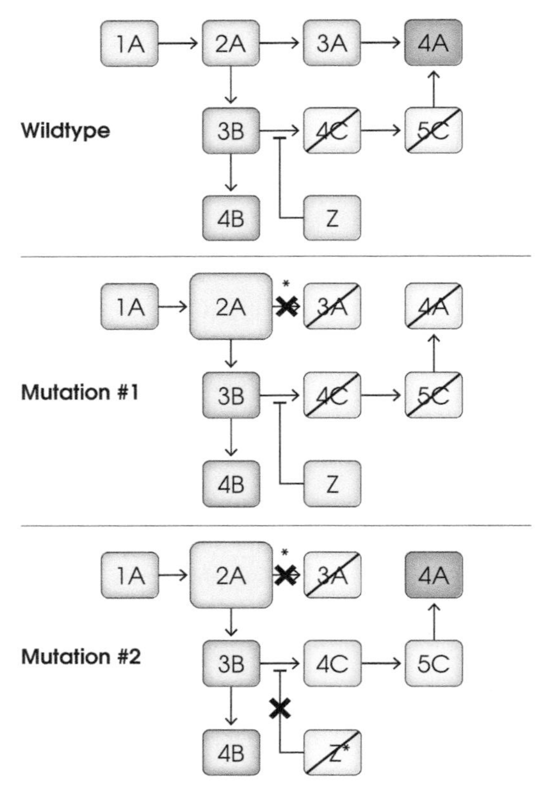

Figure 6.5 Bypass suppression. Product 4A is usually made from a series of steps beginning with substrate 1A. A shunt off this pathway produces product 4B. Another branch off *that* pathway can also make product 4A, but the use of this branch is usually blocked by protein Z. A mutation (indicated by *) that prohibits the production of intermediate compound 3A (and thus product 4A) can be suppressed by a second mutation in protein Z that allows the use of the alternative pathway to produce product 4A.

Box 6.1 Bypass Suppression of a Telomere Defect in the Yeast *S. pombe*

There is an elegant and instructive case of bypass suppression in yeast in which a mutant that blocked proper telomere function was suppressed by circularization of all three chromosomes (Baumann and Cech 2001). The mutant *pot1* (protection of telomeres) causes rapid loss of telomeres and resulting chromosome instability. Thus, sporulating a diploid heterozygous for a *pot1* mutant resulted in mutant-bearing spores that produced small colonies that grew quite slowly. For the next 10 generations, these colonies were composed primarily of very elongated cells that were unable to divide further. Cytological studies revealed that when cell division occurred, one of the two daughter cells often failed to inherit any DNA.

Nonetheless, after this period of instability, the cells in these colonies eventually stabilized and were then able to propagate and divide in the presence of the defect. Indeed, after 75 generations, the colony and cell morphology returned to normal. But these cells still lacked functional telomeres and the *pot1* mutant was very much still there. How, then, did they suppress this defect? The answer lied in the occurrence of ring chromosome derivatives resulting from chromosome breakage events near the ends of chromosomes. Ring chromosomes, by definition, do not have ends and thus do not have telomeres. Once three such events had occurred, all three chromosomes of *S. pombe* were organized as rings. The telomere functions become all but vestigial. The cell got around a telomere-defective mutant by rearranging chromosomes in such a way as to alleviate the need for telomeres at all. This has to be the defining case of bypass suppression.

physically interact with the suppressed mutant protein. But bypass suppression can identify a number of genes in related biological pathways. Thus, this technique simply has a wider net and captures related functions. Moreover, to observe conformational suppression, one needs exactly the right pair of mutants. But bypass suppression is not likely to be allele-specific and *can* involve a null mutant at either locus. Indeed, the ability to suppress a nonsense mutant is a hallmark of bypass suppression. Thus, it's possible to identify more genes in the process of a single screen.

"Push me, Pull You" Bypass Suppression

Bypass suppression is often observed when screening for suppressors of dominant antimorphic or neomorphic mutants. In this case, the second mutant need only block or compensate for the negative effects caused by the first mutation. An example of this effect is provided by studies of the antagonistic activities of two classes of motor proteins in the process of nuclear division in yeast (Saunders et al. 1997). Two kinesin-related motor proteins, Cin8 and Kip1, play an apparently redundant role in the separation of spindle poles and in spindle elongation. A double temperature-sensitive mutant in both genes displays spindle collapse during mitosis at the restrictive temperature. However, a similar mutation in a gene encoding a third kinesin-related motor, Kar3, partially suppresses the spindle collapse in *cin8 kip1* mutants. In this case, the Cin8 and Kip1 proteins are required to elongate the spindle, while the Kar3 protein is required to contract the spindle. By balancing the loss of the outward lengthening force with a compensatory loss of the inward force created by the *kar3* mutation, the phenotype is relieved.

Another example of "push me, pull you" suppression is provided by Matthies et al. (1999) regarding the suppression of a dominant effect of an α-tubulin mutation on meiotic chromosome segregation in *Drosophila*. *Drosophila melanogaster* oocytes that are heterozygous for mutations in the *α-tubulin 67C* gene display defects in centromere positioning during the early phases of meiosis I. The centromeres do not migrate toward the poles and the chromatin fails to stretch during spindle lengthening. These results suggest that the poleward forces acting at the centromeres are compromised in the *alpha-tub67C* mutants. Proper centromere orientation and chromatin elongation can be restored to oocytes bearing the α-tubulin mutant by the presence of a loss-of-function mutation in the *nod* gene, which encodes a kinesin-like protein that serves to prevent the poleward movement of the chromosomes. These results suggest that the accurate segregation of achiasmate chromosomes requires the proper balance of forces acting on the chromosomes during prometaphase. Once again, the loss of some functions required to pull chromosomes toward the poles can be compensated for by the loss of one of the functions required to retard that movement. It is all just a balance of forces, of which there are many examples (see for example Willins et al. 1995; Sharp et al. 2000).

Multicopy Bypass Suppression

We cannot conclude this chapter without at least mentioning the technique of **high-copy** (or **multicopy**) **suppression** in yeast. The idea is that you can suppress a given mutant defect by the presence of a high number of copies of some gene(s) with a related function (Bender and Pringle 1989). In this sense, it is really an example of bypass suppression. The mechanics of this technique involve creating a library of genes to be tested, using as a vector a yeast plasmid that is maintained at a high copy number within the cell.[11] You can then transform mutant cells with this library and search for transformants that can survive that appropriate restrictive condition (i.e. in which the mutant phenotype is rescued), discarding those cases where the rescuing plasmid contains the wildtype copy of the mutant gene that one is trying to suppress and focusing on the other surviving colonies. A detailed protocol for the analysis of multicopy suppression in yeast can be found in Appling (1999).

An elegant example of true bypass suppression using high-copy suppression is provided by the work of Doug Bishop and his colleagues. The yeast recombination protein DMC1 is required for the completion of meiotic recombination and of meiotic prophase. In a *dmc1* mutant, recombination intermediates cannot be resolved and thus accumulate, causing the yeast cells to arrest permanently in prophase. Bishop et al. (1999) isolated genes that, when present in high copy numbers, suppressed the meiotic arrest phenotype conferred by *dmc1* mutations. Two of the suppressors recovered, *REC114* and *RAD54*, act by altering the recombination process. In high copy numbers, *REC114* suppresses the formation of recombination intermediates. Because meiotic arrest only occurs in the presence of unresolved recombination intermediates, as long as recombination is either completed or never begun, prophase will proceed normally in yeast. High copy numbers of *RAD54* suppress meiotic arrest and promote the repair of recombination intermediates. Several lines of evidence suggest that although RAD54 is not required for the repair of recombination intermediates during meiosis, is it required for efficient repair of double-strand breaks by an alternative pathway that involves sister-chromatid exchange and operates primarily in vegetative cells (see also Schmuckli-Maurer and Heyer 2000). To quote Bishop et al. (1999), "The ability of *RAD54* to promote DMC1-independent recombination is proposed to involve suppression of a constraint that normally promotes recombination between homologous chromatids rather than sisters." Thus, the presence of a high copy number of *RAD54* genes appears to bypass the *dmc1* defect, as obviated by shunting those unrepaired recombination intermediates into an alternative pathway.

---

11 There are problems inherent in constructing such multicopy libraries. The nature of these difficulties and a possible solution are discussed by Ramer et al. (1992).

In some cases, high-copy suppression analysis *can* identify proteins that physically interact. A nice example of the power of this type of suppression is provided by the work of Susan Ferro-Novick and her collaborators on the genetics of vesicle transport between the Golgi and endoplasmic reticulum in yeast. Ferro-Novick began with a temperature-sensitive mutation in the *BET3* gene in yeast. At the restrictive temperature, these cells are defective in this process of intracellular transport and die (Jiang et al. 1998). A screen for extra-copy suppressors of this mutant yielded a new gene, *BET5*. This suppression is specific for *bet3* mutants, as *BET5* overexpression does not rescue other mutants defective in ER transport. The BET3 and BET5 proteins turn out to be components of the same protein complex. Ferro-Novick's lab also reported a similar high-copy suppression analysis that revealed an interaction between the proteins encoded by the *SEC34* and *SEC35* genes (Kim et al. 1999).

Lillie and Brown (1992) reported another case of high-copy suppression involving the physical interaction of two motor proteins in yeast. These authors demonstrated that a temperature-sensitive mutant in a myosin gene (*MYO2*) could be suppressed by high copy numbers of a kinesin-related protein encoded by the *SMY1* gene. They also demonstrated that these two proteins physically interact in the cell (Beningo et al. 2000) and observed that extra copies of *SMY1* can suppress mutations (*sec2* and *sec4*) that define components of the late secretory pathway.

The use of high-copy suppression can also be adapted to higher organisms. The application of this technique to *Drosophila* by Rorth and her collaborators was described in Section 1.1 (Rorth et al. 1998). There is also a variation of this technique in which one screens for suppressors of the effects caused by a given high-copy plasmid. This involves the isolation of suppressor mutations that obviate the defects caused by having the bait gene present in a high copy number. Sandrock et al. (1999) demonstrated that the overexpression of *SAC6* was lethal to the yeast cell and that this effect could be specifically suppressed by mutations in the *ACT1* gene. An impressive use of this technique in *S. pombe* has also been reported by Cullen et al. (2000).

6.8 Suppression of Dominant Mutations

The isolation of suppressors of dominant mutations offers an opportunity to identify proteins involved in downstream regulatory processes. For example, Karim et al. (1996) performed a screen for genes that function downstream of *Ras1* during *Drosophila* eye development. They began their screen by using a dominant *Ras1* mutant that causes the eye to become rough in appearance and then screened for dominant suppressors and enhancers of this rough-eye phenotype. From the 850,000 mutagenized flies screened, they recovered 282 dominant suppressors. Mutations in known components of the Ras signaling pathway and 577 dominant enhancers were also isolated (such as the *Drosophila* homologs of Raf, MEK, MAPK, and protein phosphatase 2A), as were mutations in several novel signaling genes. Some of these mutant genes appear to be general signaling factors that function in other Ras1 pathways, while one seems to be more specific for photoreceptor development.

6.9 Designing Your Own Screen for Suppressor Mutations

Before beginning your own suppressor screen, we suggest you consider the following issues:

1) **Carefully choose the mutant you wish to suppress.** Perhaps no step in a suppression screen is as critical as this one. A screen for conformational suppression is best done with a missense

mutant, while a bypass-suppression screen is best done with a null mutant, and so on. Everything you end up getting will depend on the bait you use.

2) **Carefully choose your mutagen.** It would be ill-advised to use transposons, X-rays, or a high-copy library to screen for conformational suppressors. If you are attempting to probe for subtle protein interactions, you will wish to change one, and only one, amino acid at a time. The choice is likely to be EMS, CRISPR, or a custom-designed transgene. But these same tools can be excellent in screens for various types of bypass suppression.
3) **Think about your selection or screen: what question are you asking?** Just what are you asking the cell to do? What phenotype are you trying to suppress? Are there obvious, easy ways the cell could ameliorate the effect (such as growing more slowly)? We urge you to choose the easiest selection you can devise but to have a way to sort among the suppressors for those that remedy the defect that is most specifically related to the process you are studying. You are going to get lots of mutants if you work hard enough. Think carefully about how to sort them out. We are partial to those suppressors that have an interesting phenotype on their own or that suppress other aspects of the original mutant phenotypes other than one used in the initial screen.

6.10 Summary

We hope to have made it clear that there are many genetic interactions that can result in one mutation suppressing another. Screening for suppressors is a powerful tool for dissecting a given biological process. The value of whatever screen you create will be determined by the kinds of secondary screens performed. More than anything else, these secondary screens will help you find the few mutations you probably want from among the many varied types of mutations you will probably get.

But we cannot leave you without a strong warning. David Botstein, quoted in Manson (2000), noted that, "When you push a genetic selection hard enough, you will almost always get what you are selecting for. However, it will usually not be what you want." Manson reinforced this point by stating, "It is imperative not to underestimate potentially bizarre and improbable consequences that can transpire when rigorous genetic selection is maintained for an appreciable length of time." Nowhere in the annals of genetics are these statements truer than in the characterization of suppression and suppressors. So, if you work hard enough, you will get the suppressor lines you seek; but remember you aren't really asking the cell, "What other proteins interact with my favorite protein?" You are asking, "What genetic tricks can the cell pull to get around a mutant in my protein of interest?" Those are very different questions. Our best advice is: Be careful out there. Crazy things abound, and they will bite you if you aren't careful.

References

Adams AE, Botstein D. 1989. Dominant suppressors of yeast actin mutations that are reciprocally suppressed. *Genetics* 121:675–683.

Alfonso A, Grundahl K, Duerr JS, Han HP, Rand JB. 1993. The *Caenorhabditis elegans unc-17* gene: a putative vesicular acetylcholine transporter. *Science* 261:617–619.

Appling DR. 1999. Genetic approaches to the study of protein–protein interactions. *Methods* 19:338–349.

Armbruster BN, Etheridge KT, Broccoli D, Counter CM. 2003. Putative telomere-recruiting domain in the catalytic subunit of human telomerase. *Mol. Cell. Biol.* 23:3237–3246.

Armbruster BN, Linardic CM, Veldman T, Bansal NP, Downie DL, *et al.* 2004. Rescue of an hTERT mutant defective in telomere elongation by fusion with hPot1. *Mol. Cell. Biol.* 24:3552–3561.

Baumann P, Cech TR. 2001. Pot1, the putative telomere end-binding protein in fission yeast and humans. *Science* 292:1171–1175.

Bender A, Pringle JR. 1989. Multicopy suppression of the *cdc24* budding defect in yeast by CDC42 and three newly identified genes including the ras-related gene RSR1. *Proc. Natl. Acad. Sci. U.S.A.* 86:9976–9980.

Beningo KA, Lillie SH, Brown SS. 2000. The yeast kinesin-related protein Smy1p exerts its effects on the class V myosin Myo2p via a physical interaction. *Mol. Biol. Cell* 11:691–702.

Betzner AS, Oakes MP, Huttner E. 1997. Transfer RNA-mediated suppression of amber stop codons in transgenic *Arabidopsis thaliana*. *Plant J.* 11:587–595.

Bishop DK, Nikolski Y, Oshiro J, Chon J, Shinohara M, *et al.* 1999. High copy number suppression of the meiotic arrest caused by a *dmc1* mutation: REC114 imposes an early recombination block and RAD54 promotes a DMC1-independent DSB repair pathway. *Genes Cells* 4:425–444.

Brachmann RK, Yu K, Eby Y, Pavletich NP, Boeke JD. 1998. Genetic selection of intragenic suppressor mutations that reverse the effect of common *p53* cancer mutations. *EMBO J.* 17:1847–1859.

Buckingham RH. 1994. Codon context and protein synthesis: enhancements of the genetic code. *Biochimie* 76:351–354.

Buvoli M, Buvoli A, Leinwand LA. 2000. Suppression of nonsense mutations in cell culture and mice by multimerized suppressor tRNA genes. *Mol. Cell. Biol.* 20:3116–3124.

Carlson M. 1997. Genetics of transcriptional regulation in yeast: connections to the RNA polymerase II CTD. *Annu. Rev. Cell Dev. Biol.* 13:1–23.

Crick FH, Barnett L, Brenner S, Watts-Tobin RJ. 1961. General nature of the genetic code for proteins. *Nature* 192:1227–1232.

Cullen CF, May KM, Hagan IM, Glover DM, Ohkura H. 2000. A new genetic method for isolating functionally interacting genes: high *plo1*$^{+}$-dependent mutants and their suppressors define genes in mitotic and septation pathways in fission yeast. *Genetics* 155:1521–1534.

Dorsett D. 1993. Distance-independent inactivation of an enhancer by the suppressor of *Hairy-wing* DNA-binding protein of *Drosophila*. *Genetics* 134:1135–1144.

Drubin D, Kobayashi S, Kellogg D, Kirschner M. 1988a. Regulation of microtubule protein levels during cellular morphogenesis in nerve growth factor-treated PC12 cells. *J. Cell Biol.* 106: 1583–1591.

Drubin DG, Miller KG, Botstein D. 1988b. Yeast actin-binding proteins: evidence for a role in morphogenesis. *J. Cell Biol.* 107:2551–2561.

Erdman SE, Chen HJ, Burtis KC. 1996. Functional and genetic characterization of the oligomerization and DNA binding properties of the *Drosophila* doublesex proteins. *Genetics* 144:1639–1652.

Fridell RA, Searles LL. 1994. Evidence for a role of the *Drosophila melanogaster suppressor of sable* gene in the pre-mRNA splicing pathway. *Mol. Cell. Biol.* 14:859–867.

Fridell RA, Pret AM, Searles LL. 1990. A retrotransposon 412 insertion within an exon of the *Drosophila melanogaster vermilion* gene is spliced from the precursor RNA. *Genes Dev.* 4:559–566.

Garza D, Medhora MM, Hartl DL. 1990. *Drosophila* nonsense suppressors: functional analysis in *Saccharomyces cerevisiae*, *Drosophila* tissue culture cells and *Drosophila melanogaster*. *Genetics* 126:625–637.

Gesteland RF, Wolfner M, Grisafi P, Fink G, Botstein D, *et al.* 1976. Yeast suppressors of UAA and UAG nonsense codons work efficiently in vitro via tRNA. *Cell* 7:381–390.

Guarente L. 1993. Synthetic enhancement in gene interaction: a genetic tool come of age. *Trends Genet.* 9(10):362–366.

Hanein D, Volkmann N, Goldsmith S, Michon AM, Lehman W, *et al.* 1998. An atomic model of fimbrin binding to F-actin and its implications for filament crosslinking and regulation. *Nat. Struct. Biol.* 5:787–792.

Hartman PE, Roth JR. 1973. Mechanisms of suppression. *Adv. Genet.* 17:1–105.

Hawthorne DC, Leupold U. 1974. Suppressors in yeast. *Curr. Top Microbiol. Immunol.* 64(0):1–47.

Hoja U, Wellein C, Greiner E, Schweizer E. 1998. Pleiotropic phenotype of acetyl-CoA-carboxylase-defective yeast cells – viability of a BPL1-amber mutation depending on its readthrough by normal tRNA(Gln)(CAG). *Eur. J. Biochem.* 254:520–526.

Hong S, Spreitzer RJ. 1997. Complementing substitutions at the bottom of the barrel influence catalysis and stability of ribulose-bisphosphate carboxylase/oxygenase. *J. Biol. Chem.* 272:11114–11117.

Honts JE, Sandrock TS, Brower SM, O'Dell JL, Adams AE. 1994. Actin mutations that show suppression with fimbrin mutations identify a likely fimbrin-binding site on actin. *J. Cell Biol.* 126:413–422.

Jarvik J, Botstein D. 1975. Conditional-lethal mutations that suppress genetic defects in morphogenesis by altering structural proteins. *Proc. Natl. Acad. Sci. U.S.A.* 72:2738–2742.

Jiang Y, Scarpa A, Zhang L, Stone S, Feliciano E, *et al.* 1998. A high copy suppressor screen reveals genetic interactions between BET3 and a new gene. Evidence for a novel complex in ER-to-Golgi transport. *Genetics* 149:833–841.

Jung KH, Spudich JL. 1998. Suppressor mutation analysis of the sensory rhodopsin I-transducer complex: insights into the color-sensing mechanism. *J. Bacteriol.* 180:2033–2042.

Kang ME, Dahmus ME. 1995. The unique C-terminal domain of RNA polymerase II and its role in transcription. *Adv. Enzymol. Relat. Areas Mol. Biol.* 71:41–77.

Karim FD, Chang HC, Therrien M, Wassarman DA, Laverty T, *et al.* 1996. A screen for genes that function downstream of Ras1 during *Drosophila* eye development. *Genetics* 143:315–329.

Kim J, Shen B, Rosen C, Dorsett D. 1996. The DNA-binding and enhancer-blocking domains of the *Drosophila* suppressor of Hairy-wing protein. *Mol. Cell. Biol.* 16:3381–3392.

Kim DW, Sacher M, Scarpa A, Quinn AM, Ferro-Novick S. 1999. High-copy suppressor analysis reveals a physical interaction between Sec34p and Sec35p, a protein implicated in vesicle docking. *Mol. Biol. Cell* 10:3317–3329.

Kopczynski JB, Raff AC, Bonner JJ. 1992. Translational readthrough at nonsense mutations in the HSF1 gene of *Saccharomyces cerevisiae*. *Mol. Gen. Genet.* 234:369–378.

Lillie SH, Brown SS. 1992. Suppression of a myosin defect by a kinesin-related gene. *Nature* 356:358–361.

Lundgren K, Walworth N, Booher R, Dembski M, Kirschner M, *et al.* 1991. mik1 and wee1 cooperate in the inhibitory tyrosine phosphorylation of cdc2. *Cell* 64:1111–1122.

Magliery TJ, Anderson JC, Schultz PG. 2001. Expanding the genetic code: selection of efficient suppressors of four-base codons and identification of "shifty" four-base codons with a library approach in *Escherichia coli*. *J. Mol. Biol.* 307:755–769.

Manson MD. 2000. Allele-specific suppression as a tool to study protein–protein interactions in bacteria. *Methods* 20:18–34.

Mathews EA, Mullen GP, Hodgkin J, Duerr JS, Rand JB. 2012. Genetic interactions between UNC-17/VAChT and a novel transmembrane protein in *Caenorhabditis elegans*. *Genetics* 192:1315–1325.

Matthies HJ, Messina LG, Namba R, Greer KJ, Walker MY, *et al.* 1999. Mutations in the *alpha-tubulin 67C* gene specifically impair achiasmate segregation in *Drosophila melanogaster*. *J. Cell Biol.* 147:1137–1144.

Mendenhall MD, Leeds P, Fen H, Mathison L, Zwick M, *et al.* 1987. Frameshift suppressor mutations affecting the major glycine transfer RNAs of *Saccharomyces cerevisiae*. *J. Mol. Biol.* 194:41–58.

therefore be prior to or at the step that makes the intermediate C because the addition of C allows the mutant to continue down the pathway and grow.

We now consider the mutant combinations *m1, m2* and *m1, m3*. The double mutant *m1, m2* grows when compound C is supplied. The double mutant *m1, m3* grows only if compound E is added. We can conclude that *m2* defines a gene that acts upstream of the synthesis of C, while *m3* defines a gene that acts downstream of the synthesis of C. We cannot yet order genes *m1* and *m2* with respect to the synthesis of C by genetic analysis alone, unless we determine whether *m1* or *m2* can grow in the presence of exogenous compound B.

A useful way to consider these data is to think of them in terms of their defects. Amending our pathway drawing by placing the genes at the appropriate steps yields:

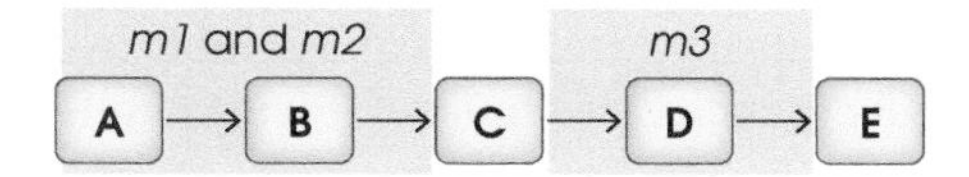

While at one level, we might say that the phenotype of each mutant is simply auxotrophy for E, in truth we now know more. The information we have allows us to order the functions of these gene products. The real phenotype of the double mutant *m1, m2* is that it cannot make C. The phenotype of *m3* is that it cannot process C into E. Thus, the real phenotype of the *m1, m3* and *m2, m3* double mutants is identical to that of *m1* or *m2* alone: the double mutants cannot make C. It does not matter how many more downstream mutants (that *could* have made compound C) you have, they still cannot make compound C. If the cell cannot make C, it does not matter if the cell can synthesize C → D or D → E, it will not be able to complete the pathway.

Epistasis describes these interactions and can be thought of as the ability of the genotype at one locus to influence the effect of an allele at another locus. Thus, with respect to the ability to produce compound C, *m1* and *m2* are said to be epistatic to *m3*. Even though *m3* could have made compound C by itself, in the presence of either mutant 1 or mutant 2, it can no longer do so.

Experimentally, we can define epistasis as follows: for two mutants that produce different but related phenotypes, mutant X is said to be epistatic to mutant Y if the X phenotype predominates in the double mutant organism. Mutant Y, therefore, is hypostatic to mutant X. A set of rules for epistasis analysis of gene order in biosynthetic pathways is listed here:

- **Rule 1:** All mutants involved must be null loss-of-function alleles.[1]
- **Rule 2:** Upstream mutants are epistatic to downstream mutants.
- **Rule 3:** Two null mutants that fail to show epistasis cannot be in the same pathway.

Nonbiosynthetic Pathways

There are many cases outside of simple biosynthetic pathways in which one mutant might be epistatic to another. An eye color mutant that blocks any pigment deposition (e.g. *white* mutants in *Drosophila*) will be epistatic to mutants that affect the color of pigment that is deposited.[2] Similarly, a wingless mutant will be epistatic to a mutant that produces curled or otherwise abnormal wings, an eyeless mutant would be epistatic to a mutant that alters eye color, etc. Although these are

1 In truth, people do perform epistasis tests with hypomorphs. It often works, but for the reasons considered above, it is risky.

2 Excellent examples of the application of epistasis analysis may be found in the study of eye color pigmentation in *Drosophila* (Phillips and Forrest 1980).

rather obvious examples, more sophisticated versions of epistasis analysis are necessary for more complicated biological questions.

Botstein and Maurer (1982) described the canonical example of pathway dissection by examining mutants that affect the assembly of phage T4. Mutants affecting head assembly have no effect on tail assembly and vice versa (so the processes are presumably distinct), but mutants that block DNA replication are epistatic to mutants in both assembly pathways. Thus, after DNA replication, the pathway would branch into two independent pathways of phage assembly, one for head and one for tail assembly.

This example is also a good warning about overinterpreting the biological mechanisms that underlie these functional dependencies. It is not the case, as might be assumed, that replicated DNA is required for head or tail assembly. Rather, DNA replication is required for the activation of the "late genes" in this phage whose products are required during head and tail assembly. All epistasis can tell you is whether step 2 requires step 1 in order to occur. It cannot tell you the nature of that requirement.

Epistasis analysis can, however, reveal the existence of multiple pathways, even when both (or all) of your mutants define but one of those pathways. An example of this type of analysis comes from the work of Jeff Sekelsky and Kim McKim on meiotic mutants in *Drosophila* (Sekelsky et al. 1995; McKim et al. 1996). There are two meiotic mutants in *Drosophila*, *mei-9* and *mei-218*, that independently reduce the frequency of meiotic recombination by more than 90%. These maximum reductions of 90% are observed even with null alleles of these genes. There are two ways to think about this observation:

1) There are two pathways of meiotic recombination in *Drosophila*, a major one that requires *mei-9* or *mei-218* and a minor one that does not. Such secondary recombination pathways have been observed in other organisms [see Whitby (2005) for a discussion of Class I vs. Class II pathways].
2) There is only one pathway, but there are redundant proteins that can substitute for either Mei-9 or Mei-218 with a low probability of success.

The two hypotheses ask a simple question: Is the small percent (10%) of exchange that remains in *mei-9* and *mei-218* mutants the result of a second, independent pathway that does not require either gene? If this is the case, then we expect the double homozygotes to be no more severe than homozygotes of either mutant alone – the *mei-9 mei-218* double mutant should still show 10% residual recombination. The other possibility is that there is a single pathway in which both the Mei-9 and Mei-218 proteins act separately, with neither mutant being strong enough to fully ablate the function of the pathway. Based on this second model, we would expect the effects of the two mutants on this pathway to be multiplicative, and thus expect only 1% of residual exchange. Sekelsky et al. (1995) observed that the frequency of residual exchange in double homozygotes was not reduced below the level observed in *mei-218* homozygotes alone. Thus, these data support the first hypothesis in which there are both major and minor pathways of meiotic recombination in *Drosophila*, similar to those observed in other organisms. It is likely that both pathways share a common set of functions because there are other mutants that completely ablate recombination in *Drosophila*.

The usefulness of this type of analysis is hard to overstate. As pointed out by Botstein and Maurer (1982), "tests of epistasis allow the provisional grouping of mutations into a pathway structure: two mutations failing to show epistasis cannot be on the same dependent pathway." Nonetheless, our ability to assess epistatic interactions often depends both on the type of mutants we have and on the process or pathway that we are trying to dissect.

7.2 Dissection of Regulatory Hierarchies

The real value of epistasis analysis is in the genetic dissection of complex regulatory pathways, such as those that determine cell fate or sex. To perform this type of analysis, you must have already identified a group of mutants that misdirect the outcome of a given developmental process (e.g. gain- or loss-of-function mutants that cause genetic males to develop as phenotypic females, or mutants that cause cells that should become bristle-formers to take on some alternative fate, such as becoming neurons). These mutants are rather special – it is not that that they simply cause some structure to fail to develop or some process to occur, but rather that they change the direction or outcome of that process.

This analysis also requires that the mutants be fully penetrant, which is to say that the phenotype is expressed in every cell or organism and that your collection includes mutants that misdirect the pathway in both directions (i.e. you need mutants to cause genetically male individuals to develop as females *and* mutants that cause genetically female individuals to develop as males). This form of epistasis analysis is somewhat less genetically restrictive than the one used for studying biosynthetic pathways in that we can now use both dominant gain-of-function and recessive loss-of-function alleles – as long as they are fully penetrant. You now need to order the functions of the genes defined by these mutants into a regulatory pathway.

Epistasis Analysis Using Mutants with Opposite Effects on the Phenotype

Let us consider a hypothetical pathway in the development of a sensory structure in the fly. This structure is built from 16 progenitor cells. During development, the localized presence of an inductive chemical signal (CS) causes the center four cells out of the 16 progenitor cells in this structure to become neurons. In the absence of that CS, the remaining 12 outer cells become bristle-forming epithelial cells. If you prevent the formation of the CS, all cells develop as bristle epithelial cells. If you flood the structure with CS, all cells become neurons. Thus, the fate of these cells is dependent on a simple binary signal: CS is present, *or* CS is absent. We refer to these two situations as signal states, following the review by Avery and Wasserman (1992).

You have isolated dominant **constitutive mutations** (mutations that cause the protein encoded by this gene to be locked into an active form, even if they do receive the CS signal) and null loss-of-function mutations that affect this differentiation process in two separate genes (*neu1* and *epi1*). Dominant *Neu1* mutants cause all 16 cells to become neurons, even those that do not receive the CS. Conversely, null alleles of *neu1* (denoted *neu1*$^{-}$) lead to a neuron-less phenotype, even in the presence of the CS. Dominant *Epi1* mutants also present a neuron-less phenotype, while loss-of-function *epi1* mutants (denoted *epi1*$^{-}$) produce the all-neuron phenotype. To summarize:

- *Neu1* or *epi1*$^{-}$ → All neurons
- *Epi1* or *neu1*$^{-}$ → No neurons

It is critical to realize that the two dominant mutants (*Neu1* and *Epi1*) are exerting their effects in two quite different fashions. *Neu1* causes cells that have not received the CS to create neurons, while *Epi1* causes cells that do receive the CS to block neuronal differentiation and become epithelial cells. Thus, each of these mutants affects only one of the two possible signal states (CS present or CS absent). In this case, the mutants have opposite effects: *Neu1* turns cells that should become epithelial cells into neurons, while *Epi1* converts cells that should become neurons into epithelial cells.

It is fortunate here that the phenotypes are clean (i.e. fully penetrant) and reciprocal. We can use these mutants to determine whether the *neu1* and *epi1* genes act in a single pathway and, if so, to determine the order in which they act. The rules for such an analysis are as follows:

1) The pathway we are dissecting must be binary, both in terms of the initiating signal (CS) and the result (neuron vs. non-neuron cell fate).
2) Sequential steps in the pathway must act as a series of binary switches.
3) All mutants tested are either complete loss-of-function alleles or fully penetrant dominants, and each mutant serves to lock a signal switch in the + or – setting.
4) The test is designed to work with two mutants that produce fully opposite effects (i.e. they affect opposite signal states).
5) When the two mutants affect opposite signal states, the setting of the last "switch" determines the phenotype. The downstream mutant will be epistatic to the upstream mutant.

Remembering that mutational ablation of the CS gene (cs^-) results in a neuron-less, all-epithelial-cell phenotype, we can envision a pathway in which the CS leads to the neuronal pattern of differentiation by acting through the Neu1 and Epi1 proteins. To determine how that pathway would be constructed, let's look at the phenotypes of two double mutant combinations:

- cs^- *Neu1* → All neurons
- *Neu1 Epi1* → No neurons

These data tell us that *Neu1* mutants are epistatic to cs^-, thus the Neu1 protein must function downstream of the CS in the regulatory cascade. Similarly, because *Epi1* mutants are epistatic to *Neu1*, the Epi1 protein must function downstream of Neu1 in the regulatory cascade.

We can confirm this by observing the phenotypes of two more double mutant combinations:[3]

- cs^- $epi1^-$ → All neurons
- $neu1^-$ $epi1^-$ → All neurons

Thus, the Epi1 protein is the critical step in this cascade. We can now understand this process of cell differentiation as consisting of three steps:

1) The function of the Epi1 protein is to prevent neuronal development. We know this because loss-of-function alleles of *epi1* (the $epi1^-$ allele in this example) allow all the cells to become neurons and gain-of-function (constitutively activated) alleles of *Epi1* cause all cells to become epithelial.
2) The function of the Neu1 protein is to repress the activity (or synthesis) of Epi1. We know this because the loss-of-function *epi1* alleles have the same phenotype as do the dominant gain-of-function alleles of *Neu1*. That is, constitutively expressing Neu1 has the same effect as removing Epi1.
3) CS activates Neu1. A cs^- mutant has the same phenotype as a $neu1^-$ mutant, and the overexpression of Neu1 is epistatic to a cs^- mutant.

In this case, we can easily diagram the regulatory hierarchy:

CS → activates Neu 1 → represses Epi1 → allows neural development

3 Note that there would be no point in constructing *Neu1* $epi1^-$ or $neu1^-$ *Epi1* double homozygotes because in each case, both mutants have the identical phenotype and so, presumably, would the double mutant.

Thus, the most proximate regulatory element is the Epi1 protein, which, when left alone, will repress the neuronal choice. Only the interference by a CS-activated Neu1 allows neuronal differentiation.

Hierarchies for Sex Determination in *Drosophila*

There are many cases of epistatic interactions leading to the construction of hierarchies. The best studied of these examples are the hierarchies for sex determination in *Drosophila* and *Caenorhabditis elegans* (Cline and Meyer 1996; Garrett-Engele et al. 2002).[4] Consider the regulatory hierarchy for sexual development in *Drosophila*:

$$\text{Two X chromosomes} \rightarrow \text{Sxl protein present} \rightarrow \text{Tra present} \rightarrow \text{DsxF}$$

$$\text{One X chromosome} \rightarrow \text{Sxl protein absent} \rightarrow \text{Tra absent} \rightarrow \text{DsxM}$$

In the third step of the sex determination hierarchy, the *dsx* mRNA is spliced to produce a transcript encoding a female-producing protein (DsxF) or a male-producing protein (DsxM). The direction of that splice is determined by the presence or absence of functional Tra protein. In the presence of Tra, the *dsx* mRNA is spliced to produce DsxF; in the absence of Tra, the default splice of the *dsx* mRNA produces DsxM. The presence or absence of functional Tra is controlled at the level of splicing by the presence or absence of Sxl protein. At the beginning of this hierarchy, the presence or absence of Sxl protein is determined by the number of X chromosomes to autosomes.

Sex determination is an ideal case to study epistasis because the initial signal (one X versus two) and the final result (male or female differentiation) are binary, and null and fully penetrant dominant alleles are available for each gene involved in the pathway. The dominant alleles of each gene affect one signal state, while the null alleles of that same gene affect the other. For example, constitutive alleles of *Sxl* will initiate the female pathway of development in the presence of a single X chromosome, while null alleles of *Sxl* will result in XX individuals developing as males.[5] The same can be said for constitutive and loss-of-function alleles of *tra*. Although null alleles of *dsx* develop as intersex individuals, dominant mutants of *dsx* (dsx^M and dsx^F) exist that lock the *dsx* gene into producing only DsxM or DsxF protein regardless of the presence or absence of Tra protein. The dominant dsx^M allele will exert a phenotypic effect only in the XX signal state, while the dominant dsx^F allele exerts an effect only in the single-X signal state.

We can now combine any two mutants that produce opposite sexual transformations – that is to say, any two mutants that exert their effects in opposite signal states. The dominant *dsx* mutants will be epistatic to either constitutive or loss-of-function mutants anywhere above them in the hierarchy of XX-induced events. Similarly, null alleles of *dsx* will differentiate as intersex offspring regardless of the genotype with respect to *Sxl* and *Tra*. In every case, it will be the phenotype of the downstream mutant (whether loss-of-function or constitutive) that will predominate. But remember, it is not possible to use two mutants that have their effects on the same signal state; both mutants will produce the same transformation.

We can use dominant constitutive mutants in these analyses as long as their effects are fully penetrant. That is, we could combine a dominant "always on" allele of *Sxl* with a loss-of-function *tra* mutant. The result (defined by the phenotype of the downstream mutant, *tra*) would be a

---

4 Perhaps the nicest exposition of epistasis analysis in any higher eukaryote is the study of fly sex determination by Baker and Ridge (1980).

5 Well, actually, they are lethal, but patches develop as male tissue.

commitment to male differentiation. All that matters in this pathway is the status of the final binary switch in this pathway, the splicing of the *dsx* gene. If there is no Tra activity, then the *dsx* gene will be spliced according to the male pathway and male differentiation can proceed.

Epistasis Analysis Using Mutants with the Same or Similar Effects on the Final Phenotype

The analysis described above requires that the two mutants act on opposite signal states, such that each tested pair of mutants exhibits opposite effects. This is obviously a serious restriction. Many screens are based on recovering mutants that fail to respond to some signal, and hence all mutants have the same phenotypes. In this case, epistasis analysis is still possible if you can meet any of three conditions:

1) You can obtain opposite-acting conditional mutants of the genes involved in the pathway.
2) You have available to you some drug or agent that allows you to block the progression of the pathway under study without aborting that pathway entirely or killing the cell.
3) Mutants in the two genes have at least slight differences in their phenotypes.

Let us discuss these three conditions further.

Using Opposite-Acting Conditional Mutants to Order Gene Function by Reciprocal Shift Experiments

The reciprocal shift method was first developed by Jarvik and Botstein (Jarvik and Botstein 1973) for use in their study of phage morphogenesis and then applied to the study of the yeast cell cycle by Hereford and Hartwell (1974). To do **reciprocal shift experiments**, you need conditional mutants at both genes that allow you to block the process in two different ways – for example, both cold-sensitive (*cs*) and heat-sensitive (*ts*) mutants for each gene.

As shown in Figure 7.2, we can learn a great deal by combining a *ts* allele of gene *A* with a *cs* allele of gene *B* (or vice versa). Assuming a method exists to initially synchronize our population of cells or organisms, we can then allow the cell or organism to go through a process of reciprocal temperature shifts: first a high-temperature phase (inactivating gene *A*), then a low-temperature phase (inactivating gene *B*). If gene *A* acts before gene *B*, this set of temperature shifts will produce a mutant phenotype. Even though gene *B* was functioning during the high-temperature period, the preceding (and presumably essential) prior function of gene *A* was inactivated. The restoration of gene *A* activity during the subsequent shift to cold was of no importance because gene *B* was no longer able to provide the essential subsequent function.

Suppose we reverse the order of temperature shifts. The initial cold treatment knocks out gene *B* but still allows gene *A* to produce its product. The subsequent shift to the high temperature stops gene *A* from functioning but frees up gene *B* to act on the substrate provided by gene *A* during the previous shift. This reversed order of temperature shift is expected to produce a normal or near-wildtype outcome. This method yields information about whether two genes act in the same or different pathways and about their order if they do function in the same pathway. The power of this method is demonstrated by the schematic diagram of the yeast cell cycle pathway (Figure 7.3) as deduced from reciprocal shift experiments (Hartwell et al. 1974; Moir and Botstein 1982).

Using a Drug or Agent that Stops the Pathway at a Given Point

Suppose you have only *ts* mutants in genes that affect the completion of the yeast cell cycle. Obviously, without some set of internal or differentiating landmarks for cell cycle progression, you can neither order these genes nor assign them to pathways. Now imagine that you had an agent that blocked the

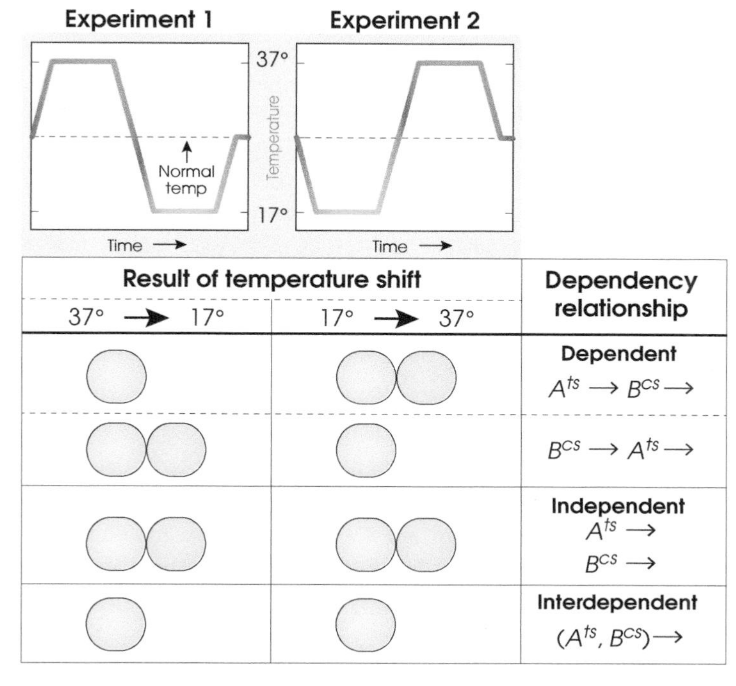

Figure 7.2 A reciprocal temperature-shift experiment. This example uses a heat-sensitive allele of gene *A* (A^{ts}) and a cold-sensitive allele of gene *B* (B^{cs}) combined with reciprocal temperature-shift experiments to determine the order in which the two genes act in a pathway. If gene *B*'s function is dependent on the prior action of gene *A* (*A*→*B*→), then experiment 2, which begins with a low-temperature phase followed by a high-temperature phase, will restore function and two cells will be observed. If gene *B*'s function precedes that of gene *A* (*B*→*A*→), then experiment 1 will restore function. This experiment can also tell us if the relationship between the two genes is independent (the order is not critical for function) or interdependent (both must be active at the same time to restore function).

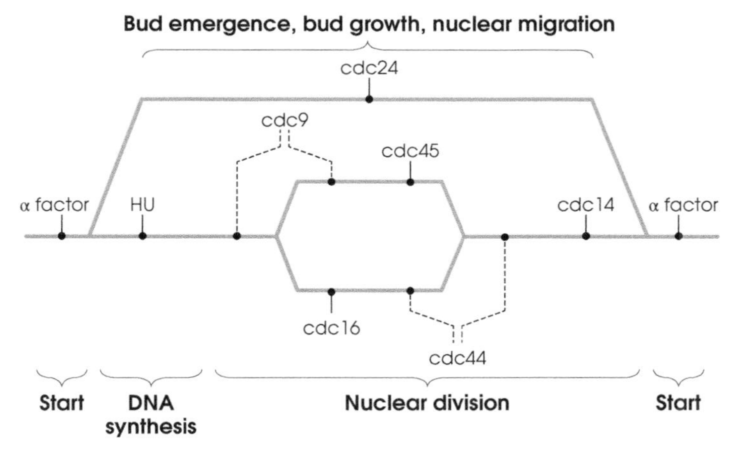

Figure 7.3 The yeast cell cycle pathway as deduced from reciprocal shift experiments. The order of gene function was determined by both cold- and heat-sensitive mutants. This demonstrates that *cdc45* and *cdc16* functions are independent of each other. The *cdc9* and *cdc44* results were inconclusive. Thus, the authors could only state that either those two genes are independent of each other or cdc9 functions before the *cdc44* step of the pathway (Adapted from Moir and Botstein 1982).

pathway midcycle without killing the cells. You could then test each mutant by comparing the effects of a regime of high-temperature/drug-arrest-and-release/permissive-temperature (HT/DAR/PT) with a cycle of permissive-temperature/drug-arrest-and-release/high-temperature (PT/DAR/HT). Heat-sensitive mutants in genes that act prior to the drug arrest will block cell cycle completion in the HT/DAR/PT cycle but allow completion in the PT/DAR/HT cycle. Mutants in genes that act after the arrest point will exhibit a reciprocal effect. Note, however, that this method only assigns a temporal order – it does not allow you to determine whether two genes act in the same pathway.

Exploiting Subtle Phenotypic Differences Exhibited by Mutants that Affect the Same Signal State
In the cell death cascade in *C. elegans*, loss-of-function mutants in two genes (*ced-3* and *ced-1*) exert different effects on the process of programmed cell death and engulfment by neighboring cells (reviewed by Avery and Wasserman 1992). Both of these mutants exert their effects in the same signal state, after the cell death pathway has already been switched on.

We can perform an epistasis analysis in this case only because the two mutants have different phenotypic effects in the same signal state. In *ced-3* mutants, cell death is completely suppressed. In *ced-1* mutants, the cell dies, but the dead cell is not engulfed by neighboring cells. In a *ced-1 ced-3* double mutant, the phenotype is identical to that of *ced-3* – cell death is completely prevented. In this process, the outcome is not a single binary state, but rather the pathway branches. That differences in the phenotypes of these mutants, or branching of the pathway, in terms of engulfment, allows the analysis.

We can understand this by diagramming the regulatory cascade:

Cell death signal → activates Ced3 → triggers cell death → activates Ced1 → triggers engulfment

In cases such as this, where both your mutants have their effects on the same signal state, the rules of epistasis analysis are different. Here, where the two null mutants tested affect the same signal state, the setting of the first switch determines the phenotype. Thus, the *ced-3 ced-1* double mutant will be phenotypically identical to *ced-3*.

This analysis becomes more complex if we use constitutive dominant mutations. A constitutively activated *ced-3* mutant would presumably trigger cell death and engulfment in the absence of a signal. However, a cell combining a constitutively activated *ced-3* along with a *ced-1* loss-of-function allele would still cause death but would not allow corpse engulfment. In this sense, the downstream mutant is epistatic to the upstream mutant in terms of the engulfment phenotype, but the *ced-3* mutant is epistatic in terms of the cell death phenotype. It is amusing to imagine the phenotype of a *ced-3 ced-1* double mutant, where *ced-1* is a constitutive dominant allele of *ced-1* – living cells trying to induce their neighbors to eat them! Thus, with respect to engulfment, when combining dominant constitutive and loss-of-function alleles, it is the downstream mutant that will be epistatic to the upstream mutant. In terms of cell death, it is the upstream phenotype (no death) that is epistatic to the downstream mutant. The issue here is that the rules have changed slightly because the pathway no longer has a simple binary outcome. A simple regulatory hierarchy now has a complex side branch. How we assess the result depends on which branch we are examining.

7.3 How Might an Epistasis Experiment Mislead You?

Can you be fooled by epistasis? The answer is yes, easily. Avery and Wasserman (1992) describe cases in which the functional orders of gene action were incorrectly obtained because the investigators used partial loss-of-function alleles. So, it's important to verify that your alleles are indeed null mutants (*see* Chapters 1 and 2). As noted by Botstein and Maurer (1982), ". . . tests of epistasis . . . require detailed knowledge of mutant phenotype." We would add to that only, ". . . and of the nature of the mutants themselves." Epistasis analysis can also fail if a given component of a regulatory hierarchy has more than two states (ON/OFF) and instead functions more like a dimmer switch. The analysis can also fail if you simply misunderstand the type of pathway you are dissecting (Ferguson et al. 1987). If, for example, the final phenotype involves the interaction of parallel or partially redundant cellular pathways, or if unknown backup or salvage pathways exist, one can easily misunderstand the result. Finally, it is not at all uncommon for the double mutant to have a phenotype quite different from that of either of the two single mutants. Such pitfalls are dangerous and perhaps not all that uncommon. We echo Avery and Wasserman (1992) in pointing out that ". . . epistasis analysis alone will not tell you in molecular detail how genes are regulated." All epistasis analysis can do is suggest who is regulating whom. We suggest that epistasis is best thought of as a tool that can implicate pairs of genes in the same process. Beyond that, one needs to rely on the tools of molecular or cell biology to elucidate the exact function of the relevant gene products in that process.

7.4 Summary

This chapter addressed different aspects of the same question: when and where do genes work? The "when" was addressed in the context of biosynthetic pathways or regulatory hierarchies and the "where" in the context of whether two or more genes act in the same or different pathways. The next chapter looks at mosaic analysis and asks similar questions, but in terms of the when and where of a developing organism. Both types of analysis can be extremely powerful if used correctly. However, they also require a thorough understanding of the nature of the mutants being used.

References

Avery L, Wasserman S. 1992. Ordering gene function: the interpretation of epistasis in regulatory hierarchies. *Trends Genet.* 8:312–316.

Baker BS, Ridge KA. 1980. Sex and the single cell. I. On the action of major loci affecting sex determination in *Drosophila melanogaster. Genetics* 94:383–423.

Botstein D, Maurer R. 1982. Genetic approaches to the analysis of microbial development. *Annu. Rev. Genet.* 16:61–83.

Cline TW, Meyer BJ. 1996. Vive la différence: males vs females in flies vs worms. *Annu. Rev. Genet.* 30:637–702.

Ferguson EL, Sternberg PW, Horvitz HR. 1987. A genetic pathway for the specification of the vulval cell lineages of *Caenorhabditis elegans. Nature* 326:259–267.

Garrett-Engele CM, Siegal ML, Manoli DS, Williams BC, Li H, *et al.* 2002. *Intersex*, a gene required for female sexual development in *Drosophila*, is expressed in both sexes and functions together with doublesex to regulate terminal differentiation. *Development* 129:4661–4675.

Hartwell LH, Culotti J, Pringle JR, Reid BJ. 1974. Genetic control of the cell division cycle in yeast. *Science* 183:46–51.

Hereford LM, Hartwell LH. 1974. Sequential gene function in the initiation of *Saccharomyces cerevisiae* DNA synthesis. *J. Mol. Biol.* 84:445–461.

Jarvik J, Botstein D. 1973. A genetic method for determining the order of events in a biological pathway. *Proc. Natl. Acad. Sci. U. S. A.* 70:2046–2050.

McKim KS, Dahmus JB, Hawley RS. 1996. Cloning of the *Drosophila melanogaster* meiotic recombination gene *mei-218*: a genetic and molecular analysis of interval 15E. *Genetics* 144:215–228.

Moir D, Botstein D. 1982. Determination of the order of gene function in the yeast nuclear division pathway using *cs* and *ts* mutants. *Genetics* 100:565–577.

Phillips JP, Forrest HS. 1980. Ommochromes and pteridines. In: Ashburner M, Wright T, editors. *The Genetics and Biology of Drosophila*, Vol. 2d. Academic Press. pp. 541–623; 83 p.

Sekelsky JJ, McKim KS, Chin GM, Hawley RS. 1995. The *Drosophila* meiotic recombination gene *mei-9* encodes a homologue of the yeast excision repair protein Rad1. *Genetics* 141:619–627.

Whitby MC. 2005. Making crossovers during meiosis. *Biochem. Soc. Trans.* 33:1451.

8

Mosaic Analysis

What are researchers to do if mutating or knocking out their gene of interest causes lethality in their model system? Or what if the mutation they are studying has a broad effect on the organism, but they are only interested in the response from a particular tissue or at a particular stage of development? The answer is mosaic analysis. While epistasis analysis can determine whether two genes function in the same pathway, **mosaic analysis** helps us examine at what times during development and in which tissues these genes interact. Using mosaic analysis, we can induce a mutation in a specific tissue or at a specific time and study its effects.

If a given mutation results in a phenotype, such as an absent antenna structure, it might seem straightforward to assume that the wildtype product of that gene acts in the tissue primordium that forms that antenna structure. Although reasonable, such assumptions are often misleading. The gene defined by your mutant may act at a distance by producing a diffusible factor that induces competent cells to take on an antenna fate, or its product might act much earlier in development by controlling the specification of some region or compartment that will eventually be required to make an antenna. How, then, do we determine when or where a given gene exerts its function? To address these questions, we need to be able to produce and analyze genetic mosaics. A **mosaic** is an animal or plant in which cells or tissues within that single organism have different genotypes and can be distinguished as having a separate origin. Indeed, all multicellular organisms, including you, are mosaics at some level (Acuna-Hidalgo et al. 2015; Campbell et al. 2015).

The basic question here is very simple: if an organism possesses both wildtype and mutant cells, which cells must be wildtype to produce a normal phenotype? More importantly, *how many* of those cells must be wildtype to produce a normal phenotype? Being able to address either question requires that we accomplish three objectives:

1) We must be able to make the appropriate mosaics at the appropriate time during development (and they must survive long enough to assess the phenotype of interest).
2) We must be able to easily distinguish mutant cells from wildtype cells.
3) We must be able to assess the phenotype in the whole organisms we produce.

Some excellent reviews of mosaic analysis include Stern (1936, 1968), Nesbitt and Gartler (1971), Hall et al. (1976), Ashburner et al. (2005), Golic (1993), Greenspan (1997), and Yochem and Herman (2005). We lack the space here to cover in detail the rather large variety of methods available for creating mosaics. Rather we will only briefly summarize the techniques themselves, focusing primarily on the types of questions that can be answered by using mosaic analysis. We begin with perhaps the oldest form of creating genetic mosaics: tissue transplantation.

Genetic Theory and Analysis: Finding Meaning in a Genome, Second Edition.
Danny E. Miller, Angela L. Miller, and R. Scott Hawley.

8.1 Tissue Transplantation

Tissue transplantation is simply the process of taking cells from one individual and transplanting them into a second individual. This is typically done from embryo to embryo, as in zebrafish or mice, where cells from one blastocyst can be injected into a second blastocyst, creating a genetic mosaic. This is an especially powerful technique when the mutant you are trying to study is lethal – a problem for in vivo analysis. Mosaics allow you to study the impact of a mutation in the background of a presumably wildtype individual.

Early Tissue Transplantation in *Drosophila*

In the early 1900s, Ephrussi and Beadle were working on the relationship between genotype and phenotype with respect to eye color in *Drosophila*. The basic question was: did the genes that produced the color of the fly's eye act in the eye cells themselves? The experiment involved transplanting the larval precursors of the eyes, known as eye imaginal discs, from one larva to another. When a disc from larva A was implanted into larva B, the transplanted disc did differentiate into an eye and could be "scored" for eye color in the next stage of fly development in the pupa. They could then ask whether transplanting an eye disc from a mutant larva into a wildtype larva could rescue the eye pigment defect. If the surrounding wildtype tissue could rescue the pigmentation of the mutant eye disc, then the necessary gene product for proper pigmentation could be (and probably is) made in cells outside the eye.

The issue we are getting here is called **cell autonomy**. We say that the phenotype of genetic function is cell autonomous if that phenotype is determined by the genotype of the cell or cells that exhibit it. So, in modern parlance, Beadle and Ephrussi (1936) were asking whether eye color was cell autonomous in *Drosophila*.

They were working with two loss-of-function eye color mutants, *vermillion* and *cinnabar*, both of which produce a fly with bright red eyes. Normal flies have dark or brick-red eyes as a result of producing both a red and a brown pigment. The wildtype *cinnabar* and wildtype *vermillion* genes are required to produce the brown pigment; hence, the eyes of flies homozygous or hemizygous for either mutant are bright red (*vermillion* is on the X chromosome, so males with only one mutant copy of *vermillion* are affected). If they transplanted an eye disc from a larva that was homozygous for either a *vermillion* or *cinnabar* mutation into a wildtype larva, the resulting supernumerary eye was normal in eye color. The wildtype host larva could provide whatever function the *vermillion* or *cinnabar* eye discs were missing.[1] Clearly, the functions of the *vermillion* and *cinnabar* genes are not cell or tissue autonomous. These genes must produce diffusible products that play some role in the synthesis of the normal pigment. When the mutant discs were placed into the wildtype host larvae, the developing discs were able to obtain these factors from the host tissue and bypass the mutant defect.

Beadle and Ephrussi also used this surgical form of mosaic analysis to ask a second question, namely, could one use mosaics to order two genes that function in a given biosynthetic pathway? They noted that the transplantation of *vermillion* discs into *cinnabar* larvae resulted in flies with normal-colored supernumerary eyes. Thus, the *cinnabar* host could provide some component

1 Think for a moment about the obvious control for this experiment: transplanting *vermillion* eye discs into *vermillion* larvae or transplanting *cinnabar* discs into *cinnabar* larvae. In both cases, you would expect to recover the same phenotype as the flies you transplanted the disc from.

8.3 Mitotic Recombination

As shown in Figure 8.3, mitotic chromosome recombination between a heterozygous mutant and its centromere *can* produce daughter cells that are each homozygous for one of the two alleles present in the parent cell. We put the word *can* in italics because only half of the two possible alignments of the two homologs at metaphase will result in homozygous daughter cells. (Remember this is mitosis – the homologs are not going to segregate from each other even though they recombined.) Since its discovery by Stern (1936), mitotic recombination has long been the favorite tool of fly geneticists both for lineage analysis (see for example Bryant and Schneiderman 1969; Wieschaus and Ghering 1976) and for studies that permit the use of small clones. The later during development the exchange event is induced, the smaller the number of divisions that daughter cells will have remaining, and thus the smaller the clone. There are also methods for increasing the proliferative capacity of a clone relative to its neighbors (thus producing bigger clones) and for genetically marking clones so they can be distinguished from the majority of parental cells. These techniques are reviewed in Ashburner et al. (2005). Although the frequency of spontaneous mitotic recombination is low, it can be increased by treating larvae with X-rays.

Initially, there were real difficulties in the use of mitotic recombination for mosaic analysis. For example, in many cases, the mutant one used to identify clones was separate from the mutant one wanted to test in that clone. Since it was not possible to control where exchange occurred, there was a real chance that the clones would not be the ones desired. Moreover, one could not control either the timing or the tissue in which the mitotic exchange events occurred. Timing matters, because a constant worry in studies of induced mosaicism is the **perdurance of gene products**. For example, suppose a protein product made before the induced exchange was able to perdure (persist) in the daughter cells long enough to create a wildtype or wildtype-appearing outcome.

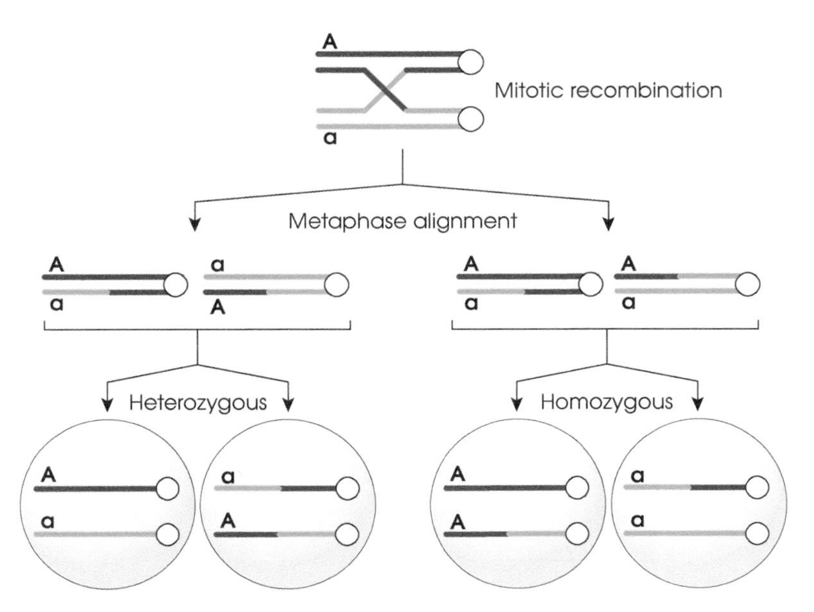

Figure 8.3 Mitotic chromosome recombination. Recombination between a heterozygous mutant and its homolog can, depending on how the chromosomes align on the metaphase plate, produce daughter cells that are each homozygous for one of the two alleles present in the parent cell.

The clone might then appear phenotypically normal, leading the investigator to conclude, incorrectly, that a wildtype gene product was not required in those cells to obtain a normal phenotype. The most useful tool would be a system in which one could control both the frequency and position of mitotic exchange while also controlling its timing and position during development.

Gene Knockout Using the FLP/FRT or Cre-Lox Systems

The utility of mitotic recombination for mosaic analysis was initially limited by its inability to specify precisely when, where along a chromosome, and in what tissues recombination occurred – it was not genetically controllable. The value of mitotic recombination in *Drosophila* was enhanced enormously by Kent Golic's use of the **FLP/FRT system** (Golic 1991; Golic and Golic 1996). This system involves the introduction into *Drosophila* (or a number of other organisms, for that matter) of a two-component site-specific recombination system derived from the yeast 2-μm plasmid. In this system, the recombinase enzyme FLP is uniquely and solely capable of inducing high frequencies of mitotic recombination at a specific site, *FRT* (for *FLP recombinase target*) (Figure 8.4). One can thus use transformation to integrate the 34-base pair *FRT* element into the fly genome (at one of any number of sites) and a *FLP* gene whose expression you can control. For example, if your *FLP* construct is controlled by a **heat-shock promoter**, you can induce a high frequency of site-specific exchanges at any time in development that you choose, depending only on when you expose the developing flies to heat shock. The advantages of this system are enormous:

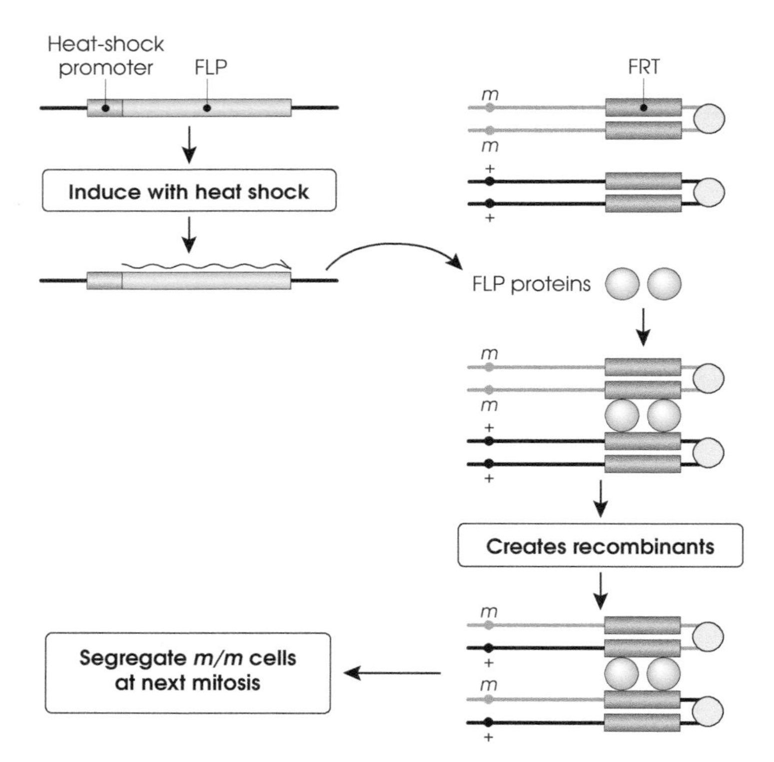

Figure 8.4 The FLP/FRT system. The FLP enzyme induces high frequencies of mitotic recombination between *FRT* sites. Using an inducible or conditional promoter, such as a heat-shock promoter, ensures the FLP protein will be expressed only at the desired time during development or in a particular tissue.

1) The exchanges you induce occur only at the sites where an *FRT* element is homozygous. Thus, all markers distal to the *FRT* will co-segregate at the ensuing mitotic division. You can use any visible marker distal to the *FRT* site to mark your clones.
2) You do not need to use radiation to obtain high levels of exchange.
3) The active form of FLP is short-lived, so all the induced events are likely to have occurred at or close to the time of heat shock.
4) By using a tissue-specific regulatory element (rather than a heat-shock regulator), one can control where in the fly the exchanges occur (rather than when). The value to this technique is that one can use intrachromatid exchanges between *FRT* elements flanking a gene of interest to excise that gene in a specific tissue (Figure 8.5).[4]

A similar technique, the **Cre-Lox system**, was derived from bacteriophage P1 for use in yeast, plants, and mice (Hoff et al. 2001; Liu et al. 2001; Yamanishi & Matsuyama 2012). Like *FRT*, *loxP* is a 34-base pair site that is recognized by a recombinase (in this case Cre), and the Cre-Lox system can be used to induce site-specific mitotic recombination in a manner similar to FLP/FRT (Lewandoski 2001). The goal of using either method is the same – to delete a segment of DNA either at a specific time or in a specific tissue to affect the function or expression of a gene of interest. The key is to express the Cre or FLP recombinase when you want to, which will then target the *loxP* or *FRT* sites, respectively, within each cell (Scharfenberger et al. 2014).

To assess the functionality of both Cre and FLP recombinase in *Drosophila*, Frickenhause and colleagues (2015) tested cell- and time-specific knockouts of the *cabeza* gene, a homolog of a human gene implicated in several neurodegenerative disorders such as amyotrophic lateral sclerosis (ALS) and frontotemporal dementia. Either Cre or FLP was selectively expressed in neurons or

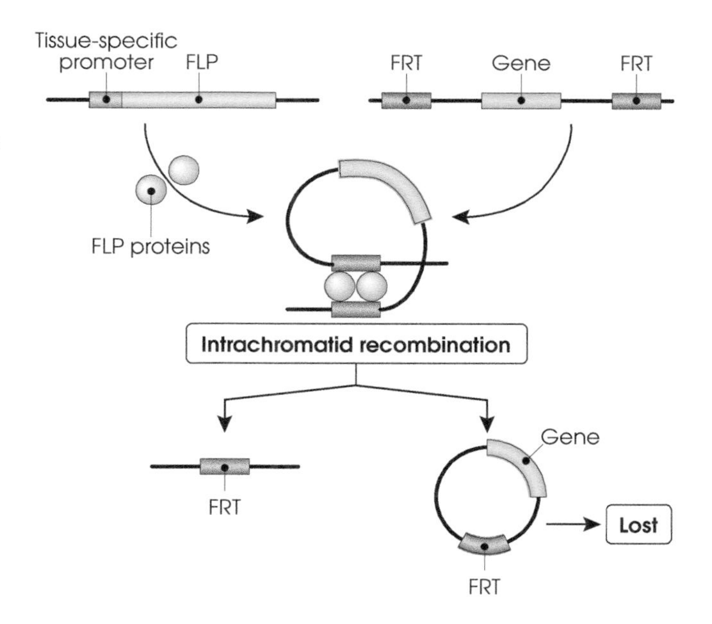

Figure 8.5 Using the FLP/FRT system to excise a gene of interest. *FLP* expression can mediate intrachromatid exchange between two *FRT* elements surrounding a gene of interest. This results in excision both of the sequence between the two *FRT* regions (which includes your gene) and of one *FRT* region. If a tissue-specific promoter is used, the excised gene will be lost only from that tissue type.

4 We can also use this system in a slightly different way – during mutant screens (Page et al. 2007). By placing two *FRT* sites at proximal locations on the chromosome, we can induce recombination in females heterozygous for a mutation. This allows us to recover homozygous mutagenized chromosome arms, saving a step in the screening process and allowing us to quickly filter out new homozygous lethal mutants.

muscle cells and effectively removed several introns from *cabeza* between either the *loxP* or *FRT* sites that the researchers had engineered into the endogenous gene. Strikingly, they found that using a pan-neuronal promoter to drive either *Cre* or *FLP* resulted in 100% knockout of *cabeza* in the adult flies, resulting in adults with the expected phenotypes. Thus, both the Cre-Lox and FLP/FRT systems appear to be robust.

8.4 Tissue-Specific Gene Expression

Lucky for us, technology only continues to improve, and we now have multiple methods of mosaic analysis that permit us to study mutations in specific tissues, at specific times, or under specific conditions. A tissue-specific promoter, for example, can be used to express a construct that produces a product designed to knock down a particular gene. This method is capable of remarkable specificity, and it allows the study of mutations that may be lethal to the organism as a whole or whose phenotype may vary from tissue to tissue or among developmental stages. Tissue-specific drivers are available for most model organisms for nearly any tissue or stage necessary.

An elegant example of tissue-specific expression comes from work done by Huang and Stern (2004) to understand fibroblast growth factor (FGF) signaling in *C. elegans*. The investigators were interested in understanding how the receptor tyrosine phosphatase CLR-1 attenuates expression of EGL-15, the FGF receptor. Worms with overactive FGF signaling (e.g. *clr-1* mutants) accumulate fluid in the hypodermis – a phenotype known as Clear – and those with underactive FGF signaling (e.g. *egl-15* mutants) arrest during early development, making it difficult to study how CLR-1 and EGL-15 interact.

To understand how fluid balance is regulated in *C. elegans*, Huang and Stern turned to tissue-specific promoters to drive the expression of both *egl-15* and *clr-1* transgenes in different tissues. Using promoters from two different hypodermal genes, two neuronal genes, and one from a gene expressed in body-wall muscles, they expressed *egl-15* or *clr-1* transgenes in either *egl-15* or *clr-1* mutants, respectively. The researchers found that expression of the transgenes in the hypodermis fully rescued their respective mutant phenotypes, but body-wall muscle expression did not. And although the neuronal promoters conferred some rescue for both genes, further cytological testing suggested that was not the primary location of action for either protein. Thus, mosaic analysis allowed the researchers to conclude that the two proteins likely both function predominantly in the hypodermis (Huang 2004).

Gene Knockdown Using RNAi

Sometimes it is not tissue-specific *rescue* of a gene you are interested in, but tissue-specific or stage-specific *knockdown* of a gene. This type of knockdown may be achieved in a number of ways, including knockdown by RNAi. As an example, Zeng et al. (2015) were interested in identifying genes involved in intestinal stem cell (ISC) differentiation in *Drosophila*. Understanding this process, which regulates the rapid turnover of the intestinal epithelium, has obvious clinical correlations to humans in areas such as response to chemotherapy or tumor development.

Zeng and colleagues screened 16 562 publically available transgenic flies carrying RNAi constructs (either dsRNA or shRNA constructs). This collection targeted about 90% of the nearly 14,000 protein-coding genes in *Drosophila*. These RNAi constructs were driven using the yeast GAL4 transcriptional activator, which targets a specific sequence called the upstream activation sequence (UAS) (Brand and Perrimon 1993). Flies carrying just the RNAi construct plus the UAS would not express the RNAi unless GAL4 was present. To express the construct, all the investigators had to do was cross an RNAi-carrying fly with a GAL4-carrying fly and collect those flies that carried both constructs.

The investigators first wanted to identify the lines that were lethal, reasoning that those genes would be essential for stem cell maintenance. To screen for lethality, they crossed the stock carrying the UAS+RNAi construct to a stock carrying GAL4 under the control of the *actin* promoter. In these flies, the RNAi would be expressed at all times and in all tissues because *actin* is a constitutively expressed gene. This revealed 7,429 lines targeting 6,170 genes that were lethal. The investigators then wanted to drive expression of the lethal RNAi constructs only in the intestine. To do this they crossed the lines to a gut-specific *esg-Gal4* line, which would only express GAL4 in the intestine (Micchelli and Perrimon 2006).

As a useful side note, the investigators used a neat trick to keep their stocks healthy. You can imagine that keeping a stock with the gut-specific *esg-Gal4* driver and the RNAi construct could be a problem for those stocks where the knockdown in the gut was still lethal or made a very sick fly. To alleviate this problem, they included a temperature-sensitive inhibitor of the *esg-Gal4* driver, *tub-Gal80^{ts}*, which is driven off the tubulin promoter and is only active at 18 °C. Keeping the stocks at 18 °C prevents GAL4 from activating the RNAi construct. To turn the RNAi on all they had to do was move their flies from 18 to 29 °C, the *tub-Gal80^{ts}* would stop being expressed, and the GAL4 would be free to activate the UAS and express the RNAi construct. To identify ISC-interacting genes, the investigators examined the midguts of each of those 7,429 lines and identified interesting ISC phenotypes in 478 of them (Zeng et al. 2015). They then studied the interaction of those 478 genes to better understand how they influence things like ISC maintenance, growth, and differentiation – a useful resource for anyone studying stem cell maintenance in the gut.

It is important to keep in mind that RNAi does not always completely knock down a gene, and it may have off-target effects, meaning you may not be targeting the gene you think you are targeting or you may end up targeting multiple genes with a single construct. When performing tissue- or stage-specific RNAi knockdown of a gene, it is also important to be wary of either the persistence of the RNAi construct or its effect on neighboring tissues or descendent cells. Bosch et al. (2016) reported persistent RNAi knockdown of genes after shRNAs were produced that targeted GFP. Importantly, the same "shadow" effect was not observed when using dsRNA targeting GFP, suggesting that the type of construct may affect the degree of persistent knockdown. The fact that it may take time to reverse the effects of RNAi knockdown is not unexpected. As pointed out by Bosch and colleagues, there have been similar reports in mouse (Bartlett and Davis 2006) and mammalian cell culture experiments (Baccarini et al. 2011). These kinds of reports suggest that observations based on RNAi knockdown should be carefully interpreted and the persistence of transcripts or off-target effects should always be considered.

Tissue-Specific Gene Editing Using CRISPR/Cas9

Using the CRISPR/Cas9 system, researchers have begun to make precise modifications to the genome at chosen times or within specific tissues. The details of the CRISPR/Cas9 system are described in Section 2.2: Mutagenesis and mutational mechanisms in the context of introducing new mutations. Here, the concept is the same, except the expression of the Cas9 endonuclease is under the control of a tissue- or drug-inducible promoter. In addition, a guide RNA (gRNA) that targets the Cas9 enzyme to specific regions of the genome must be either constitutively available or also under the control of the same promoter as the Cas9 enzyme.

There are many reports of successful gene editing using this system, with the obvious goal being to edit human somatic DNA to correct single-gene diseases (see Box 8.1). A wonderful example of this is a group of studies that used CRISPR/Cas9 to edit the dystrophin gene in mouse models of human muscular dystrophy (Long et al. 2015; Nelson et al. 2015; Tabebordbar et al. 2015). Specifically, researchers used adenovirus to deliver cDNA encoding the CRISPR/Cas9 machinery along with gRNAs that target a specific exon (exon 23) carrying a deleterious mutation that prevents proper expression of the dystrophin gene. Deletion of this exon eliminates the mutation and results in proper expression of the gene. Notably, one group of researchers was able to induce this mutation in nearly half of the muscular and cardiac cells of postnatal mice and ameliorate many of the phenotypes typically seen with this disease (Long et al. 2015).

To perform complete gene excisions similar to the Cre and FLP systems, you would need to use two gRNAs targeting different locations within the gene. Although tissue-specific knockouts using CRISPR/Cas9 are being pursued and demonstrated successfully, there are some concerns that anyone planning this type of experiment should keep in mind.

1) **The CRISPR/Cas9 system is not perfect.** You will not get a cut and excision in 100% of targeted cells. Instead, a number of outcomes may be observed. You may get cutting at only one gRNA position rather than both, which might not result in ablation or excision of the gene you are working on.
2) **The cell repairs the double-strand breaks introduced by Cas9 using the error-prone nonhomologous end-joining system.** This may result in very small (or very large) deletions that you had not planned on. If your goal is to ablate the function of a gene and you are targeting an exon, then this is probably fine. But, if your goal is to remove only a single exon or group of exons, then you might encounter problems.

Box 8.1 The Ethics of Targeted Gene Editing in Humans

The power of the CRISPR/Cas9 system to selectively and specifically edit DNA has been heralded as one of the greatest scientific discoveries in modern history. It is now possible to imagine making directed genetic changes in living humans or in human embryos. The potential for this technique to treat, and perhaps even cure, myriad genetic diseases is immeasurable. Indeed, human trials are underway using CRISPR to modify patient cells (http://ClinicalTrials.gov Identifier: NCT04774536) or to treat specific conditions such as viral keratitis (http://Clinicaltrials.gov Identifier: NCT04560790). Countless other studies have been proposed or are actively being planned. This means that what previously was a theoretical debate about where and how to use gene-editing techniques is no longer theoretical.

This brings us to the most important question we must address: How far do we go? Obviously, using CRISPR to correct deleterious somatic mutations is an exciting prospect with the potential to improve or save the lives of thousands of people. But is it acceptable to make germline alterations in humans as well? We submit that at this time the answer to this question is an emphatic "No." Not only does germline alteration push us down the slippery slope toward human eugenics, but it also assumes that we fully understand the genetics of all the diseases we may wish to treat, which we do not. Will it be acceptable in the future to make directed germline changes in human embryos? Our sense is that yes, this will eventually be acceptable; we only hope these techniques remain focused on reducing the burden of disease, not directed at other traits such as height, personality, or any other number of unique human characteristics that create the rich diversity we are so fortunate to have among us as a species.

3) **If you are trying to couple cutting and repair by homologous recombination with a template, it is not clear how successful you will be.** Several problems persist, including how to deliver the template at the time and in the tissue you are interested in (keep in mind that you repair off of DNA, so you cannot use a driver to make that DNA available at the same time your Cas9 and gRNA(s) are made – it always has to be there). For example, Xue et al. (2014) reported knock-in of a homologous donor template at a rate of 4.5% in *Drosophila* embryos, a number not high enough to permit reasonable interpretation of findings from tissue-specific knock-in experiments.
4) **There is lingering concern about the off-target effects of the CRISPR/Cas9 system.** You are releasing an endonuclease into the genome and asking it to cut only at a very specific location – a stringent request for any biological system. Off-target effects have been reported in a number of studies and they should always be in the back of your mind when interpreting the results from these kinds of experiments.

An example of the limited efficacy of the CRISPR/Cas9 system is the tissue-specific ablation of the *urod* gene, which is required for heme biosynthesis in Zebrafish (Ablain et al. 2015). Importantly, Ablain et al. demonstrate ablation of this gene only in erythroid precursor cells. But they also found that the knockout resulted in mosaic individuals with deletions within *urod* of varying lengths, ranging from single-nucleotide deletions to complete gene excision. Thus, although the CRISPR/Cas9 system did produce tissue-specific results, those results were not uniform.

Another example comes from a study by Maresch et al. (2016), who attempted to edit multiple genes in pancreatic cells of adult mice using a transfection-based approach.[5] The major finding of Maresch and colleagues was the extensive heterogeneity created by delivering CRISPR/Cas9 via transfection, meaning that only a small percentage of the cells were successfully edited. This is actually kind of handy if you are interested in modeling something like a solid tumor, which is typically very genetically heterogeneous, but not so useful if you are interested in studying how the knockout of a specific gene affects the function of an organ.

Although CRSPR/Cas9 is a relatively new technology, it seems to have limitless potential (Sánchez-Rivera and Jacks 2015; Shalem et al. 2015). We suspect that its issues (e.g. incomplete cutting, off-target effects) will be reduced with improvements to the structure of the Cas9 enzyme itself and with improved techniques for delivering gRNA and designing targets. The problem of delivering a repair template to cells in a tissue-specific manner will be challenging but may be overcome using strategies such as tissue-specific viral introduction of the donor template. Regardless of the methods employed, it will be exciting to watch this technology develop.

8.5 Summary

Mosaic analysis is a key tool when working on a gene or process that may be lethal in your model system or when you are worried about the effects of your mutant on other tissues in your organism. In this chapter, we discussed some of the classic approaches to generating mosaic individuals along with some newer and usually more precise methods. We saw that a mosaic may be an organism expressing a specific gene at a specific time or in a specific tissue, or it may be the knockdown of a specific gene. Systems that allow for tissue- and time-specific expression or knockdowns allow a whole new set of questions to be addressed in a variety of model systems.

---

5 Note that this is not driving expression off the genome, so it is unclear how well this method directly compares to a transgene-based approach like that described above.

References

Ablain J, Durand EM, Yang S, Zhou Y, Zon LI. 2015. A CRISPR/Cas9 vector system for tissue-specific gene disruption in zebrafish. *Dev. Cell* 32:756–764.

Acuna-Hidalgo R, Bo T, Kwint MP, van de Vorst M, Pinelli M, *et al.* 2015. Post-zygotic point mutations are an underrecognized source of de novo genomic variation. *Am. J. Hum. Genet.* 97:67–74.

Ashburner M, Golic K, Hawley RS. 2005. *Drosophila: A Laboratory Handbook*, 2nd ed. Cold Spring Harbor Laboratory Press.

Baccarini A, Chauhan H, Gardner TJ, Jayaprakash AD, Sachidanandam R, *et al.* 2011. Kinetic analysis reveals the fate of a microRNA following target regulation in mammalian cells. *Curr. Biol.* 21:369–376.

Baker BS. 1975. *Paternal loss* (*pal*): a meiotic mutant in *Drosophila melanogaster* causing loss of paternal chromosomes. *Genetics* 80:267–296.

Bartlett DW, Davis ME. 2006. Insights into the kinetics of siRNA-mediated gene silencing from live-cell and live-animal bioluminescent imaging. *Nucleic Acids Res.* 34:322–333.

Beadle GW, Ephrussi B. 1936. The differentiation of eye pigments in *Drosophila* as studied by transplantation. *Genetics* 21:225–247.

Bosch JA, Sumabat TM, Hariharan IK. 2016. Persistence of RNAi-mediated knockdown in *Drosophila* complicates mosaic analysis yet enables highly sensitive lineage tracing. *Genetics* 203(1):109–118.

Brand AH, Perrimon N. 1993. Targeted gene expression as a means of altering cell fates and generating dominant phenotypes. *Development* 118:401–415.

Bryant PJ, Schneiderman HA. 1969. Cell lineage, growth, and determination in the imaginal leg discs of *Drosophila melanogaster*. *Dev. Biol.* 20:263–290.

Campbell IM, Shaw CA, Stankiewicz P, Lupski JR. 2015. Somatic mosaicism: implications for disease and transmission genetics. *Trends Genet.* 31:382–392.

Frickenhaus M, Wagner M, Mallik M, Catinozzi M, Storkebaum E. 2015. Highly efficient cell-type-specific gene inactivation reveals a key function for the *Drosophila FUS* homolog *cabeza* in neurons. *Sci. Rep.* 5:9107–9110.

Garcia-Bellido A, Merriam JR. 1969. Cell lineage of the imaginal discs in *Drosophila* gynandromorphs. *J. Exp. Zool.* 170:61–75.

Gelbart WM. 1974. A new mutant controlling mitotic chromosome disjunction in *Drosophila melanogaster*. *Genetics* 76:51–63.

Golic KG. 1991. Site-specific recombination between homologous chromosomes in *Drosophila*. *Science* 252:958–961.

Golic KG. 1993. Generating mosaics by site-specific recombination. In: Hartley DA, editor. *Cellular Interactions in Development: A Practical Approach*. Oxford: IRL Press. pp. 1–32.

Golic MM, Golic KG. 1996. Engineering the *Drosophila* genome: chromosome rearrangements by design. *Genetics* 144:1693.

Greenspan RJ. 1997. *Fly Pushing: The Theory and Practice of Drosophila Genetics*. New York: Cold Spring Harbor Laboratory Press.

Haldi M, Ton C, Seng WL, McGrath P. 2006. Human melanoma cells transplanted into zebrafish proliferate, migrate, produce melanin, form masses and stimulate angiogenesis in zebrafish. *Angiogenesis* 9:139–151.

Hall JC. 1979. Control of male reproductive behavior by the central nervous system of *Drosophila*: dissection of a courtship pathway by genetic mosaics. *Genetics* 92:437–457.

Hall JC, Gelbart WM, Kankel DR. 1976. Mosaic systems. In: Ashburner M, Novitski E, editors. *The Genetics and Biology of Drosophila*. 1a London: Academic Press. pp. 265–314; 50 p.

Herman RK. 1984. Analysis of genetic mosaics of the nematode *Caneorhabditis elegans*. *Genetics* 108:165–180.

Hinton CW. 1955. The behavior of an unstable ring chromosome of *Drosophila melanogaster*. *Genetics* 40:951–961.

Hinton CW. 1957. The analysis of rod derivatives of an unstable ring chromosome of *Drosophila melanogaster*. *Genetics* 42:55–65.

Hoff T, Schnorr KM, Mundy J. 2001. A recombinase-mediated transcriptional induction system in transgenic plants. *Plant Mol. Biol.* 45:41–49.

Hoppe PE, Greenspan RJ. 1986. Local function of the *Notch* gene for embryonic ectodermal pathway choice in *Drosophila*. *Cell* 46:773–783.

Huang P. 2004. FGF signaling functions in the hypodermis to regulate fluid balance in *C. elegans*. *Development* 131:2595–2604.

Kemp HA, Carmany-Rampey A, Moens C. 2009. Generating chimeric zebrafish embryos by transplantation. *J. Vis. Exp.* 1–5. https://doi.org/10.3791/1394.

Kimble J, Austin J. 1989. Genetic control of cellular interactions in Caenorhabditis elegans development. *Ciba Foundation Symposium* 144:212–20– discussion 221–6– 290–5.

Koehler KE, Boulton CL, Collins HE, French RL, Herman KC, *et al.* 1996. Spontaneous X chromosome MI and MII nondisjunction events in *Drosophila melanogaster* oocytes have different recombinational histories. *Nat. Genet.* 14:406–414.

Lange J, Skaletsky H, van Daalen SKM, Embry SL, Korver CM, *et al.* 2009. Isodicentric Y chromosomes and sex disorders as byproducts of homologous recombination that maintains palindromes. *Cell* 138(855):869.

Lewandoski M. 2001. Conditional control of gene expression in the mouse. *Nat. Rev. Genet.* 2:743–755.

Li P, White RM, Zon LI. 2011. Transplantation in zebrafish. *Methods Cell Biol.* 105:403–417.

Liu P, Jenkins NA, Copeland NG. 2001. Efficient Cre-*loxP*–induced mitotic recombination in mouse embryonic stem cells. *Nat. Genet.* 30:66–72.

Long C, Amoasii L, Mireault AA, McAnally JR, Li H, *et al.* 2015. Postnatal genome editing partially restores dystrophin expression in a mouse model of muscular dystrophy. *Science*:1–7. https://doi.org/10.1126/science.aad5725.

Maresch R, Mueller S, Veltkamp C, Öllinger R, Friedrich M, *et al.* 2016. Multiplexed pancreatic genome engineering and cancer induction by transfection-based CRISPR/Cas9 delivery in mice. *Nat. Commun.* 7:10770.

Micchelli CA, Perrimon N. 2006. Evidence that stem cells reside in the adult *Drosophila* midgut epithelium. *Nature* 439:475–479.

Miller LM, Waring DA, Kim SK. 1996. Mosaic analysis using a *ncl-1* (+) extrachromosomal array reveals that *lin-31* acts in the Pn.p cells during *Caenorhabditis elegans* vulval development. *Genetics* 143:1181–1191.

Morgan TH. 1914. Mosaics and gynandromorphs in *Drosophila*. *Exp. Biol. Med.* 11:171–172.

Nelson CE, Hakim CH, Ousterout DG, Thakore PI, Moreb EA, *et al.* 2015. in vivo genome editing improves muscle function in a mouse model of Duchenne muscular dystrophy. *Science*:1–8. https://doi.org/10.1126/science.aad5143.

Nelson CR, Szauter P. 1992. Timing of mitotic chromosome loss caused by the *ncd* mutation of *Drosophila melanogaster*. *Cell Motil. Cytoskeleton* 23:34–44.

Nesbitt MN, Gartler SM. 1971. The applications of genetic mosaicism to developmental problems. *Annu. Rev. Genet.* 5:143–162.

Page SL, Nielsen RJ, Teeter K, Lake CM, Ong S, *et al.* 2007. A germline clone screen for meiotic mutants in *Drosophila melanogaster*. *Fly* 1:172–181.

Reinke R, Zipursky SL. 1988. Cell-cell interaction in the *Drosophila* retina: the *bride of sevenless* gene is required in photoreceptor cell R8 for R7 cell development. *Cell* 55:321–330.

Sánchez-Rivera FJ, Jacks T. 2015. Applications of the CRISPR–Cas9 system in cancer biology. *Nat. Publ. Group* 15:387–395.

Scharfenberger L, Hennerici T, Király G, Kitzmuller S, Vernooij M, *et al.* 2014. Transgenic mouse technology in skin biology: generation of complete or tissue-specific knockout mice. *J. Investig. Dermatol.* 134:e1–e5.

Seydoux G, Greenwald I. 1989. Cell autonomy of *lin-12* function in a cell fate decision in *C. elegans*. *Cell* 57:1237–1245.

Shalem O, Sanjana NE, Zhang F. 2015. High-throughput functional genomics using CRISPR–Cas9. *Nat. Rev. Genet.* 16(5):299–311.

Simpson L, Wieschaus E. 1990. Zygotic activity of the *nullo* locus is required to stabilize the actin-myosin network during cellularization in *Drosophila*. *Development* 110:851–863.

Stern C. 1936. Somatic crossing over and segregation in *Drosophila melanogaster*. *Genetics* 21:625–730.

Stern G. 1968. *Genetic Mosaics and Other Essays*. Cambridge, MA: Harvard University Press.

Tabebordbar M, Zhu K, Cheng JKW, Chew WL, Widrick JJ, *et al.* 2015. in vivo gene editing in dystrophic mouse muscle and muscle stem cells. *Science* 351(6271):407–411.

Villeneuve AM, Meyer BJ. 1990. The role of *sdc-1* in the sex determination and dosage compensation decisions in *Caenorhabditis elegans*. *Genetics* 124:91–114.

White RM, Sessa A, Burke C, Bowman T, LeBlanc J, *et al.* 2008. Transparent adult zebrafish as a tool for in vivo transplantation analysis. *Cell Stem Cell* 2:183–189.

Wieschaus E, Gehring W. 1976. Clonal analysis of primordial disc cells in the early embryo of *Drosophila melanogaster*. *Dev. Biol.* 50:249–263.

Xue Z, Ren M, Wu M, Dai J, Rong YS, *et al.* 2014. Efficient gene knock-out and knock-in with transgenic Cas9 in *Drosophila*. *G3: Genes Genom. Geneti.* 4:925–929.

Yamanishi M, Matsuyama T. 2012. A modified Cre-*lox* genetic switch to dynamically control metabolic flow in *Saccharomyces cerevisiae*. *ACS Synth. Biol.* 1: 172–180.

Yochem J. 2005. *Genetic Mosaics*. WormBook. 1–6.

Zalokar M, Erk I, Santamaría P. 1980. Distribution of ring-X chromosomes in the blastoderm of gynandromorphic *D. melanogaster*. *Cell* 19:133–141.

Zeng X, Han L, Singh SR, Liu H, Neumüller RA, *et al.* 2015. Genome-wide RNAi screen identifies networks involved in intestinal stem cell regulation in *Drosophila*. *Cell Rep.* 10:1226–1238.

Zhang P, Hawley RS. 1990. The genetic analysis of distributive segregation in *Drosophila melanogaster*. II. Further genetic analysis of the *nod* locus. *Genetics* 125:115–127.

9

Meiotic Chromosome Segregation

Because the ability of chromosomes to divide reductionally at meiosis I is the physical basis of Mendelian inheritance, we must address the mechanism by which that separation takes place. The most critical issues to understand about meiotic chromosome segregation are the following:

1) **At the first meiotic division (MI), homologs separate to opposite poles of the spindle; MII is simply a haploid mitosis.** For the most part, the second meiotic division simply "processes" the complement of chromosomes delivered by the first meiotic spindle. However, chromosomes that have misbehaved at the first meiotic division are more likely to misbehave at the second meiotic division and in subsequent mitotic divisions.
2) **Proper segregation is mediated by structures (usually crossovers or chiasmata) that act to hold bivalents together and *not* by the homology of centromeres.** Thus, two chromosomes that have undergone exchange will usually segregate from each other even if their centromeres are not homologous. Similarly, in the absence of a backup segregation system, nonexchange homologs will not properly segregate from each other even though their centromeres are homologous.
3) **A single crossover event involves two nonsister chromatids within the homolog pair.** Two or more crossovers can (and often do) occur between the same pair of homologs; when they do, the choice of nonsister chromatid is random for each crossover event.
4) **The position of crossovers is not random.**[1] Crossing over does occur in the heterochromatin, and most single crossover events occur in the middle of the euchromatin. There are polar forces along paired chromosomes that can diminish the frequency of crossing over in those regions. These include the **centromere effect**, which strongly suppresses crossing over in the vicinity of the centromere, and **crossover interference**, which reduces the frequency that two crossovers will occur in the same vicinity.
5) **The primary function of meiotic recombination is to ensure homolog segregation** by creating chiasmata that hold homologs together with their centromeres oriented toward opposite poles of the metaphase I spindle. The beginning of anaphase I triggers the loosening of the chiasmata, allowing the homologs to separate and move to opposite poles.

1 If your professor says it is or, even worse, that the frequency of crossing over is proportional to the physical length of DNA in the region, please tell them that they are wrong and to call me (RSH).

Genetic Theory and Analysis: Finding Meaning in a Genome, Second Edition.
Danny E. Miller, Angela L. Miller, and R. Scott Hawley.

6) **Thus, the failure of two homologs to undergo crossing over usually increases the likelihood that they will fail to properly segregate from each other at anaphase I** (i.e. they will **nondisjoin**). Recombination accomplishes this goal by using sister chromatid cohesion to create interhomolog adhesion.
7) **Some organisms, such as *Drosophila*** (Hawley 1993; Hawley et al. 1993), **yeast** (Guacci and Kaback 1991; Molnar et al. 2001), **Lepidopteran females** (Rasmussen 1977a, b), **and *Caenorhabditis elegans*** (Riddle et al. 1997) **have systems that ensure the segregation of those homolog pairs that fail to recombine**. However, there are also organisms that appear to lack such backup systems, such as higher plants (Ines et al. 2014). In those organisms, achiasmate chromosomes dissociate from each other prior to the first meiotic division and segregate at random with respect to each other at that division, often resulting in nondisjunction.

9.1 Types and Consequences of Failed Segregation

The failure of proper meiotic segregation is heralded by **nondisjunction**. We define nondisjunction as either the movement of both homologs to the same pole at meiosis I (**MI nondisjunction**) or the presence of two separated sister chromatids at the same pole at the end of anaphase II (**MII nondisjunction**) (Figure 9.1). We distinguish nondisjunction from chromosome loss events in which one homolog or one chromatid simply fails, for whatever reason, to reach a pole. Regardless of their origin, nondisjunction can produce gametes, and thus zygotes, carrying too many or too few chromosomes, a condition known as **aneuploidy**.

The result of nondisjunction at either division is a daughter cell carrying two copies of the nondisjoining chromosome (called a **diplo**-cell) or no copy of the chromosome in question (a **nullo**-cell). The results of the two nondisjunctional events differ in terms of the genetic composition of the diplo-products. In the case of MI nondisjunction, the two chromosomes may be from two different homologs and thus carry different centromeres; in the case of MII nondisjunction, both chromosomes will have come from the same homolog, and thus the two chromatids were once sisters. Keep in mind, however, that although MII nondisjunction may result in a single cell containing two sister chromatids, one or both of the sisters might have undergone recombination with a homolog and so could slightly differ from one another genetically. The preservation or loss of centromere heterozygosity thus becomes the defining tool for determining the division at which a diplo-exception arose (Figure 9.1).

There is no method for determining absolutely when a nullo-exception arises. Indeed, nullo-exceptions can arise as simple chromosome loss, without nondisjunction. We distinguish between processes that result in nondisjunction and those that result in chromosome loss by looking at the ratio of diplo- to nullo-exceptions. In the absence of viability effects that favor one type of exceptional progeny or gamete over the other, nondisjunction will produce equal frequencies of nullo- and diplo-exceptions. Chromosome loss will produce only nullo-exceptions.

In general, MI nondisjunction will be followed by a regular MII that produces two diplo-daughter cells and two nullo-daughter cells. In other words, the two chromosomes that nondisjoined at MI usually will behave properly at MII, each segregating one sister chromatid to each of the two daughter cells. However, there are cases where nondisjunction at MI predisposes a chromosome to loss or subsequent nondisjunction at MII (Hawley 1988; Koehler et al. 1996).

Table 9.1 Meiotic segregation of minichromosomes.

			Distribution in tetrads of genetic marker on minichromosomes (%)					Test for centromere linkage of marker on minichromosome		
Cross	**Minichromosome**	**Marker**	**4+ : 0−**	**3+ : 1−**	**2+ : 2−**	**1+ : 3−**	**0+ : 4−**	**PD**	**NPD**	**T**
1	pYe(*CDC10*)1	*TRP*	1 (6)	0	10 (63)	0	5 (31)	2	8	0
2	pYe(*CDC10*)1	*CDC10*	1 (8)	0	11 (92)	0	0	2	8	1
3	pYe(*CDC10*)1	*TRP*	1 (7)	0	11 (79)	0	2 (14)	4	7	0
4	pYe(*CDC10*)1	*TRP*	4 (21)	0	11 (58)	0	4 (21)	ND	ND	ND
5	pYe(*CEN3*)11	*TRP*	2 (13)	0	9 (60)	0	4 (27)	4	4	0

Box 9.1 Identifying Genes that Encode Centromere-Binding Proteins in Yeast

Doheny et al. (1993) described two assays to identify genes that affect kinetochore function in yeast: one assay detects the relaxation of a transcriptional block that is observed at centromeres and the other detects an increase in the mitotic stability of dicentric chromosomes. Doheny and her collaborators used these assays to sort through a set of mutants (called *ctf* mutants) recovered by Forrest Spencer on the basis of impaired transmission of minichromosomes.

The first assay is based on screening for mutants that allow transcriptional readthrough of a centromere, presumably by removing a centromere protein that serves to block polymerase progression. When transcription from a strong promoter is initiated toward a *CEN* DNA sequence, the mitotic segregational function of the centromere is destroyed. However, the chromatin structure of the kinetochore remains undisrupted, indicating that at least some of the centromere remains intact. Moreover, the majority of the transcripts terminate at the border of the *CEN* sequence. The active polymerase is therefore sufficient to kill centromere activity, but the remaining centromere structure is sufficient to impede the polymerase.

Doheny et al. exploited this standoff by screening for mutants that allowed readthrough across such an inactive centromere construct that had been integrated into one of the arms of a yeast chromosome. (Because that centromere is inactive, we can integrate such constructs into a normal yeast chromosome without creating a dicentric). One such construct is shown in Figure B9.1. The promoter carried by this construct (GAL) is inducible and very strong. Moreover, lying beyond the *CEN* sequence is a transcription mutant capable of changing the color of the yeast colony, meaning that readthrough equals color change. But inserting the 165 bp of *CEN6* knocks down transcription by 100X, blocking that color change. To show that the transcriptional block is due to at least partial kinetochore assembly, Doheny replaced the normal *CEN* sequence with a single mutant in CDEIII. This mutant reduces minichromosome transmission by 250X, meaning that it clearly impairs kinetochore function. The use of this mutant insertion elevated levels of transcriptional readthrough to 20% of normal. This change was sufficient to create a color change in colonies, providing a rapid tool for screening through the *ctf* mutants – just insert the construct carrying the normal *CEN6* construct into yeast carrying each of the *ctf* mutants. This test identified seven possible candidate mutants from among the available collection of *ctf* mutants.

Of course, this screen tested only for mutants in genes whose products bound to the *CEN* sequence. The second assay was a lovely, if counterintuitive, screen for those mutants affecting centromere function. The basic idea is that a mutant that impairs centromere function ought

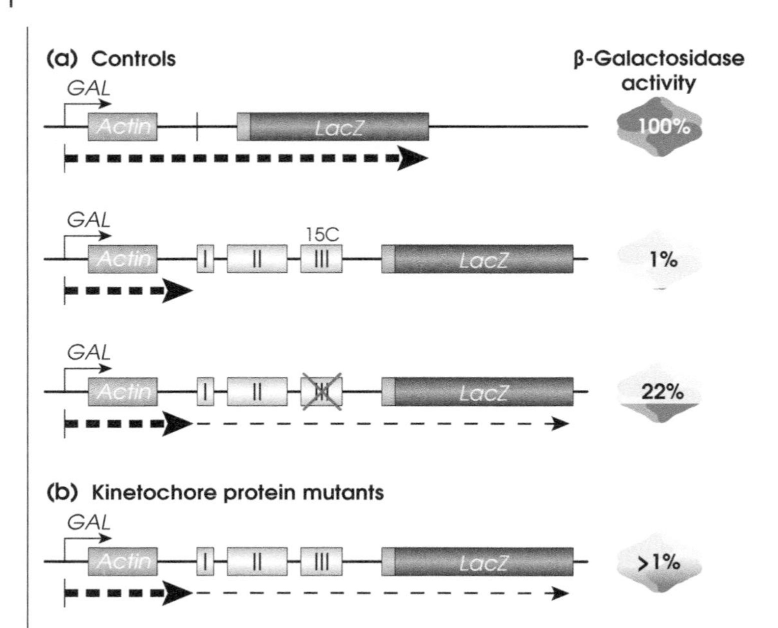

Figure B9.1 Transcriptional readthrough assay. (a) Constructs carrying the wildtype *CEN* sequence (CDEI, CDEII, and CDEIII) blocked nearly all transcription, preventing the production of B-galactosidase. CDEIII-C15, a mutant known to affect kinetochore function, allowed some transcriptional readthrough and resultant B-galactosidase production, visualized as color change of the colony. (b) Doheny et al. (1993) predicted that other genes affecting kinetochore function could be identified by screening for those that produced color change in colonies. *Source:* Adapted from Doheny et al. (1993).

to restabilize a dicentric chromosome by making one of the two centromeres "weaker." To perform this experiment, Doheny inserted her regulatable centromere construct into a dispensable chromosome fragment (a partial disome) carrying both a functional *CEN* sequence and a useful reporter gene. She could then activate that centromere by repressing the activity of the *GAL* promoter. The result was a dicentric chromosome that was quickly lost, but mutants that disrupted *CEN* function weakened the expression of one or both of those centromeres and stabilized the chromosome. Indeed, these two assays together identified two genes of interest, at least one of which is now a known component of the centromere.

Unfortunately, the simplicity of the budding yeast centromere would not be a harbinger of things to come. The centromere of the fission yeast *Schizosaccharomyces pombe* would turn out to be much larger and composed of arrays of repeated elements. The basic elements of that story can be found in Allshire and Ekwall (2015). The centromeres of higher eukaryotes would turn out to be rather more complex, and their understanding would require a major paradigm shift in modern biology. The story, described in the next section, begins in *Drosophila*.

The Isolation and Analysis of the *Drosophila* Centromere

Gary Karpen began the assault on the fly centromere using the smallest known stable chromosome, *Dp(1;f)1187* (Le et al. 1995; Murphy and Karpen 1995; Sun et al. 1997). As shown in Figure 9.3, this small chromosome was derived from a normal sequence X chromosome by an

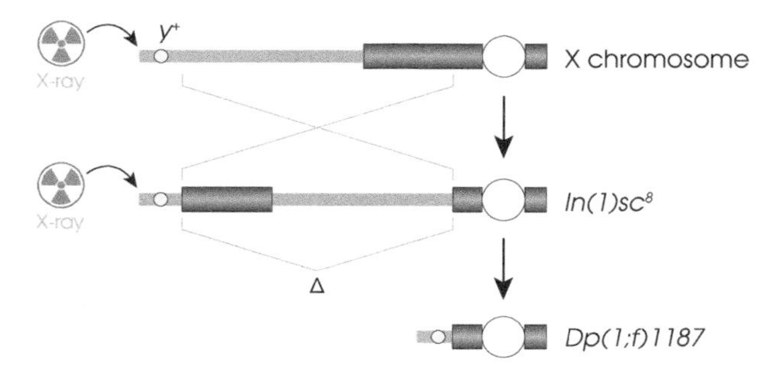

Figure 9.3 The origin of the smallest known stable chromosome. The *In(1)sc*^{8} chromosome was created by an inversion resulting from X-irradiation of a *D. melanogaster* X chromosome carrying a visible y^+ marker. A second round of X-irradiation induced a large deletion, resulting in *Dp(1;f)1187*, the smallest known stable chromosome, which was then used to investigate the *Drosophila* centromere.

X-ray-induced inversion and a subsequent X-ray-induced deletion. Despite its small size, the *Dp(1,f)1187* chromosome can easily be followed in crosses because it carries a useful visible marker (y^+) at its left tip. Even though this chromosome is only approximately 1 Mb in length, it segregates faithfully in both meiosis and mitosis. To be able to create deletion derivatives of this chromosome, Karpen first created a variant, known as *Dp8-23*, that carries two marked (*rosy*$^+$) transposon insertions distal to the y^+ marker. (One needed to have at least a second scorable marker here in order to screen for further deletions of this chromosome.) Karpen then screened for both terminal (loss of the *rosy*$^+$ marker) and internal (loss of the y^+ markers) deficiency derivatives on this chromosome.

To further define the centromere sequence, Karpen created a set of gamma-radiation-induced deletion derivatives of the *Dp8-23* chromosome (Murphy and Karpen 1995). He then assessed the stability of these derivatives by testing their ability to be properly transmitted through meiosis in males or females when present as a single copy – the **monosome transmission test**. (Remember that to even be recovered, these derivatives had to be at least mitotically stable.) The monosome transmission test looks at the fraction of progeny that receive the derivative from a male or female parent carrying a single copy of that chromosome. A fully functional chromosome would yield a value close to 50%. The authors also characterized the structure of these deletion derivatives by pulsed-field gel electrophoresis.

Surprisingly, deletions removing nearly half of the DNA content of this chromosome were fully meiotically and mitotically stable. To quote Murphy and Karpen (1995), "The normal behavior of all the *Dp1187* derivatives ..., including the 620 kb [deletion] derivative *γ1230*, indicates that the entire euchromatin, subtelomeric heterochromatin and telomere (−290 to 0), and over one half of the *Dp1187* centric heterochromatin (0 to +580) are dispensable for chromosome stability." Thus, sequences necessary for centromere function must reside in the 620-kb region. But the real find obtained by creating this set of deletion derivatives was not a deletion at all. Rather, it was an inversion, referred to as *γ238*, in which the y^+ and *rosy*$^+$ markers marked separate ends of the chromosome. Now that both ends were marked, a second screen for gamma-ray-induced deletions would allow Karpen and his collaborators to obtain a series of deficiencies extending into the chromosome from both ends. The structure and transmissibility of those derivatives are described in Murphy and Karpen (1995).

A look at Murphy and Karpen (1995) suggests that full stability of a deletion derivative does require a small region of the original *Dp(1;f)1187* chromosome that Murphy and Karpen refer to as the *Bora Bora* island. All derivatives (Group A and Group B) that retain that region are fully stable regardless of what else they have lost. Moreover, deficiencies that have lost some fraction of the

Bora Bora interval (Group C) show a reduced level of transmission, especially in females. All of this would lead Murphy and Karpen to conclude that "completely normal chromosome stability requires *Bora Bora* plus flanking heterochromatin."

That argument was fine and not the least heretical. At the time the Murphy and Karpen paper was published, the scientific community was still wrapped in a view that could be summarized as, "in yeast, as in all creatures great and small." Everyone assumed flies had to have a yeast-like discrete centromere; it just might be harder to find. The problem with that argument was the recovery of the Group D derivatives. These elements are at least weakly transmissible; after all, Karpen could recover them, keep them in stock, and test them. But they do not contain the *Bora Bora* island. Indeed, some of them seem to contain only material that was telomeric on the original *Dp(1;f)1187* chromosome and that could be deleted without consequence for transmission. (Compare the top two Group A derivatives with the bottom five Group D derivatives.) Karpen explained these outliers by suggesting that they corresponded to transmissible acentric fragments, which might be maintained by some other means. This is *not* to say that they don't have some centromere activity, but rather just to say that they do not contain the normal centromere.

Yet a subsequent paper, Williams et al. (1998), would show that these odd acentric fragments possessed many of the properties expected of centromere-bearing chromosomes. These presumably acentromeric fragments bound the centromere-specific protein ZW10 and associated with the spindle poles at anaphase. Because these derivatives contain DNA normally found near the tip of the X chromosome, the authors suggest that these sequences have *acquired* centromere function. This is important because Williams et al. (1998) demonstrated that these fragment chromosomes carry only sequences that are usually found only near the X chromosome telomere. Such fragments of telomeric DNA do not bind to ZW10, and if they are simply "broken off" the end of the X, they only acquire this ability if they were derived from a region near the centromere. Indeed, Maggert and Karpen (2001) would show elegantly that a usually noncentromeric sequence needed to be in close proximity to a centromere sequence to *acquire* centromeric activity.

These findings have led to a critically important model of centromere structure and function in which the centromere is a state of chromatin rather than a specific sequence (Karpen and Allshire 1997). In this model, the centromeric state becomes stable and heritable by virtue of the duplication of specific chromatin structures during each cell cycle. Rarely, sequences that do not normally function as centromeres can acquire centromere function if they lie close to functioning centromeric intervals. Centromere function can also presumably be lost at a low frequency, but this is usually a lethal event. Further data in support of this model and its implications are described next.

The Concept of the Epigenetic Centromere in *Drosophila* and Humans

The finding by Karpen and his collaborators that sequences born by the Group D derivative chromosomes might have *acquired* centromere activity was stunning. The only reasonable interpretation of these data was that being a "centromere" in flies was less about possessing a specific sequence than it was about possessing a certain state. The idea was strongly supported by four types of evidence:

1) The acentric chromosomes are transmitted at high frequency in several types of cell divisions. However, they are quite unstable in other types of divisions. For example, they are transmitted moderately well in male meiosis but fare quite poorly in female meiosis. They seem to do well in pre-blastoderm mitoses but are frequently lost in neuroblast meiosis.

2) These acentric chromosomes bind a centromeric marker protein (ZW10).
3) Female transmission is greatly reduced in *nod*/+ heterozygotes, demonstrating that segregation is mediated by movement on the spindle. Similarly, transmission is enhanced in females carrying extra doses of nod^{+}. (The *nod* gene produces a chromosomal protein required to hold chromosomes on the developing spindle by opposing the poleward forces exerted by kinetochores.)
4) Molecular analysis demonstrates that these chromosomes have not acquired new centromeric sequences.

To explain these observations, Karpen suggested that these chromosomes reflected "neocentromeric activity" and that centromeric activity is not itself the direct property of some particular DNA sequence. Rather, centromeric activity reflects an epigenetic state that is stably passed on with each cell division. He further imagined, and later demonstrated (Maggert and Karpen 2001), that centromeric activity can spread to nearby sequences. The concept of an **epigenetic centromere** is fully presented in a seminal review by Karpen and Allshire (1997). These authors noted that viewing the centromere as a state, instead of a sequence, allowed one to understand rather a lot of previously confusing observations regarding the centromeres of other higher eukaryotes. For example, the centromeres of *S. pombe* are much larger and more complex than their *S. cerevisiae* counterparts (Clarke et al. 1993; Clarke 1998). Moreover, when centromere-repeat-bearing sequences are transformed back into *S. pombe* cells, they acquire centromere competence only rarely but without sequence change (Steiner and Clarke 1994). Similarly, the analysis of small human marker chromosomes suggested that such chromosomes can acquire centromere function without acquiring the alpha-satellite DNA sequences present at normal centromeres (Karpen and Allshire 1997).

The model proposed by Karpen and Allshire (1997) is simple: centromeric activity in higher eukaryotes reflects a state of chromatin, not just a DNA sequence. That state is self-perpetuating at the time of DNA replication. This is not to say that the content of certain regions might favor the assembly or maintenance of that chromatin state but only that such sequences are not absolutely required. More recent discussions of the epigenetic centromere may be found in Allshire and Karpen (2008), Sullivan and Sullivan (2020), and Hoffman et al. (2020).

Understanding the concept of epigenetic centromeres helps us understand a curious type of rearrangement found in human cells – chromosomes that appear to possess two centromeric regions. Although such dicentric chromosomes might be expected to be unstable, they can propagate if one of the two centromeres is inactive (Stimpson et al. 2012; McNulty and Sullivan 2017).

Holocentric Chromosomes

Being holocentric does not pose a problem for chromosomes in mitosis – microtubules simply attach along all the lengths of both sister chromatids. Presumably, some mechanism prevents longer chromosomes from twisting in the middle to result in a chromatid attaching to two poles. However, chiasmate meiosis poses additional problems. The solution in several organisms appears to be having the holocentric bivalent attach to the spindle at only one point on each chromosome, usually near an end – basically mimicking the canonical process (Girard et al. 2007). That said, there are systems (for example, *Bombyx mori* females) where crossing over doesn't occur at all, and bivalents are held together by proteinaceous structures derived from the synaptonemal complex. In these cases, during the first meiotic division, bivalents still attach all along their length to the spindle (again, presumably avoiding twists that could create bipolar attachment).

Several insects and plants with holocentric chromosomes undergo a process called **inverted meiosis** in which a meiosis II-like division that includes sister chromatid segregation occurs first and then is followed by a second division that segregates homologous nonsister chromatids. The complexity of these processes exceeds the scope of this book, but we direct interested readers to two excellent reviews published by Li and He (2020) and Hofstatter et al. (2021).

9.4 Chromosome Segregation Mechanisms

The world would be a much simpler place if there were but one method of ensuring proper homolog separation at meiosis I.[2] However, there are many ways that evolution has decided to accomplish this single task. Nonetheless, in the vast majority of organisms, meiotic segregation is mediated by chiasma function. For that reason, we shall divide the universe of segregational mechanisms into two camps: chiasmate segregation and achiasmate segregation. Both types are reviewed briefly here.

Chiasmate Chromosome Segregation

In Chapter 4, when discussing meiotic recombination, we cited the work of Bruce Nicklas (1974, 1977) as demonstrating that exchanges ensure homolog separation at the first meiotic division. Because sister chromatids display tight cohesion during meiotic prophase and metaphase I, chiasmata hold the paired chromosomes together and thus commit the two centromeres physically linked by the exchange to orient toward opposite poles of the spindle. However, the exact mechanism by which this is achieved does differ among meiotic systems. In plants and most animal male meioses, the spindle is assembled by centrosomes. In such meioses, each kinetochore captures (or is captured by) microtubule fibers emanating from the kinetochore immediately after nuclear envelope breakdown. Thus, each centromere of the bivalent attaches to the spindle and begins to move toward one of the two poles. Most bivalents achieve a bipolar orientation immediately because at the start of prometaphase, the two homologous centromeres are usually oriented in opposite directions such that if one centromere is pointed at one pole, the other centromere is pointed at the opposite pole (Nicklas and Staehly 1967). Once the two centromeres have oriented toward opposite poles, the progression of the two centromeres toward the poles is halted at the metaphase plate by the chiasma. This represents a stable position in which the bivalent will remain until anaphase I. If the bivalent does not immediately acquire a bipolar orientation, centromeres are capable of breaking and reforming spindle attachments until a stable position at the metaphase plate is achieved (Nicklas 1967; Nicklas and Staehly 1967).

In females of many organisms, meiosis is acentriolar – the chromosomes themselves organize the spindle. Initially, each pair of homologs establishes two oppositely oriented half-spindles that then coalesce into a bipolar spindle (Theurkauf and Hawley 1992). Perhaps because it is opposed centromeres that facilitate bipolar assembly in female meiosis, observations of initial chromosome malorientation are rare.

The ability of chiasmata to link centromeres together does *not* require that the centromeres themselves be homologous. Homologous exchanges that conjoin (link) heterologous centromeres together will ensure that those chromosomes segregate from each other. As shown in Figure 9.4,

2 Or of regulating gene expression, for that matter.

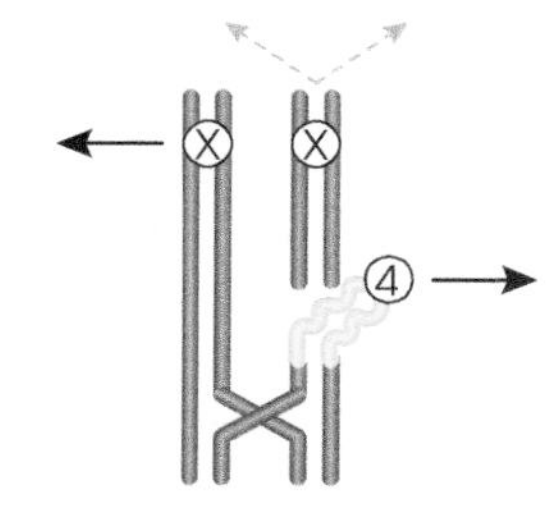

Figure 9.4 Chiasmate nonhomologous segregation. The physical link created by crossing over between paired chromosomes – even nonhomologous ones – causes them to orient toward opposite poles of the meiotic spindle. Shown here, exchange in a *Drosophila* X–4 translocation heterozygote commits nonhomologous centromeres to segregate away from each other.

the exchange between two homologous euchromatic regions of an X–4 translocation in *Drosophila* will commit the X and 4th centromeres to segregate from each other quite faithfully (Parker and Williamson 1976). Similarly, centromere "replacement" studies in yeast (Clarke and Carbon 1985) have shown that crossover bivalents segregate faithfully even when their centromeres are originally derived from different chromosomes.

And if chiasmata do not form? Some organisms possess segregational backup systems to ensure the segregation of achiasmate chromosomes (Wolf 1994), but many do not. In such organisms, achiasmate bivalents fall apart at diplotene–diakinesis, resulting in the premature dissociation of the bivalent into two univalents prior to prometaphase I. The two prematurely separated homologs align separately on the spindle, and stability at the metaphase plate is never achieved. Rather, both chromosomes move up and down the spindle, frequently reorienting. Their disjunctional fate at anaphase I is determined solely by their orientation on the spindle at the time anaphase begins. In this instance, normal disjunctional events (in which homologs proceed to opposite poles) and nondisjunctional events (in which both homologs proceed to the same pole) will occur with equal frequency. Work in a variety of organisms has also shown that such univalents are frequently unstable, such that chromosome loss and breakage are not uncommon events at both meiotic divisions and in subsequent mitotic divisions (for review, see Hawley 1993). Thus, chiasmata provide an essential device for stabilizing bivalents at the metaphase plate during prometaphase.

Sex chromosome segregation in mammals: the role of the PAR region. X–Y segregation in mammalian males requires a set of processes that ensure a high-frequency set of crossover events in a short region of homology known as the **pseudoautosomal region** (PAR) (Kauppi et al. 2012). Numerous processes ensure that such PAR exchanges occur at least once in meiosis, thus creating the chiasmata required to ensure proper chromosome segregation. Failure of PAR recombination often results in sex chromosome nondisjunction (May et al. 1990; Thomas and Hassold 2003).

Segregation Without Chiasmata (Achiasmate Chromosome Segregation)

Many organisms employ elaborate mechanisms to ensure the segregation of achiasmate chromosomes. Here we discuss only those mechanisms that have been described in the most commonly studied model organisms.

Achiasmate Segregation in *Drosophila* Males

There is no recombination in wildtype *Drosophila melanogaster* males.[3] How, then, do these males carry out meiotic chromosome segregation?

3 We mean zero, none, zip, as in, not any. Nor is a synaptonemal complex constructed.

Sex chromosome segregation in *Drosophila* males is mediated by two chromosomal sites, known as **collochores** (the word collochore is Latin for sticky spot) (Cooper 1964). The collochores, which map to the tandemly repeated rDNA present in the X heterochromatin and on the short arm of the Y chromosome (McKee and Karpen 1990), are in fact comprised of many copies of a 240-bp intergenic spacer sequence within the rDNA repeats (McKee et al. 1992; Ren et al. 1997). In this curious case, the remnants of a once common nucleolus seem sufficient to function as a pseudo-chiasma to hold the homologs together until they separate at anaphase I.

The achiasmate autosomes clearly pair along their lengths at metaphase, and a cytogenetic study of the effects of duplications and deletions on this process led McKee and his collaborators to conclude that this pairing was based on widespread homology, and not on the function of specific pairing sites (McKee et al. 1993). At least one mutant, known formerly as *mei-S8* and now as *teflon*, specifically impairs autosomal pairing in *Drosophila* males (Sandler et al. 1968). An excellent study by Vazquez et al. (2002) describes the live analysis of meiotic pairing in *Drosophila* and has been extended by Vernizzi and Lehner (2022), with both studies showing that each bivalent occupies a separate territory within the prophase nucleus. Vernizzi and Lehner (2022) also showed that homologs comprising each autosomal bivalent are connected at one or two sites along the arms by aggregates of a protein called MNM, which they refer to as "spot welds," that stably link the homologs together in a fashion sufficient to ensure their segregation at anaphase I. They propose that these junctions substitute for chiasmata, providing a mechanism for a process they call alternative homolog conjunction.

Thus, for both the autosomes and sex chromosomes, *Drosophila* males use protein aggregation to link homologs together and ensure proper segregation without creating chiasmata. While it seems a high price to avoid X–Y recombination, it appears to be the path that evolution has chosen.

Achiasmate Segregation in *D. melanogaster* Females

Even though exchange is fully sufficient to ensure segregation in *Drosophila* females, it is nonetheless clear that exchange is not always required to ensure the segregation of homologous chromosomes. For example, the obligately achiasmate 4th chromosome segregates normally (Sturtevant and Beadle 1936). Three lines of evidence strongly suggest that these achiasmate segregations are mediated by heterochromatic pairings. The first evidence that heterochromatic homology plays a crucial role in achiasmate segregation came from the analysis of the segregation of two normal 4th chromosomes in the presence of a homologous duplication (Figure 9.5). These six duplications, known as *Dp(1;4)s*, all carry some or all of the heterochromatic base of the 4th chromosome capped by X chromosome euchromatin. Each was tested for its ability to ensure 4th chromosome

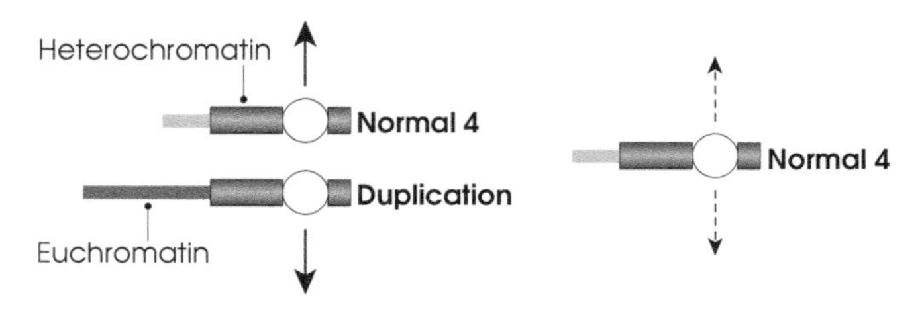

Figure 9.5 Achiasmate homologous segregation. In the absence of crossing over, the heterochromatin mediates homologous segregation. As shown here, the 4th chromosomes in *Drosophila* are mostly heterochromatic and do not undergo exchange. A duplication carrying 4th chromosomal heterochromatin plus distally located X chromosomal euchromatin will thus compete with a normal 4th chromosome for its partner. The unpaired 4th chromosome segregates at random.

nondisjunction as a result of *Dp(1;4)* ⟷ 4^4 segregation events (4^4 is an attached 4th chromosome, also known as a compound 4 chromosome). Basically, the more 4th chromosome heterochromatin carried by a given *Dp(1,4)*, the higher the observed level of induced 4th chromosome nondisjunction (Hawley et al. 1992).

Karpen et al. (1996) presented stronger evidence that achiasmate segregation is dependent on heterochromatic homology. They showed that the frequency with which two achiasmate deletion derivatives of *Dp(1;f)1187* segregate from each other in female meiosis is proportional to the amount of centric heterochromatic homology. Normal segregation requires 800 kb of overlap in the heterochromatin surrounding the centromere, whereas nearly random disjunction is observed with only 300 kb of overlap. A linear correlation between the amount of heterochromatic homology and segregation efficiency was observed in the range of 800–300 kb. They concluded that sequences found throughout the centric heterochromatin of *Dp(1;f)1187* act additively to ensure achiasmate meiotic segregation. Karpen's work here is critical for two reasons. First, it provides the strongest evidence that centromere homology alone is not sufficient to guarantee segregation. Second, it demonstrates the direct role of heterochromatic homology in ensuring achiasmate segregation.

Although heterochromatic pairings were known to exist as late as the end of pachytene (Carpenter 1975, 1979), it was not clear whether they might persist during metaphase, the time at which segregational orientation is presumably determined and critical. Dernburg et al. (1996) used three-dimensional fluorescent hybridization to investigate the physical associations between achiasmate homologs from the end of pachytene until the onset of achiasmate segregation at prometaphase. Although euchromatic pairings dissolve following pachytene, heterochromatic pairings are preserved within the karyosome until prometaphase. Thus, in *Drosophila* meiosis, heterochromatic pairings persist beyond the dissolution of the synaptonemal complex at pachytene until centromere co-orientation at prometaphase.

In most organisms, paired but achiasmate bivalents would precociously dissociate as a consequence of homolog–homolog repulsion at diplotene–diakinesis. However, in *Drosophila* female meiosis, there is no stage comparable to diplotene–diakinesis. Instead, after pachytene, the chromosomes condense into the karyosome, where they remain until nuclear envelope breakdown and spindle formation during prometaphase. Karyosome formation occurs in stage 3 oocytes and the karyosome persists until the beginning of prometaphase in stage 13 oocytes. It seems likely then, as a consequence of this curious meiotic detour into the karyosome, that the pairings that existed at the end of pachytene, and most especially heterochromatic pairings, may persist until the beginning of prometaphase. This would allow paired, but achiasmate, bivalents to maintain centromere apposition until spindle assembly at prometaphase. *Drosophila* also possess a second, very different achiasmate segregation system that allows the faithful segregation of two nonrecombining nonhomologous chromosomes, which is discussed in Box 9.2.

Achiasmate Segregation in *S. cerevisiae*

Dean Dawson and his colleagues provided the first report of achiasmate segregation in budding yeast (Dawson et al. 1986). Using circular minichromosomes with limited homology, they observed that the two chromosomes segregated faithfully, without benefit of chiasmata, in 100 out of 112 cases. The remaining 12 meioses included eight cases of meiosis I nondisjunction and four cases of nondisjunction at meiosis II. They also demonstrated that this process was independent of homology; that is, in a diploid with three minichromosomes, only two of which were homologous, the two homologs segregated from each other no more than 33% of the time. Guacci and Kaback (1991) went on to show that "authentic" full-length yeast chromosomes

Box 9.2 Achiasmate Heterologous Segregation in *Drosophila* Females

The curious pattern by which *Drosophila* oocytes build and organize their spindle may also explain perhaps the most vexing of all meiotic events in *Drosophila*, namely the ability of the oocyte to segregate any two nonhomologous chromosomes that would otherwise lack a partner. For example, a compound *X* chromosome will virtually always segregate from a compound 4th chromosome. Indeed, two compound autosomes will virtually always segregate from each other regardless of their identity. The participation of such compound chromosomes in achiasmate disjunctions tells us two important things. First, the ability of any two compound chromosomes to segregate from each other, regardless of their genetic content, demonstrates that this process does not require homology. Second, the fact that even compounds that have undergone inter-arm exchanges participate in achiasmate segregations differentiates between the requirement for recombination *per se* and the requirement for chiasmata; only the latter are sufficient to prevent a compound chromosome from undergoing heterologous segregations. Like achiasmate homologous segregations, these nonhomologous events require the function of the NOD protein. However, unlike achiasmate homologous segregations, these segregation events are not preceded by physical interactions between the partner chromosomes either in pachytene or during prometaphase (Dernburg et al. 1996).

Hawley and Theurkauf (1993) presented a model for heterologous segregation in which heterologous chromosomes remain randomly oriented until the spindle is assembled and then orient such that a given chromosome orients toward the least crowded pole. This model is based on two well-documented properties of the meiotic spindle in *Drosophila* females. First, the movement toward the pole is size-dependent, in that smaller chromosomes leave the metaphase plate earlier and move farther toward the pole than larger achiasmate chromosomes do (Theurkauf and Hawley 1992). According to this model, a given nonexchange chromosome will move to the least crowded pole. Given the narrowness of the meiotic spindle in *Drosophila* females relative to the width of a chromosome, the physical basis of this blockage by other chromosomes might be nothing more than a physical obstruction of the ability of kinetochore microtubules to reach the poles. Accordingly, when there are only two nonexchange chromosomes, they always segregate from each other. That this model may indeed be correct is suggested by live studies of achiasmate segregation in *S. pombe* (Molnar et al. 2001).

could be segregated by this method as well and confirmed that neither size nor homology was critical to ensure segregation (Kaback 1989; Ross et al. 1996). Curiously enough, however, these nonhomologous segregations are preceded by nonhomologous pairings. In diploids monosomic for chromosomes I and III, the eventual segregation of chromosome I from chromosome III in almost 90% of the meioses is preceded by the pairing or tight association of these chromosomes at pachytene (Loidl et al. 1994).

Dawson would go on to show that these segregations require centromere pairing that appears to be mediated by a synaptonemal complex protein known as Zip1 (Kurdzo et al. 2017, 2018). Zip1 can be thought of as the "teeth of the zipper" in the synaptonemal complex. The use of the synaptonemal complex and some of its components seems to be a common theme in achiasmate segregational systems, finding fulfillment in the fully achiasmate male meiosis of the Australian praying mantis where the entire SC, appearing structurally normal, persists until metaphase 1 (Gassner 1969) and, presumably, is the primary means of homolog disjunction.

Achiasmate Segregation in *S. pombe*

Several studies of recombination-deficient mutants in *S. pombe* have revealed the existence of an achiasmate segregation system that may well be homology dependent. The evidence for this assertion is that achiasmate homologs appear to segregate from each other faithfully in Rec$^-$ cells in more than 75% of the meioses (Davis and Smith 2001). This is true even though two small minichromosomes do not segregate faithfully in this organism (Niwa et al. 1989). Achiasmate segregation can indeed be observed cytologically in this organism and appears to function by a crowded pole-like mechanism (Molnar et al. 2001).

Achiasmate Segregation in Silkworm Females

As previously noted, Lepidopteran females possess an unusual system in which the synaptonemal complex is transformed into a much larger but rather amorphous structure that holds the achiasmate bivalents together until homolog separation at anaphase I. This process is best described in the silkworm *B. mori* (Rasmussen 1977a, b).

9.5 Meiotic Drive

We have said before and will say again, meiosis is the physical basis of Mendelian inheritance. For example, the fact that an *Aa* heterozygote will produce *A*- and *a*-bearing gametes at equal frequency is no more than a statement that the two homologs segregate to opposite poles at anaphase I. Similarly, the intrinsic statistical fairness of Mendel's Law of Independent Assortment depends on the random orientation of two bivalents at metaphase I followed by their segregation at anaphase I. (There is also an often-unstated assumption that gametes will function at equal frequencies.) And fortunately, meiosis is intrinsically fair.

However, in any biological or human system, some genes, chromosomes, or people will cheat. In the case of meiosis, cheating often implies cold-blooded murder – gametes carrying one of the two alleles are eliminated by the other. Imagine that some allele of the *A* gene could destroy gametes bearing other alleles of this gene. In male meiosis, this could confer a huge advantage to this killer allele in terms of its transmission to the next generation. In female meiosis, where only one of the four meiotic products becomes an oocyte, suppose a gene or chromosome could increase its chance of ending up in the oocyte pronucleus, and not in a dead-end polar body, by biasing segregation to one pole or another at anaphase I. If you are shaking your head and saying, "Yeah, but cheating in such cases would be so difficult that it wouldn't be worth it," remember what the people who run gambling casinos, the folks at the IRS, and college professors already know – if it is possible to bias a process or an event, it will happen. Indeed, it has evolved many times in different meiotic systems.

Taken together, these processes of making meiosis unfair are known as **meiotic drive**. There are too many such processes for us to list here, but such lists can be found in reviews by Bravo Núñez et al. (2018a), Kruger and Mueller (2021), and Arora and Dumont (2022). We will focus on two kinds of drive: (i) spore/sperm killing in meioses that produce four functional gametes and (ii) directed segregation at either meiosis I or II in oocytes.

Meiotic Drive Via Spore Killing

An Example in *Schizosaccharomyces pombe*

We begin with spore killing in the fission yeast *S. pombe* by the *wtf* system, the name of which is derived from a family of genes with *Tf* transposons flanking them. Some *wtf* elements can function as powerful meiotic drivers. If one mates a *wtf*-containing strain to a strain missing such an element, only the *wtf*-bearing spores will survive; the spores lacking *wtf* elements will die an ugly death. The basis for this phenomenon has been elegantly elucidated by Bravo Núñez et al. (2018b) and Nuckolls et al. (2017). In the meiotic ascus (a structure containing the four spores produced by meiosis), spores bearing *wtf* elements produce a diffusible poison that reaches all four spores. Conveniently, they also produce a nondiffusible antidote that saves themselves while their sister spores perish.

An Example in *D. melanogaster*

A similar case of fratricidal drive can be found in *D. melanogaster* males in the form of the *Segregation Distorter* (*SD*) chromosome (Ganetzky 1977). Males heterozygous for the neomorphic *Sd* mutant will transmit the *SD* chromosome to 99% of their progeny. They do so by ensuring the destruction of the sperm that lack *Sd*. After the completion of meiosis, the *Sd* allele alters nuclear transport in a fashion that causes nearly all sperm bearing a type of heterochromatic repeat sequence known as *Responder* (*Rsp*) to fail to condense their chromatin. This failure blocks the maturation of those cells into mature sperm. However, the *Sd* allele is linked by an inversion to a deletion of the *Rsp* element called *Responder-insensitive* (Rsp^i). This means that the *SD* chromosome, which carries both *Sd* and Rsp^i, can induce a severe defect in chromosome condensation – a defect to which it is itself immune. As such, primarily *SD*-bearing sperm survive to maturity.

Meiotic Drive Via Directed Segregation

There are also mechanisms that function by biasing segregation itself, such as in systems like female meiosis, in which only one of the four products of meiosis becomes a gamete. If one can bias which chromatid ends up in the gamete pronucleus, the result can be powerful meiotic drive. For example, in some wild populations of *D. melanogaster*, a mechanism exists that allows paracentric inversions to be maintained in the population. As Figure 9.6 illustrates, in females heterozygous for a paracentric inversion on the *X* chromosome and a normal sequence *X* chromosome, crossing over within the inversion will create disaster in the form of both a dicentric and an acentric chromosome, neither of which is transmissible. How, then, can such inversions be maintained in natural populations? The answer to this dilemma was published by Sturtevant and Beadle (1936) and summarized by Ganetzky and Hawley (2016). In *Drosophila* females, the four products of meiosis are arranged as an ordered line of pronuclei that lies perpendicular to the inner surface of the egg (there is no cell division, just nuclear division). The four products reflect the two sequential meiotic divisions, but only the innermost of the four nuclei can be fertilized when the sperm enters the egg. After meiosis I, the crossover products (the dicentric chromosome and the centromere-less acentric fragment) must remain in the two middle pronuclei; the noncrossover chromatids are free to reach the nuclei at the ends of the line and thus participate in fertilization.[4,5]

---

4 What would happen to a double crossover that involved all four chromatids? As shown in Figure 9.6, such an event creates a double dicentric ring-like chromosome, which is also captured only by the two middle pronuclei and two very stuck-in-the middle acentrics. The outermost pronuclei will receive no *X* chromosomes and thus look like products of nondisjunction. However, two-strand doubles (in which both crossovers involve the same two chromatids) are perfectly transmissible. This paradox of recovering no singles, but plenty of double crossover chromatids and some nullo-X eggs, was solved by Sturtevant and Beadle (1936) and reviewed by Ganetzky and Hawley (2016). That solution is the basis of the explanation provided in text.

5 If you are inclined to ponder the real drama of meiotic crossing over in paracentric inversions in an organism like yeast, in which all the products of meiosis are functional, we gleefully refer you to Haber and Thorburn (1984) and Haber *et al.* (1984). It's a great story, and these two papers make perfect companions for a cold day with a thermos of hot chocolate.

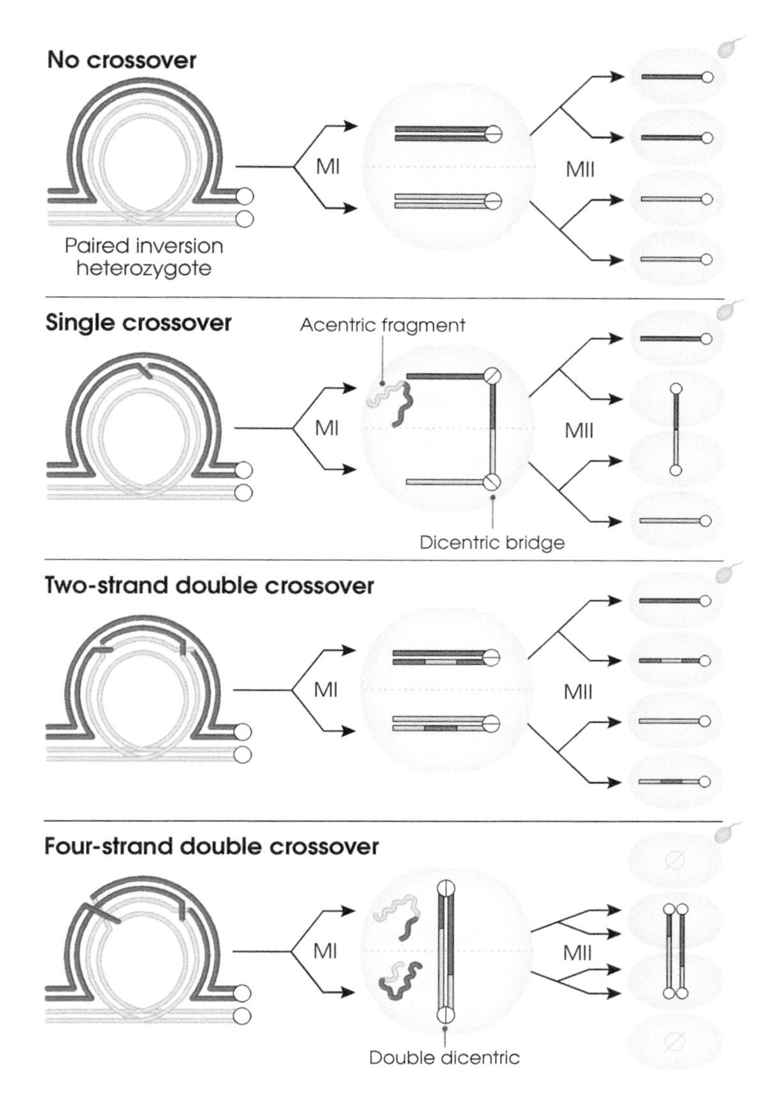

Figure 9.6 Directed segregation. Although inversions are often deleterious to an organism, in some wild *Drosophila* populations, X-chromosome inversion heterozygotes are maintained in the population. In these populations, single crossovers and four-strand double crossovers that occur within the inversion will not segregate properly during meiosis because they produce dicentric bridges that are directed into the middle two nuclei, which do not participate in fertilization. The only viable gametes are those that either do not cross over or those with a two-strand double crossover within the inverted segment. Because half of these gametes carry the inversion – and are equally as likely as the wildtype gametes to make it into the oocyte nucleus – the inversion is able to persist in the population.

The type of meiotic drive that involves using the orientation of the metaphase I and metaphase II spindles to place recombinant or nonrecombinant chromatids in the single functional pronucleus explains two other well-known drive systems: nonrandom disjunction in *Drosophila* oocytes and the chromosomal knob on the maize Ab10 chromosome (Novitski 1951; Mark and Zimmering 1977; Buckler et al. 1999; Dawe 2022). Other examples of the meiotic drive have been reviewed by Clark and Akera (2021). Examples of other drive systems in plants, animals, and fungi are numerous and the mechanisms are almost limitless. Fortunately, there are excellent reviews in plants (Chen et al. 2022), mice (Arora and Dumont 2022), and fungi (Zanders and Johannesson 2021; Komluski et al. 2022; Vogan et al. 2022).

One could think of the drive systems presented here as an example of a process referred to as genome conflict. Genomic elements or chromosomes evolve ways to cheat, while the organism evolves means to suppress this type of selfish misbehavior. An example of such a genome–selfish element interaction might be found in the behavior of B chromosomes in *D. melanogaster* (Bauerly et al. 2014). These chromosomes were discovered in a strain that was heterozygous for a mutant in the *matrimony* (*mtrm*) gene, the product of which acts to block the activity of the powerful cell regulator Polo kinase during female meiosis. In a *mtrm/+* background, B chromosome number increases in oocytes to approximately 15 copies per cell and is maintained at that level. However, in the absence of a *mtrm* mutant, B chromosomes are quite quickly lost from the genome in a few generations (Bauerly et al. 2014). Perhaps one function of Mtrm's regulation of Polo in female meiosis is to block the accumulation of such small selfish chromosomes, but the Bs exploited the more permissible environment found in a *mtrm/+* strain. It is a most intriguing view of genome conflict.

9.6 Summary

In this chapter, we have given you a picture of the mechanisms that ensure proper chromosome segregation during meiosis. It is segregation that explains Mendelian assortment – and that brings us back to the beginning. Perhaps that is a very good place to end.

References

Allshire RC Ekwall K. 2015. Epigenetic regulation of chromatin states in *Schizosaccharomyces pombe. Cold Spring Harb. Perspect. Biol.* 7:a018770.

Allshire RC Karpen GH. 2008. Epigenetic regulation of centromeric chromatin: old dogs, new tricks? *Nat. Rev. Genet.* 9:923–937.

Arora UP Dumont BL. 2022. Meiotic drive in house mice: mechanisms, consequences, and insights for human biology. *Chromosom. Res.* 30:165–186.

Bauerly E, Hughes SE, Vietti DR, Miller DE, McDowell W, *et al.* 2014. Discovery of supernumerary B chromosomes in *Drosophila melanogaster. Genetics* 196:1007–1016.

Bonner AM, Hughes SE, Hawley RS. 2020. Regulation of polo kinase by matrimony is required for cohesin maintenance during *Drosophila melanogaster* female meiosis. *Curr. Biol.* 30:715–722. e3

Bravo Núñez MA, Nuckolls NL, Zanders SE. 2018a. Genetic villains: killer meiotic drivers. *Trends Genet.* 34:424–433.

Bravo Núñez MA, Lange JJ, Zanders SE. 2018b. A suppressor of a *wtf* poison-antidote meiotic driver acts via mimicry of the driver's antidote. *PLoS Genet.* 14(11):e1007836.

Bridges CB. 1914. Direct proof through non-disjunction that the sex-linked genes of *Drosophila* are borne by the X-chromosome. *Science* 40:107–109.

Bridges CB. 1916. Non-disjunction as proof of the chromosome theory of heredity. *Genetics* 1 (1):53.

Buckler ES 4th, Phelps-Durr TL, Buckler CS, Dawe RK, Doebley JF, *et al.* 1999. Meiotic drive of chromosomal knobs reshaped the maize genome. *Genetics* 153(1):415–426.

Carpenter AT. 1973. A meiotic mutant defective in distributive disjunction in *Drosophila melanogaster. Genetics* 73:393–428.

Carpenter AT. 1975. Electron microscopy of meiosis in *Drosophila melanogaster* females: II: the recombination nodule – a recombination-associated structure at pachytene? *Proc. Natl. Acad. Sci. U.S.A.* 72:3186–3189.

Carpenter AT. 1979. Synaptonemal complex and recombination nodules in wild-type *Drosophila melanogaster* females. *Genetics* 92:511–541.

Chen J, Birchler JA, Houben A. 2022. The non-Mendelian behavior of plant B chromosomes. *Chromosom. Res.* 30:229–239.

Clark FE, Akera T. 2021. Unravelling the mystery of female meiotic drive: where we are. *Open Biol.* 11:210074.

Clarke L. 1998. Centromeres: proteins, protein complexes, and repeated domains at centromeres of simple eukaryotes. *Curr. Opin. Genet. Dev.* 8:212–218.

Clarke L, Carbon J. 1980. Isolation of a yeast centromere and construction of functional small circular chromosomes. *Nature* 287:504–509.

Clarke L, Carbon J. 1985. The structure and function of yeast centromeres. *Annu. Rev. Genet.* 19(29):55.

Clarke L, Baum M, Marschall LG, Ngan VK, Steiner NC. 1993. Structure and function of *Schizosaccharomyces pombe* centromeres. *Cold Spring Harb. Symp. Quant. Biol.* 58:687–695.

Cooper KW. 1964. Meiotic conjunctive elements not involving chiasmata. *Proc. Natl. Acad. Sci. U.S.A.* 52:1248–1255.

Davis L, Smith GR. 2001. Meiotic recombination and chromosome segregation in *Schizosaccharomyces pombe*. *Proc. Natl. Acad. Sci. U.S.A.* 98:8395–8402.

Dawe RK. 2022. The maize abnormal chromosome 10 meiotic drive haplotype: a review. *Chromosom. Res.* 30(2–3):205–216.

Dawson DS, Murray AW, Szostak JW. 1986. An alternative pathway for meiotic chromosome segregation in yeast. *Science* 234:713–717.

Dernburg AF, Sedat JW, Hawley RS. 1996. Direct evidence of a role for heterochromatin in meiotic chromosome segregation. *Cell* 86:135–146.

Doheny KF, Sorger PK, Hyman AA, Tugendreich S, Spencer F, *et al.* 1993. Identification of essential components of the *S. cerevisiae* kinetochore. *Cell* 73:761–774.

Ganetzky B. 1977. On the components of segregation distortion in *Drosophila melanogaster*. *Genetics* 86:321–355.

Ganetzky B, Hawley RS. 2016. The centenary of genetics: bridges to the future. *Genetics* 202:15–23.

Gassner G. 1969. Synaptonemal complexes in the achiasmatic spermatogenesis of *Bolbe nigra* Giglio-Tos (Mantoidea). *Chromosoma* 26:22–34.

Girard LR, Fiedler TJ, Harris TW, Carvalho F, Antoshechkin I, *et al.* 2007. WormBook: the online review of *Caenorhabditis elegans* biology. *Nucleic Acids Res.* 35:D472–D475.

Gruhn JR, Hoffmann ER. 2022. Errors of the egg: the establishment and progression of human aneuploidy research in the maternal germline. *Annu. Rev. Genet.* 56:369–390.

Gruhn JR, Zielinska AP, Shukla V, Blanshard R, Capalbo A, *et al.* 2019. Chromosome errors in human eggs shape natural fertility over reproductive life span. *Science* 365:1466–1469.

Guacci V, Kaback DB. 1991. Distributive disjunction of authentic chromosomes in *Saccharomyces cerevisiae*. *Genetics* 127:475–488.

Haber JE, Thorburn PC. 1984. Healing of broken linear dicentric chromosomes in yeast. *Genetics* 106:207–226.

Haber JE, Thorburn PC, Rogers D. 1984. Meiotic and mitotic behavior of dicentric chromosomes in *Saccharomyces cerevisiae*. *Genetics* 106:185–205.

Harris D, Orme C, Kramer J, Namba L, Champion M, *et al.* 2003. A deficiency screen of the major autosomes identifies a gene (*matrimony*) that is haplo-insufficient for achiasmate segregation in Doocytes. *Genetics* 165:637–652.

Hassold TJ, Hunt PA. 2021. Missed connections: recombination and human aneuploidy. *Prenat. Diagn.* 41(5):584–590.

Hassold T, Merrill M, Adkins K, Freeman S, Sherman S. 1995. Recombination and maternal age-dependent nondisjunction: molecular studies of trisomy 16. *Am. J. Hum. Genet.* 57:867–874.

Hassold T, Maylor-Hagen H, Wood A, Gruhn J, Hoffmann E, *et al.* 2021. Failure to recombine is a common feature of human oogenesis. *Am. J. Hum. Genetics* 108:16–24.

Hawley RS. 1988. Exchange and chromosomal segregation in eukaryotes. In: Kucherlapati R, Smith G, editors. *Genetic Recombination*. Washington, DC: American Society of Microbiology. pp. 497–527.

Hawley RS. 1993. Meiosis as an "M" thing: twenty-five years of meiotic mutants in *Drosophila*. *Genetics* 135:613–618.

Hawley RS, Theurkauf WE. 1993. Requiem for distributive segregation: achiasmate segregation in *Drosophila* females. *Trends Genet.* 9:310–317.

Hawley RS, Irick H, Zitron AE, Haddox DA, Lohe A, *et al.* 1992. There are two mechanisms of achiasmate segregation in *Drosophila* females, one of which requires heterochromatic homology. *Dev. Genet.* 13:440–467.

Hawley RS, McKim KS, Arbel T. 1993. Meiotic segregation in *Drosophila melanogaster* females: molecules, mechanisms, and myths. *Annu. Rev. Genet.* 27:281–317.

Hoffmann S, Izquierdo HM, Gamba R, Chardon F, Dumont M, *et al.* 2020. A genetic memory initiates the epigenetic loop necessary to preserve centromere position. *EMBO J.* 39:e105505.

Hofstatter PG, Thangavel G, Castellani M, Marques A. 2021. Meiosis progression and recombination in holocentric plants: what is known? *Front. Plant Sci.* 12:658296.

Hyman AA, Sorger PK. 1995. Structure and function of kinetochores in budding yeast. *Annu. Rev. Cell Dev. Biol.* 11:471–495.

Ines OD, Gallego ME, White CI. 2014. Recombination-independent mechanisms and pairing of homologous chromosomes during meiosis in plants. *Mol. Plant* 7:492–501.

Kaback DB. 1989. Meiotic segregation of circular plasmid-minichromosomes from intact chromosomes in *Saccharomyces cerevisiae*. *Curr. Genet.* 15:385–392.

Karpen GH, Allshire RC. 1997. The case for epigenetic effects on centromere identity and function. *Trends Genet.* 13:489–496.

Karpen GH, Le MH, Le H. 1996. Centric heterochromatin and the efficiency of achiasmate disjunction in *Drosophila* female meiosis. *Science* 273:118–122.

Kauppi L, Jasin M, Keeney S. 2012. The tricky path to recombining X and Y chromosomes in meiosis. *Ann. NY. Acad. Sci.* 1267:18–23.

Koehler KE, Hawley RS, Sherman S, Hassold T. 1996. Recombination and nondisjunction in humans and flies. *Hum. Mol. Genet.* 5:1495–1504.

Komluski J, Stukenbrock EH, Habig M. 2022. Non-Mendelian transmission of accessory chromosomes in fungi. *Chromosom. Res.* 30:241–253.

Kruger AN, Mueller JL. 2021. Mechanisms of meiotic drive in symmetric and asymmetric meiosis. *Cell. Mol. Life Sci.* 78:3205–3218.

Kurdzo EL, Obeso D, Chuong H, Dawson DS. 2017. Meiotic centromere coupling and pairing function by two separate mechanisms in *Saccharomyces cerevisiae*. *Genetics* 205:657–671.

Kurdzo EL, Chuong HH, Evatt JM, Dawson DS. 2018. A ZIP1 separation-of-function allele reveals that centromere pairing drives meiotic segregation of achiasmate chromosomes in budding yeast. *PLoS Genet.* 14:e1007513.

Lamb NE, Freeman SB, Savage-Austin A, Pettay D, Taft L, *et al.* 1996. Susceptible chiasmate configurations of chromosome 21 predispose to non-disjunction in both maternal meiosis I and meiosis II. *Nat. Genet.* 14:400–405.

Le MH, Duricka D, Karpen GH. 1995. Islands of complex DNA are widespread in *Drosophila* centric heterochromatin. *Genetics* 141:283–303.

Lee JY, Orr-Weaver TL. 2001. The molecular basis of sister-chromatid cohesion. *Annu. Rev. Cell Dev. Biol.* 17:753–777.

Li W, He X. 2020. Inverted meiosis: an alternative way of chromosome segregation for reproduction. *Acta Biochim. Biophys. Sin.* 52:702–707.

Loidl J, Scherthan H, Kaback DB. 1994. Physical association between nonhomologous chromosomes precedes distributive disjunction in yeast. *Proc. Natl. Acad. Sci. U.S.A* 91:331–334.

Maggert KA, Karpen GH. 2001. The activation of a neocentromere in *Drosophila* requires proximity to an endogenous centromere. *Genetics* 158:1615–1628.

Mark HFL, Zimmering S. 1977. Centromeric effect on the degree of nonrandom disjunction in the female *Drosophila melanogaster*. *Genetics* 86:121–132.

May KM, Jacobs PA, Lee M, Ratcliffe S, Robinson A, *et al.* 1990. The parental origin of the extra X chromosome in 47, XXX females. *Am. J. Hum. Genet.* 46:754–761.

McKee BD, Karpen GH. 1990. *Drosophila* ribosomal RNA genes function as an X–Y pairing site during male meiosis. *Cell* 61:61–72.

McKee BD, Habera L, Vrana JA. 1992. Evidence that intergenic spacer repeats of *Drosophila melanogaster* rRNA genes function as X–Y pairing sites in male meiosis, and a general model for achiasmatic pairing. *Genetics* 132:529–544.

McKee BD, Lumsden SE, Das S. 1993. The distribution of male meiotic pairing sites on chromosome 2 of *Drosophila melanogaster*: meiotic pairing and segregation of 2-Y transpositions. *Chromosoma* 102:180–194.

McNulty SM, Sullivan BA. 2017. Centromeres and kinetochores, discovering the molecular mechanisms underlying chromosome inheritance. *Prog. Mol. Subcell. Biol.* 56:233–255.

Miller DE, Smith CB, Kazemi NY, Cockrell AJ, Arvanitakis AV, *et al.* 2016. Whole-genome analysis of individual meiotic events in *Drosophila melanogaster* reveals that noncrossover gene conversions are insensitive to interference and the centromere effect. *Genetics* 203:159–171.

Molnar M, Bähler J, Kohli J, Hiraoka Y. 2001. Live observation of fission yeast meiosis in recombination-deficient mutants: a study on achiasmate chromosome segregation. *J. Cell Sci.* 114:2843–2853.

Murphy TD, Karpen GH. 1995. Interactions between the *nod+* kinesin-like gene and extracentromeric sequences are required for transmission of a *Drosophila* minichromosome. *Cell* 81:139–148.

Nicklas RB. 1967. Chromosome micromanipulation. II. Induced reorientation and the experimental control of segregation in meiosis. *Chromosoma* 21:17–50.

Nicklas RB. 1974. Chromosome segregation mechanisms. *Genetics* 78:205–213.

Nicklas RB. 1977. Chromosome distribution: experiments on cell hybrids and in vitro. *Philos. Trans. R. Soc. B Biol. Sci.* 277:267–276.

Nicklas RB, Staehly CA. 1967. Chromosome micromanipulation. I. The mechanics of chromosome attachment to the spindle. *Chromosoma* 21:1–16.

Niwa O, Matsumoto T, Chikashige Y, Yanagida M. 1989. Characterization of *Schizosaccharomyces pombe* minichromosome deletion derivatives and a functional allocation of their centromere. *EMBO J.* 8:3045–3052.

Novitski E. 1951. Non-random disjunction in *Drosophila*. *Genetics* 36:267–280.

Nuckolls NL, Bravo Núñez MA, Eickbush MT, Young JM, Lange JJ, *et al.* 2017. *wtf* genes are prolific dual poison-antidote meiotic drivers. *elife* 6:e26033.

Orr-Weaver T. 1996. Meiotic nondisjunction does the two-step. *Nat. Genet.* 14:374–376.

Parker DR, Williamson JH. 1976. Aberration induction and segregation in oocytes. *Genetics and biology of Drosophila.* 1:1251–1268.

Rasmussen SW. 1977a. The transformation of the synaptonemal complex into the "elimination chromatin" in *Bombyx mori* oocytes. *Chromosoma* 60:205–221.

Rasmussen SW. 1977b. Meiosis in *Bombyx mori* females. *Philos. Trans. R. Soc. Lond B Biol. Sci.* 277:343–350.

Rasooly RS, New CM, Zhang P, Hawley RS, Baker BS. 1991. The *lethal(1)TW-6cs* mutation of *Drosophila melanogaster* is a dominant antimorphic allele of *nod* and is associated with a single base change in the putative ATP-binding domain. *Genetics* 129:409–422.

Ren X, Eisenhour L, Hong C, Lee Y, McKee BD. 1997. Roles of rDNA spacer and transcription unit-sequences in X–Y meiotic chromosome pairing in *Drosophila melanogaster* males. *Chromosoma* 106:29–36.

Riddle DL, Blumenthal T, Meyer BJ, Priess JR, Albertson DG, *et al.* 1997. *Chromosome Organization, Mitosis, and Meiosis.* Cold Spring Harbor Laboratory Press.

Ross LO, Rankin S, Shuster MF, Dawson DS. 1996. Effects of homology, size and exchange of the meiotic segregation of model chromosomes in *Saccharomyces cerevisiae. Genetics* 142:79–89.

Sandler L, Lindsley DL, Nicoletti B, Trippa G. 1968. Mutants affecting meiosis in natural populations of *Drosophila melanogaster. Genetics* 60:525–558.

Steiner NC, Clarke L. 1994. A novel epigenetic effect can alter centromere function in fission yeast. *Cell* 79:865–874.

Stimpson KM, Matheny JE, Sullivan BA. 2012. Dicentric chromosomes: unique models to study centromere function and inactivation. *Chromosom. Res.* 20:595–605.

Sturtevant AH, Beadle GW. 1936. The relations of inversions in the X chromosome of *Drosophila melanogaster* to crossing over and disjunction. *Genetics* 21:554–604.

Sullivan LL, Sullivan BA. 2020. Genomic and functional variation of human centromeres. *Exp. Cell Res.* 389:111896.

Sun X, Wahlstrom J, Karpen G. 1997. Molecular structure of a functional *Drosophila* centromere. *Cell* 91:1007–1019.

Theurkauf WE, Hawley RS. 1992. Meiotic spindle assembly in *Drosophila* females: behavior of nonexchange chromosomes and the effects of mutations in the *nod* kinesin-like protein. *J. Cell Biol.* 116:1167–1180.

Thomas NS, Hassold TJ. 2003. Aberrant recombination and the origin of Klinefelter syndrome. *Hum. Reprod. Update* 9:309–317.

Vazquez J, Belmont AS, Sedat JW. 2002. The dynamics of homologous chromosome pairing during male *Drosophila* meiosis. *Curr. Biol.* 12:1473–1483.

Vernizzi L, Lehner CF. 2022. Dispersive forces and resisting spot welds by alternative homolog conjunction govern chromosome shape in *Drosophila* spermatocytes during prophase I. *PLoS Genet.* 18:e1010327.

Vogan AA, Martinossi-Allibert I, Ament-Velásquez SL, Svedberg J, Johannesson H. 2022. The spore killers, fungal meiotic driver elements. *Mycologia* 114:1–23.

Williams BC, Murphy TD, Goldberg ML, Karpen GH. 1998. Neocentromere activity of structurally acentric mini-chromosomes in *Drosophila. Nat. Genet.* 18:30–37.

Wolf KW. 1994. How meiotic cells deal with non-exchange chromosomes. *Bioessays* 16:107–114.

Zanders S, Johannesson H. 2021. Molecular mechanisms and evolutionary consequences of spore killers in Ascomycetes. *Microbiol. Mol. Biol. Rev.* 85:e00016–e00021.

Appendix A

Model Organisms

The cultivation of various model organisms is the foundation of modern genetic research. Because of their generally short lifespans, their amenability to genetic manipulation, and the relative ease of maintaining them, model organisms are particularly useful for studying genetic structure, function, mutations, or diseases. There are several model organisms available to researchers today, including – but not limited to – yeast, plants, worms, fish, flies, and mice.

The choice of a model organism is important, as some are more appropriate for particular studies than others. For example, researchers may find organisms such as *Drosophila* preferable because of their relatively short generation time and the ability to obtain hundreds of progenies from a single individual. Researchers looking for a closer evolutionary relative to humans may choose to work with mice or rats. Others prefer organisms such as yeast or nematodes because of the ability to freeze samples for long periods of time, eliminating the need to maintain stocks and reducing the risk of a useful stock unexpectedly dying.

There are too many model organisms to cover them all here. Instead, we will focus on the model organisms discussed in this text, considering their basic culturing techniques, nomenclature, and chromosome biology, as well as providing other, more detailed sources of useful information regarding each organism.

A.1 Budding Yeast: *Saccharomyces cerevisiae*

Because budding (or baker's) yeast is a single-celled eukaryote that can be handled with the same facility as *Escherichia coli*, this organism has been the workhorse for much of the development of modern eukaryotic genetics. The ease of culture and ability to grow as either a stable haploid or diploid make this organism ideal for mutant screens. Indeed, the ability to grow this yeast as a haploid at various temperatures and to replicate plate colonies makes it straightforward to isolate both loss-of-function and conditional mutations at a large number of loci. Complementation tests are easily done by mating haploids to diploids, and meiosis can be studied by inducing the diploid culture to undergo meiosis (sporulation). Sporulation results in the production of a walled ascus containing four spores – the four products of meiosis. The ability to recover all four products of meiosis greatly facilitates the analysis of the meiotic process.

Genetic Theory and Analysis: Finding Meaning in a Genome, Second Edition.
Danny E. Miller, Angela L. Miller, and R. Scott Hawley.

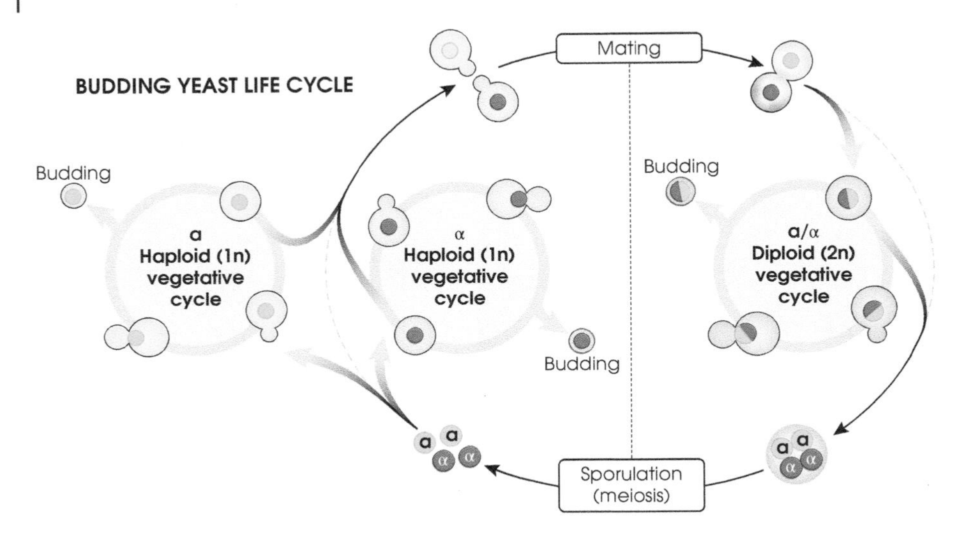

Basic Culture Techniques

Yeast can be grown in a standard broth media and on petri dishes. The haploids exist as two genetically determined mating types (a and alpha). Haploids of opposite mating type can be mated to produce stable a/alpha diploids that can be sporulated by nitrogen starvation.

Nomenclature

Genes are normally named by a three-letter symbol describing some aspect of their isolation, their phenotype, their required function, or the protein they produce. Thus, a gene isolated from an arginine auxotroph mutant would be symbolized as ARG, a gene encoding actin as ACT, and a gene required for the cell division cycle as CDC. In cases where two or more different genes have the same letter symbol, they are distinguished by numbers, such as URA1 and URA2. Alleles of a given gene are indicated by a number following the gene system and separated from the symbol by a hyphen. For example, *cdc2-1* and *cdc2-2* are different alleles of the CDC2 gene. Dominant alleles are symbolized by uppercase italics (e.g. *CDC2*), while lowercase italics denote recessive alleles (e.g. *cdc2-1*). Wildtype genes are written in uppercase, nonitalic text, and can be designated by a superscript plus sign (e.g. $ACT2^{+}$). Protein products are designated by nonitalic text with only the first letter capitalized (e.g. Spo11). It is also common to see the letter "p" following the protein name (e.g. Spo11p).

Chromosome Biology

The yeast genome is relatively small at 12.1 Mb, spread across 16 well-characterized chromosomes. Standard metaphase chromosome analysis methods allow one to visualize chromosomes, and bivalents can be visualized during meiosis.

Useful Guides and Manuals

The best guide to the genetics and biology of this organism was written by Fred Sherman at the University of Rochester, who unfortunately passed away in 2013. He published this guide in *Methods in Enzymology* (Sherman 2002), and a modified version may be found at https://instruct.

uwo.ca/biology/3596a/startedyeast.pdf. The central repository for current information regarding yeast genes and gene sequences, the Saccharomyces Genome Database (SGD), is maintained by Stanford University and can be found at www.yeastgenome.org. SGD also maintains a virtual library of information related to yeast and yeast culture, as well as yeast nomenclature information.

A.2 Plants: *Arabidopsis thaliana*

This small mustard weed plant has become one of the premier genetic organisms. A combination of a small chromosome number, a sequenced genome, and a small genome size makes it attractive as a genetic tool. Add to that a short generation time, the ability to grow under defined conditions, a very high fecundity, and the relative ease of propagation, and the attraction becomes irresistible. Efficient transformation systems exist, and there are both many available mutants and several stock centers. Moreover, the roots of young seedlings are relatively translucent – an appealing quality for microscopy work.

Mutagenesis in Arabidopsis can be done using using chemical (e.g. ethyl methanesulfonate) or physical mutagens (neutrons) but is more often done using transposable elements or random T-DNA integration following exposure of the plant to Agrobacterium. An excellent review of these methodologies may be found in Page and Grossniklaus (2002). Indeed, EMS mutagenesis of seeds works sufficiently well that after mutagenizing seeds, screens of 2000–3000 plants in the next generation usually yield two or three new mutants at most loci.

Basic Culture Techniques

At 25 cm tall, Arabidopsis is a relatively small plant, with most strains having a rapid six-week life cycle. It is flexible in that it grows well in petri plates, hydroponic systems, or under direct light. Its size and rapid development time therefore make it an ideal plant for laboratory work. In addition, each plant is self-pollinating and can produce up to 10 000 seeds per plant.

Nomenclature

Wildtype genes are denoted by three uppercase italic letters, followed by a number to differentiate genes with a similar symbol (i.e. *ARB1* or *ARB2*). Mutants are denoted by lowercase italics (*arb1* or *arb2*). Different alleles at the same locus are differentiated by a hyphen and a number (*arb3-3* would be the third allele of the *arb3* gene). Some workers denote dominant alleles by adding a *D* to the end of the symbol. Proteins are denoted using the wildtype gene symbol entirely in uppercase, no italics (e.g. ARB3). Phenotypes are usually denoted by fully writing out the gene name in nonitalic with an initial capital, followed by a superscript + to denote wildtype or – to denote mutant.

Chromosome Biology

Arabidopsis is a diploid (unusual for plants) and has a 115-Mb genome (small for a plant) distributed over five well-studied chromosomes. It is a well-annotated genome with a wide range of available tools that have been developed and validated over many years of laboratory research. The facility of genetic analysis in this organism allows screens for enhancer and suppressor mutants as well as for sophisticated epistasis analysis. Until 2001, the analysis of meiotic chromosome behavior had been hindered by the lack of useful protocols for meiotic cytology. Thanks largely to the work of Gareth Jones and his collaborators (Moran et al. 2001), this difficulty was fully overcome.

Useful Guides and Manuals

The best sources for printed information are *Arabidopsis*, edited by Meyerowitz and Somerville (1994) and Weigel and Glazebrook's *Arabidopsis: A Laboratory Manual* (2002). Web-based information can be found on the Arabidopsis Information Resource at www.arabidopsis.org. This website also contains a guide to mutant genes and their nomenclature.

A.3 Worms: *Caenorhabditis elegans*

The worm *C. elegans* is one of the most important model organisms in use today. This tiny worm lives on *E. coli* (of all things – don't you just love the symbolism of that) and is most easily cultured on petri dishes. *C. elegans* has two sexes, hermaphrodites (XX) and males (X0). Hermaphrodites can self-fertilize or mate with males, but one hermaphrodite cannot mate with another hermaphrodite. This capacity of hermaphrodites is arguably the most useful genetic feature of this system. A zygote resulting from a mutagenized gamete will develop as a heterozygote. If that heterozygote is a hermaphrodite, it will produce homozygotes in the next generation. Thus, one can go from mutagenesis to homozygotes for mutagenized chromosomes with but one intervening generation.

Basic Culture Techniques

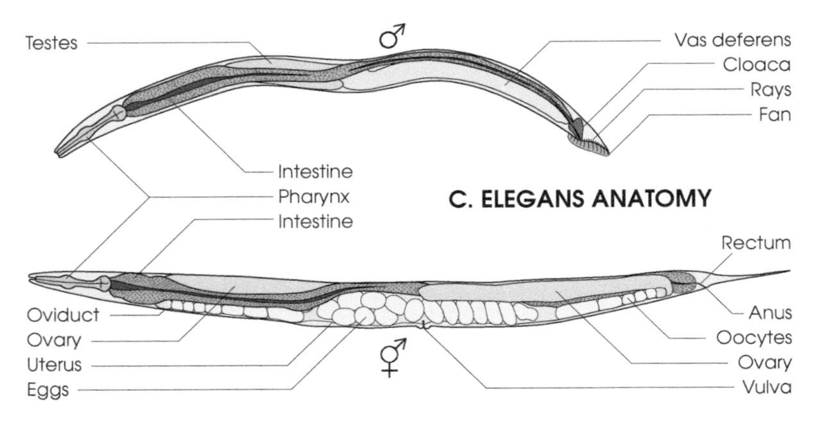

Development from an egg to a fertile adult hermaphrodite or male takes only three days, and a hermaphrodite can produce about 300 offspring using self-fertilization alone. In the presence of males, even greater numbers of progeny are easily obtained. When hermaphrodites mate with males, 50% of the progeny will be males and 50% will be hermaphrodites. The organism is transparent and contains an invariant number of somatic cells (959). The lineage relationships that produced the adult animals are fully described. Best yet, stocks of any given genotype can be stored frozen, avoiding the time-consuming process of stock keeping.

Nomenclature

As is the case in yeast, genes are normally named by a three-letter symbol describing some aspect of their isolation, their phenotype, their required function, or the protein they produce, followed by a dash and then a number. For example, the fifth gene defined by an *uncoordinated movement* mutant (*unc*) would be named *unc-5*. Gene names and symbols are italicized. Alleles are denoted by single or double letters, identifying the lab that isolated the mutant, followed by a number. So, an *unc-5* allele isolated in Kansas City might be called *unc-5(kc001)*. Wildtype alleles are denoted by a + symbol, and the protein product of a given gene is symbolized in all capital letters without italics (i.e. UNC-5).

Chromosome Biology

The *C. elegans* genome is approximately 100 Mb distributed over five pairs of autosomes and one pair of sex chromosomes. Chromosomes are holokinetic (diffuse centromeres) during mitosis but unikinetic (single centromere) during meiosis (Albertson and Thomson 1993). *C. elegans* is somewhat unique among eukaryotes in that it has operons, similar to bacteria (Blumenthal et al. 2002).

Useful Guides and Manuals

The book *C. elegans II* by Riddle et al. (1997) is available online at www.ncbi.nlm.nih.gov/books/NBK19997/ and provides an invaluable reference manual for *C. elegans* biology. An online database with more detailed information on a given strain or mutant, as well as nomenclature information, may be found on WormBase at www.wormbase.org/. Its companion site WormBook (http://www.wormbook.org/) provides a collection of peer-reviewed chapters on *C. elegans* biology and protocols.

A.4 Fruit Flies: *Drosophila melanogaster*

D. melanogaster was the organism used by Calvin Bridges to verify the chromosome theory of heredity (Bridges 1914, 1916). Since then, it has remained a central player in genetic research. Although there are other species in the *Drosophila* group (Ashburner, 1989), when we use the terms *Drosophila*, fruit fly, or simply flies in this narrative, we will always mean *D. melanogaster*.

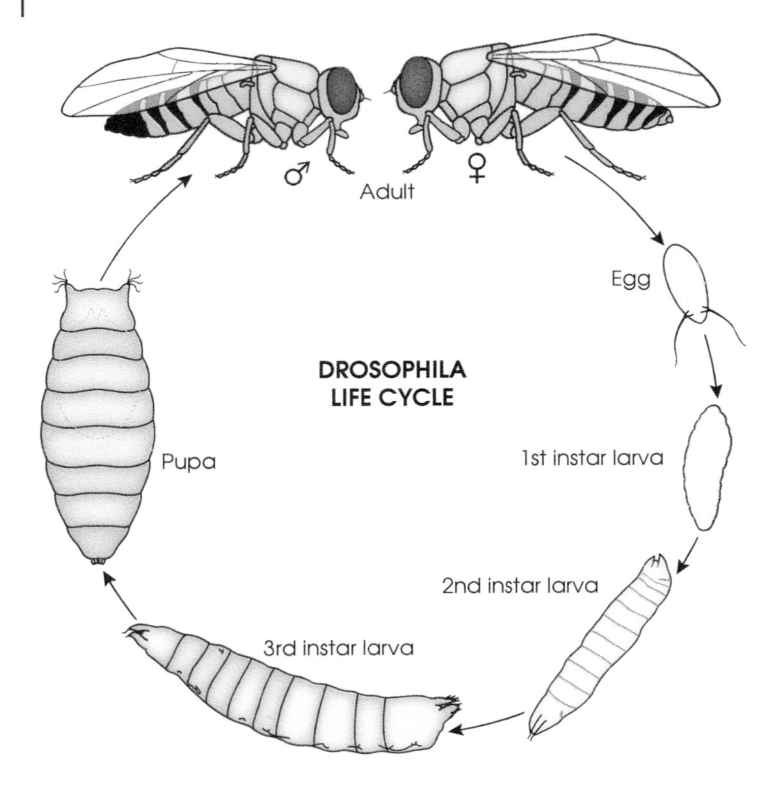

Basic Culture Techniques

Drosophila are easily raised on a cornmeal-molasses media (Ashburner 1989). They are neither harmful to humans nor do they carry parasites that are harmful to people. Under ideal conditions, an egg laid by a female fly will develop into an adult fly in nine days. This includes three larval instars and several days of pupation before hatching. Females are competent to mate within 8–12 hours after hatching. Unfortunately, they also store sperm from each mating they experience and will use that sperm throughout their lives. Thus, to set up controlled matings, one needs to collect virgin females (i.e. females who have had no previous opportunities to mate with males of a different genotype). This can be done easily by clearing bottles of hatching flies in the morning and then collecting recently hatched females in the afternoon. Once mated, a single female can produce several hundred eggs in a matter of one to two weeks.

Nomenclature

Fly nomenclature is decidedly complex. For our purposes, and oversimplifying greatly, we need only note that gene names and symbols are always printed in italics. For recessive mutants, the name and the symbol are fully in lowercase (e.g. *white*, denoted *w*, or *bithorax*, denoted *bx*). In the case of dominant mutants, the first letter of the name or symbol is capitalized (e.g. *Bar*, denoted *B*, or *Abnormal X segregation*, denoted *Axs*). A wildtype, or normal, allele is denoted either simply by a + symbol in a heterozygote (e.g. *nod*/+) or by a superscript following the gene symbol (e.g. nod^{+}). Protein products are nonitalicized, full protein names have an initial capital, and protein symbols (which match the gene symbol) are in all capital letters. For example, the protein product of the *corona* gene could be correctly denoted as the *cona* protein, Corona, or CONA.

When genes on more than one chromosome are represented in a genotype, the markers on each chromosome are separated by semicolons in the order *X; Y; 2; 3; 4*. Gene symbols are separated by a space. Homologs are separated by a forward slash mark (/). Homozygous chromosomes are defined only once, and + implies +/+. Thus, a female heterozygous for *yellow* (an X chromosome gene) and for *cinnabar* (a 2nd chromosome gene) would be denoted *y/+; cn/+*, while females homozygous at both loci would be denoted *y; cn*.

Chromosome Biology

D. melanogaster has four chromosome pairs – three autosomes and one pair of sex chromosomes. Although males are XY and females are XX, sex is essentially determined by the number of X chromosomes and not by the presence or absence of the Y chromosome. The Y chromosome is required only for male fertility. Thus, X0 animals are sterile males and XXY individuals are fertile females. The well-known polytene chromosomes are found most easily in the salivary glands of third instar larvae. Only the euchromatic regions of the chromosomes (approximately the distal two-thirds of each arm) amplify in the polytene chromosomes. The heterochromatic region appears at best as dense threads at the base of each chromosome. Mitotic chromosomes are easily found in early embryos and in dividing cells of the 3rd instar larval brain. Techniques for visualizing meiotic chromosomes in both sexes are now well developed.

There are a few interesting and useful genetic idiosyncrasies about *Drosophila* that are helpful to know:

- There is no recombination during meiosis in *Drosophila* males – none.
- There is no recombination on the small fourth chromosome in *Drosophila* female meiosis.
- The X, 2nd, and 3rd chromosomes do recombine during female meiosis, and on average, there is slightly more than one recombination event per chromosome arm.
- The Y chromosome does not determine sex but does carry genes essential for male fertility.
- X0 flies are sterile males. XXY flies are fertile females. XXX females survive but are sterile.
- Flies that carry only one 4th chromosome do survive, but they are sterile. They also exhibit a strong minute phenotype (thin bristles, delayed emergence). Flies with three copies of chromosome 4 are fully viable and fertile.
- Flies carrying either one or three copies of chromosome 2 or 3 are not viable. The exception to this rule is full triploids, which do survive.
- Multiply-inverted X, 2nd, and 3rd chromosomes that suppress recombination and contain easily identifiable visual markers are available to assist with stock construction and to simplify maintaining recessive mutations in stocks; these are known as balancer chromosomes.

Useful Guides and Manuals

A basic introduction to working with *Drosophila* can be found at www.ceolas.org/fly/intro.html. The basics of fly nomenclature are clearly spelled out on the Bloomington *Drosophila* Stock Center's website at https://bdsc.indiana.edu/information/nomenclature.html, and the full nomenclature guidelines are on FlyBase at http://flybase.org/wiki/FlyBase:Nomenclature. There are literally tens of thousands of genetically marked strains of *Drosophila*. The best guides to the available mutants are the so-called Redbooks, Lindsley and Grell (1968) and Lindsley and Zimm (1992). One can also get constantly updated information on mutants, chromosomes, and strains from several websites such as FlyBase (http://flybase.org) or the Bloomington *Drosophila* Stock Center at Indiana University (https://bdsc.indiana.edu). Information about other *Drosophila* species can be found at the website of the National *Drosophila* Species Stock Center at Cornell.

A.5 Zebrafish: *Danio rerio*

Concomitant with all the benefits of doing genetics in *Drosophila* or yeast, there are limitations. While the small size of flies makes it possible for a single researcher to easily maintain hundreds of thousands of *Drosophilae*, it also eliminates the need for *D. melanogaster* to have systems such as an endoskeleton or a closed circulatory system. When we want to study how the vertebrate heart works, for example, we are forced to turn to other organisms. One such organism that has become quite popular is the zebrafish *D. rerio*. Zebrafish offer us many of the same advantages that we enjoy when working with *Drosophila* that systems such as mouse lack. For instance, zebrafish are easy to maintain, can produce thousands of embryos in a week, and perhaps most importantly, the zebrafish embryo is transparent. This affords us the possibility of studying development unencumbered in embryogenesis. All these things, in combination with the ever-increasing collection of genetic markers and molecular methods, have quickly made zebrafish one of the most popular vertebrate model organisms.

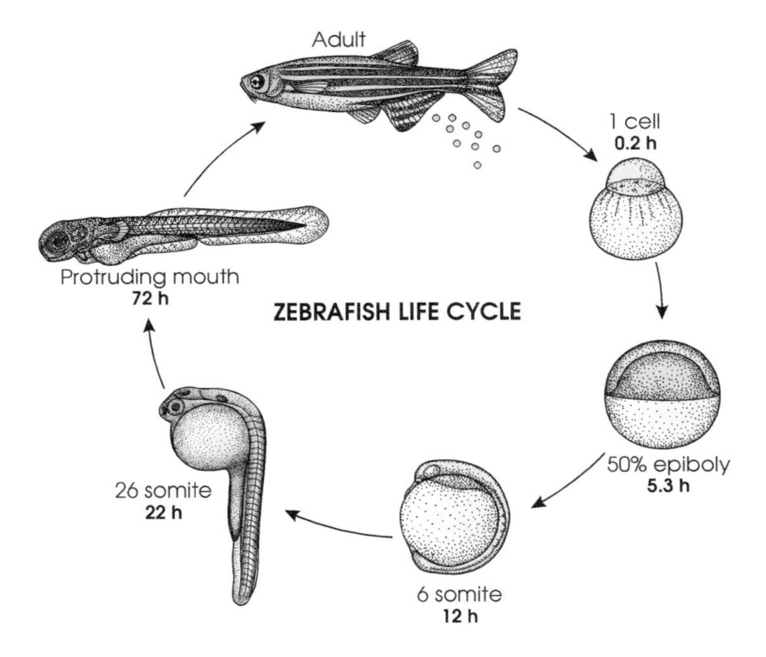

Basic Culture Techniques

Zebrafish are easily maintained in tanks ranging from 1 to 40 l in size and can be fed many types of food, including spirulina flakes and even fruit flies (Westerfield 2000). This is a handy way of feeding your fish if you work in the same building as a fly lab; fly geneticists are all too willing to relinquish old bottles of flies for which they no longer have need. The best food, however, if you want to keep your fish happy, is live brine shrimp. Like graduate students, zebrafish seem to survive better when kept on a regular day–night schedule of 14 hours of light for every 10 hours of dark. This means that if you don't have an entire room to devote to zebrafish stocks, you will have to get lighting for the tanks and covers to block out the ambient light created by your fellow scientists who might be working late.

Nomenclature

Gene names in zebrafish are written lowercase in italics. In general, gene symbols are three-letter, lowercased abbreviations. For example, the gene *albino* is abbreviated *alb*. Wildtype alleles of genes are given by the three-letter abbreviation followed by a superscript +. To continue with our example, the wildtype allele of the gene *albino* would be designated alb^{+}. Mutant alleles are also given by a three-letter designation but followed by a superscript code that generally gives an allele number and line designation of where the mutant was isolated. For example, the allele alb^{b4} is named as such because *b* is the code for Eugene, Oregon, where this allele was isolated. Dominant mutant alleles are named with a *d* as the first letter in the superscript. For example, nba^{da10} is a dominant allele of the gene *night blindness a*. Protein products of genes are simply written as the three-letter gene code, nonitalicized, with the first letter capitalized (e.g. Nba).

Chromosome Biology

The zebrafish genome is approximately 1.4 Gb distributed over 25 chromosomes. Zebrafish males and females appear to have the same complement of chromosomes (i.e. no sex chromosomes), unlike flies or mammals. Interestingly, the zebrafish genome underwent a duplication event recently that today is manifested as a large number of paralogs with similar sequences. This has implications for the knockdown of gene expression using RNAi and for genome editing using CRISPR/Cas9 or TALENs, since it may be difficult to target one specific gene if its paralog is not sufficiently different at the nucleotide level.

Useful Guides and Manuals

One of the most useful guides in working with zebrafish is *The Zebrafish Book* (Westerfield 2000). This volume covers everything from general fish care to histological, genetic, and molecular methods. The book is available online through http://zfin.org, a website that also contains information about available mutant and wildtype strains, molecular markers, recent publications, and laboratory contacts, as well as a complete list of codes for gene isolation location.

A.6 Mice: *Mus musculus*

Let us never forget that it all started with a mouse. –Walt Disney

In many ways, early genetics had its beginnings in the work of so-called mouse fanciers who helped extend the findings of Mendel. The mouse has continued to be a preeminent system for genetics, and indeed, taking the problem to the level of the mouse is a primary goal of many model organism studies. Some years ago, an administrator at the National Institutes of Health was overheard to remark, "Say what you want about all of these model organisms, they just aren't mice." Like it or not, the mouse has become the premier model organism for human biology, both because its mammalian biology closely approximates that of human beings and because of the facility for genetic manipulation that is possible in this organism.

There are hundreds of naturally occurring mouse mutant strains, but the relative ease of targeted mutagenesis in the mouse has led to the availability of thousands of mutant lines. Mutagenesis can be accomplished using either chemical or physical mutagens but is now most often achieved by targeted gene disruption. Although mouse geneticists continually claim to be on the verge of possessing respectable balancer chromosomes (Yu and Bradley 2001), by *Drosophila* standards, there is quite a long way to go in this effort.

Basic Culture Techniques

The mouse system benefits greatly from a relatively short generation time (about nine weeks), the availability of inbred strains, and the relatively low cost of rearing.[1] Laboratory mice are generally fed commercially available pelleted food. They thrive in quiet spaces with regular day/night cycles and sufficient nesting and enrichment materials. Females are often mated around 6–12 weeks of age, producing around three to 10 pups per litter, and good breeders are productive for seven months or more. Care should be taken when combining males from different litters or cages to prevent aggression. Mice live about two to five years, making them good candidates for studies involving mammalian lifespan or ageing.

Nomenclature

In mice, gene names are usually three to five, but no more than 10, characters long. Genes should be italicized with an initial capital letter, unless they're recessive, in which case the first letter should be lowercase. Protein names should be nonitalic and in all caps.

Chromosome Biology

At 2.7 Gb, the mouse genome is similar in size to the human genome and consists of 21 chromosomes, including a pair of sex chromosomes. Humans and mice diverged approximately 75 million years ago, which is reflected by the fact that about 90% of both genomes can be divided into regions of conserved synteny. In addition, about 99% of mice genes have a homolog in humans (Guénet 2005).

---

1 Having written that sentence, we can hear every mouse geneticist we have ever known screaming about the high costs of raising mice. True, they are more expensive to keep than flies, worms, or zebrafish, but they are cheaper than rats, hamsters, or primates, and incomparably less expensive to rear than larger mammals (e.g. blue whales).

B.1 Intragenic Mapping (Then)

The First Efforts Toward Finding Structure *Within* a Gene

Early intellectual ancestors viewed the gene as an indivisible particle that comprised the fundamental units of mutation, recombination, and function. Looking back at these early concepts of the gene, they seem hard to imagine. Students often ask, "Didn't they realize that genes are made out of DNA, and that DNA has length?" No, they did not. Remember, these early geneticists worked before the structure of DNA was known (Watson and Crick 1953). More critically, no one had performed the right experiment to challenge the view that genes were particulate in nature.

Such experiments were first done by Pete Oliver and Mel Green using alleles of the *lozenge* (*lz*) gene on the X chromosome of *Drosophila* (*for review, see* Green 1990). Like many great experiments, the result was obtained as the consequence of a carefully performed experiment done for another purpose. Mutations in the *lz* gene are notable for their effects on the shape and texture of the *Drosophila* eye. Many *lz* alleles are also female sterile. However, some *lz* alleles or combinations of *lz* alleles do allow the production of a few progeny (Oliver 1940; Oliver and Green 1944). Oliver and Green were examining various double-heterozygote combinations of *lz* alleles to study interactions between various alleles and their effect on female fertility. Much to their surprise, lz^+ (wildtype) male progeny emerged from several types of double heterozygotes. Such progeny were quite rare – approximately one per thousand – but the experiment was clearly reproducible.

Three lines of evidence argued that these lz^+ male progeny did not arise by simple reversion of one of the two *lz* alleles. First, wildtype progeny were not produced by homozygotes for the semi-fertile alleles (i.e. lz^x/lz^x). Second, Green used a clever set of genetic tricks to demonstrate that the reciprocal product (the double mutant chromosome $lz^x\ lz^y$) was also produced (Green and Green 1949; Green 1990). Third, production of lz^+ or double mutant combinations was often associated with the recombination of flanking markers. Thus, for the genotype $A\ lz^1\ B/a\ lz^2\ b$, where A and B are flanking genes, the lz^+ wildtype recombinant products were usually aB, while the double mutant-bearing chromosomes ($lz^1\ lz^2$) were Ab.

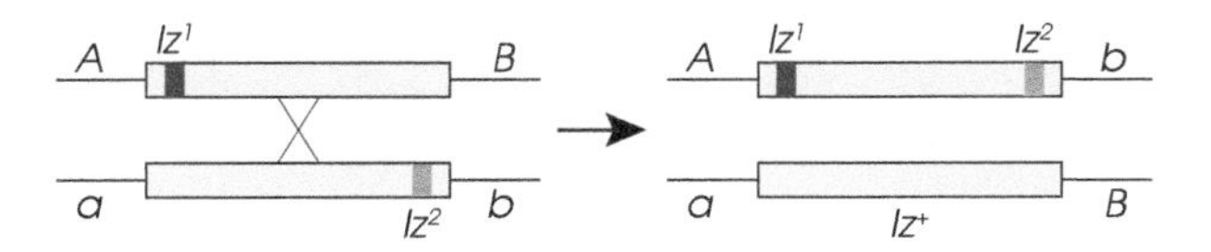

The only reasonable explanation for these data was that both types of chromosomes arose by recombination. Similar observations for another locus were obtained by Ed Lewis at Caltech (1948, 1952).

These experiments demonstrated that the gene is divisible by recombination. The importance of this result cannot be understated. As noted by Pontecorvo (1958), "The most obvious wrong idea is that of the particulate gene, i.e. of the genetic material as beads on a string in which each bead is the ultimate unit of crossing over, of mutation and of (function)." If recombination can occur between two different mutations within a gene, then the gene must have length and the two mutations must have occurred at different sites within the gene. Not only is the gene no longer the unit of recombination, but it is also no longer the unit of mutation. Different mutations cannot be thought of as different quantum states of an atom-like gene but must be viewed as different lesions along a linear gene.

Still, Mendel's view of the gene as a particle was difficult to dislodge despite the existence of the necessary evidence. Change would be created by a rather brash set of young biologists who viewed the world not through the Mendelian inheritance of flies and corn but rather through the most unusual genetics of bacteria and bacterial viruses. One of the most influential of those geneticists was Seymour Benzer. Beginning his genetic studies after the elucidation of the structure of DNA, Benzer was able to dissect the anatomy of the gene.

The Unit of Recombination and Mutation is the Base Pair

Following the discovery that DNA was the hereditary material (Avery et al. 1944) and the elucidation of its structure (Watson and Crick 1953), the concept of a linear gene with many mutable sites seemed quite reasonable. Work by Seymour Benzer on the **phage T4** allowed two adjacent genes to be dissected in a fashion that identified hundreds of separately mutable sites within each gene, each of which could be recombinationally separated from the others. Benzer was able to accomplish this feat both because of the ease with which phage recombination studies can be performed and because of his cleverness in choosing mutants that were amenable to selecting wildtype recombinants.

Benzer (1955, 1962) exploited the discovery (by others) of *rII* mutants in phage T4. In wildtype T4, phage particles multiply and destroy the bacterial cells on the surface of a culture in a glass dish, producing a normal **plaque**, or clear region, on three different *Escherichia coli* strains: S, B, and K. The ability of wildtype and *rII* mutant T4 phages to produce plaques on these three strains is shown in this table.

	Strain S	Strain B	Strain K
Wildtype	Normal plaque	Normal plaque	Normal plaque
rII	Normal plaque	Large, round plaque	No plaque

Understanding the value of these three strains is critical to understanding Benzer's analysis. The unusual plaque phenotype on strain B allows *rII* mutants to be easily isolated. (Other types of rough mutants will also be isolated by this type of screen, but only *rII* mutants fail to produce any plaques on strain K.) The inability of *rII* mutants to produce plaques on strain K allows the easy identification and collection of wildtype revertants or recombinants. Finally, under conditions where it is important to easily grow *rII* mutants in a nonselective fashion, one can use strain S.

Benzer identified *rII* $\longrightarrow$ wildtype revertants by first growing phage in strain S and then plating the phage on strain K. The various *rII* mutants tested showed a wide range of revertant frequencies, ranging from one in 10^3 progeny to less than one in 10^8 progeny phage. These mutants were presumed to be single base changes, or point mutants, that could mutate back to wildtype by processes similar to the mutational event that induced them.

Crosses of different *rII* alleles often produced wildtype recombinants at a frequency of a few percent. To obtain these recombinants and measure their frequency, strain B cells were coinfected with mutant *rII* (X)-bearing phage and mutant *rII* (Y)-bearing phage. The resulting lysate was then plated on strain K and on strain B. The number of wildtype phage produced by recombination could be determined by the number of plaques on strain K. The total number of plaques on strain B is an estimate of the total progeny, and the ratio of strain K/strain B plaques is an estimate of the recombination frequency. Using such crosses, Benzer was able to create a linear map for eight spontaneously arising *rII* mutants.

The discovery of nonreverting *rII* mutations, which correspond to deletions, enhanced the ease of mapping *rII* mutants (i.e. no revertants of these mutations were observed in more than 10^{10} progeny phage). In an important step toward fine-structure mapping, Benzer mapped these deletions relative to each other by determining which deletion-bearing phages could recombine to form a wildtype phage. To do this, strain B cells were coinfected with *rII* deletion (X)-bearing phage and *rII* deletion (Y)-bearing phage. The resulting lysate is then plated on strain K and on strain B. Two overlapping deficiencies cannot recombine to produce a wildtype recombinant, but two non-overlapping deficiencies can recombine at some frequency to produce a wildtype. An example of a deficiency map and the recombination data obtained from pairwise crosses is presented here:

Positions of the four deficiencies

rII Region

Deletion 1

Deletion 2

Deletion 3

Deletion 4

Production of wildtype recombinants

Parent X \ Parent Y	1	2	3	4
1	–	–	+	+
2		–	+	+
3			–	–
4				–

\+ Wildtype recombinants recovered

– No wildtype recombinants recovered

It is not surprising that each deletion failed to recombine with a defined set of revertible mutants (point mutations). Point mutations will not be able to recombine to produce wildtype progeny with a deletion that removes the base that is altered by that mutant. In the diagram here, the dashed line indicates the lengths of the two different deletions (A and B) and the asterisk denotes a point mutant.

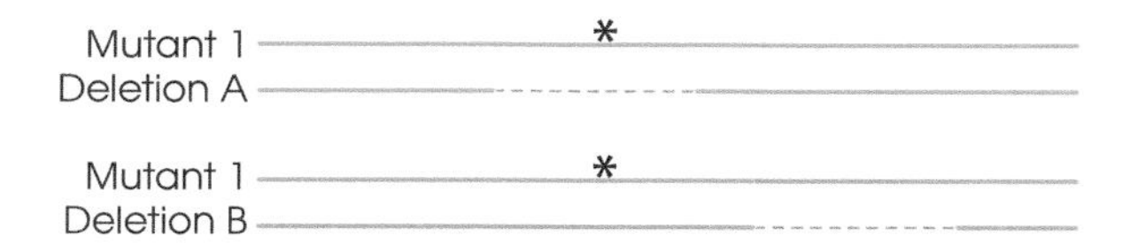

There is no way a phage carrying mutant 1 and deletion A can recombine to create a wildtype phage because there is no wildtype base pair corresponding to mutant 1. However, a phage carrying mutant 1 and deletion B can recombine to produce a wildtype because the wildtype base pair is present. By mapping deficiencies relative to each other, Benzer was able to divide the *rII* region of page T4 into 80 distinct subintervals.

Most deficiencies of the *rII* region failed to recombine with a number of revertible point mutants. Furthermore, these point mutants recombined with each other at a very low frequency. To explain these two phenomena, Benzer reasoned that all of these point mutants fell within one deletion. As shown by the drawings here, one can easily map a number of point mutants by allowing them to recombine with a set of overlapping deficiencies:

Map of 6 point mutants (A–F) and 4 deletions in the *rII* region of T4

A* B* C* D* E* F*

Deletion 1

Deletion 2

Deletion 3

Deletion 4

Production of wildtype recombinants

Deletion	Mutant A	B	C	D	E	F
1	–	–	+	+	+	+
2	+	–	–	+	+	+
3	+	+	–	–	–	+
4	+	+	+	+	–	–

+ Wildtype recombinants recovered

– No wildtype recombinants recovered

Using this method, Benzer mapped over 1000 *rII* point mutants within the 80 subintervals defined by the deletions. The point mutants within each subinterval could then be mapped against each other. Those pairs of point mutants that could recombine to produce a wildtype phage were said to define separate sites within the interval, while those mutants that could not recombine to produce a wildtype phage were said to define the same site. Based on this analysis, the *rII* region could be divided into approximately 350 separately mutable sites. Benzer argued, on statistical grounds, that at least 100 or more sites remained to be discovered, raising the number of separately mutable sites to at least 450. Most of these sites (~375) mapped into one complementation group, which Benzer named the *rIIa* gene. The others mapped in a second complementation group denoted *rIIb*. Benzer estimated that the *rIIa* and *rIIb* genes were each 4000 base pairs in length. These base pair estimations would argue that the units of mutation could be no longer than approximately 10 base pairs and that recombination occurs in intervals of 10 base pairs or less in length. (In fact, subsequent DNA sequencing of phage T4 would reveal that Benzer overestimated the physical lengths of these genes: *rIIa* is only 2178 base pairs in length and *rIIb* is only 939 base pairs long.)

The final demonstration that the unit of mutation was the base pair, and that the unit of recombination was the space between two adjacent base pairs, was accomplished by Charles Yanofsky and his collaborators (Yanofsky 1958; Yanofsky and Stadler 1958; Stadler and Yanofsky 1959; Yanofsky and Crawford 1959). While working on the tryptophan synthetase gene in *E. coli*, Yanofsky identified mutations at each of three bases within a single codon (#211) that altered the amino acid specified by that codon. Specifically, these mutations altered the gene to produce the mRNA codons shown here:

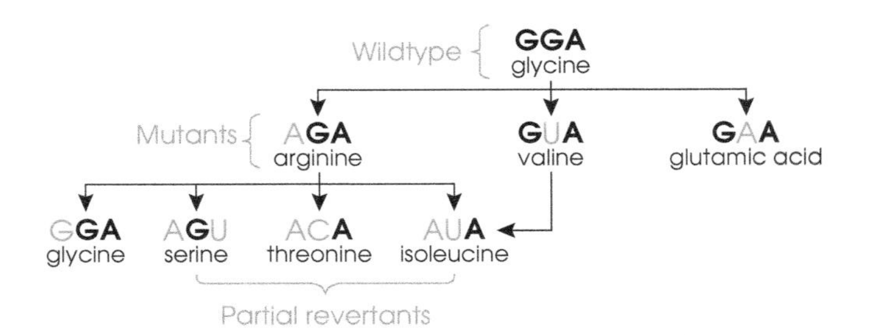

These data demonstrate that each base pair of the codon is separately mutable. More critically, Yanofsky demonstrated that two mutations, in separate but adjacent base pairs, could recombine to produce a wildtype tryptophan synthetase gene, but two mutations in the same base pair could not. For example, a bacterial cell carrying a copy of the AGA (arginine) codon and the GAA (glutamic acid) codon at position 211 could have a recombination event between the first two base pairs of this codon to yield the wildtype GGA (glycine) codon:

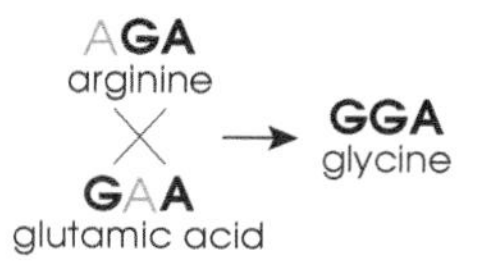

Taken together, the studies performed by Benzer and Yanofsky reduced the unit of mutation and recombination from the gene to the base pair. But if the gene is not the unit of mutation or recombination, then just how do we define a "gene?" Benzer chose to address this issue by formalizing the concept of the gene as the unit of function. Benzer noted that coinfection of strain K with wildtype and *rII* phages followed by plating on strain B produced both mutant and normal plaques. (Remember, *rII* mutants cannot survive in strain K.) The wildtype phage must provide the necessary function to the *rII* phage in *trans*, allowing the *rII* phage to replicate in the strain K cell. When strain K cells were coinfected with different pairs of *rII* mutants, two different results were obtained. For some combinations of alleles, no progeny phage was produced (no lysis), while for some combinations, billions of progeny phages were created (lysis).

These observations allowed Benzer to develop a test for gene function that he referred to as the ***cis-trans* test** (Figure B.1). If two mutations define the same gene, then coinfection of an *m1* phage and an *m2* phage (the *trans* test) should fail to produce lysis in strain K; infection of a + + phage and an *m1 m2* phage (the *cis* test) should show good lysis in strain K. When two mutations are in different genes, the results of the *cis* and *trans* tests should be identical. (Realize that what Benzer has just described is, in fact, a rather rigorous variant of the standard complementation test described in some detail in Chapter 3.)

Note that these experiments are quite different from the recombination experiments described earlier. In an experiment designed to assess the capacity of two mutants to produce wildtype phage by recombination, the two viruses were coinfected in a semipermissive (strain B) or permissive (strain S) host, and the presence of wildtype recombinants was detected by plating the resulting progeny phage on strain K. In this case, where we are assessing function, phage bearing the two different *rII* alleles are coinfected into the highly restrictive (strain K) host. Unless the two phages can themselves provide the necessary lytic functions without the need to produce a recombinant, infection will virtually always be abortive.

Benzer observed that the revertible point mutants could be divided into two groups (*rIIa* and *rIIb*) by this experiment. Coinfection of strain K with either two *rIIa* alleles or two *rIIb* alleles produced no progeny (because both phages were deficient for the same function). But coinfection with an *rIIa* and an *rIIb* allele did produce phage. Quoting Benzer, "If both mutants belong to the same segment, mixed infection on K gives the mutant phenotype (very few cells lyse). If the two mutants belong to different segments, extensive lysis occurs with liberation of both infecting types (and recombinants). ... Thus, on the basis of this test, the two segments of the *rII* region correspond to independent functional units." Fortunately, the *rIIa* mutants mapped on one side of Benzer's *rII* map, while the *rIIb* mutants mapped to the other. These results divided the *rII* locus into two adjacent, but functionally separate, segments, *rIIa* and *rIIb*.

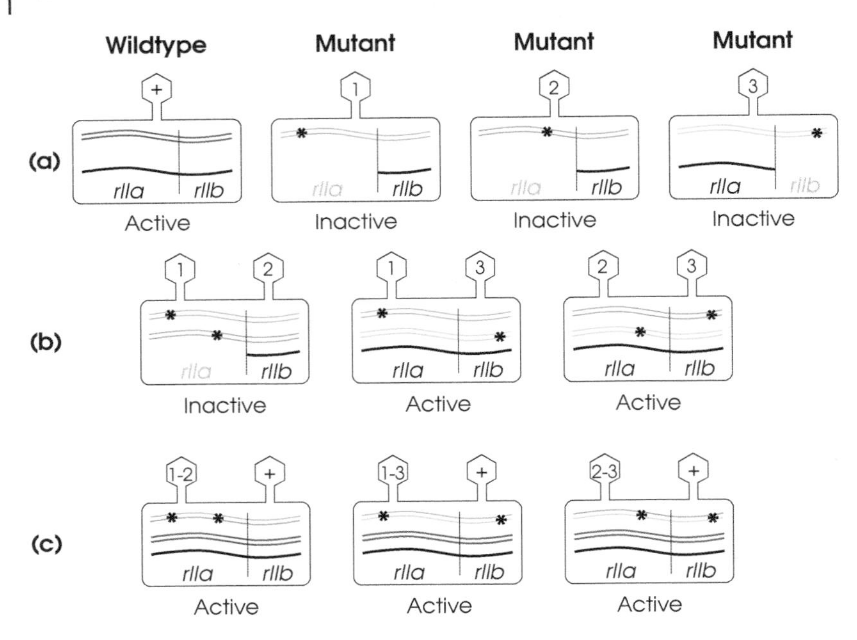

Figure B.1 The *cis-trans* test. This test determines whether two mutants map within the same gene. (a) When wildtype T4 phages infect bacteria, they will produce phage progeny. Mutant phages fail to produce progeny (or lyse the bacteria). (b) When two different T4 phages (both carrying an *rII* mutant) infect the same bacterial cell from strain K, the infection will produce phage progeny only if the mutants lie in the two different cistrons (genes) *rIIa* and *rIIb*. (c) A test of each mutant with the wildtype phage provides an important control. *Source:* Adapted from Benzer (1962).

Benzer redefined the concept of the gene (or *cistron* as he called it) as being the unit of function. He further showed that each of these functional units could be further divided into a number of separately mutable sites, each of which was separable by recombination.

Although the pioneering studies of Green and Oliver (Oliver 1940; Oliver and Green 1944) emboldened eukaryotic geneticists to attempt to fine-structure map eukaryotic genes, intragenic recombination frequencies are often quite low in higher organisms and strong selection techniques had to be applied to recover recombinants. We will not discuss the details of any of these studies for two reasons. First, geneticists no longer position mutants by intragenic recombination and have not done so for several decades. It is more straightforward and efficient to sequence even a large collection of mutants. Second, the analysis of the process of recombination itself using intragenic recombination is considered in Chapter 4. But the process of aligning the map with complex complementation patterns, or a diversity of phenotypes exhibited by mutants in a given gene, is a problem that still very much vexes the modern geneticist. It is that problem that we shall consider next.[1]

1 For those students curious enough to want to see how such intragenic maps were constructed prior to the availability of DNA sequencing, the following papers might prove useful. Methods for the construction of genetic fine-structure maps in *Drosophila* are reviewed by Chovnick (1989), Finnerty (1976), and Judd (1976). Rand (1989), Ruvkun et al. (1989), and McKim et al. (1988) have described fine-structure mapping in Caenorhabditis elegans. Finally, intragenic recombination in maize has also been studied at several loci in maize, and the paper by Okagaki and Weil (1997) provides an excellent introduction to this literature.

B.2 Intragenic Complementation Meets Intragenic Recombination: The Basis of Fine-Structure Analysis

Genetic fine-structure analysis combines intragenic mapping with intragenic complementation studies and has been done in many eukaryotes. These studies have allowed investigators to observe three primary types of intragenic complementation:

1) cases involving genes that encode multifunctional proteins,
2) cases involving transcriptional regulatory elements that do not depend on somatic pairing, and
3) cases involving transcriptional regulatory elements that *do* depend on somatic pairing.

Researchers can then use those findings to deduce various aspects of gene structure and function.

The first class of interactions involves those genes that encode multifunctional proteins. It is often possible to obtain mutants in such genes that disrupt only one of the two or more functional domains of those proteins. Two such mutants can display intragenic complementation if they fall in different domains. They will obviously not be able to complement each other if they fall in the same domain. We can thus group the mutants together on the basis of their pattern of intragenic complementation. In the case where we can map mutants that ablate the same function to a specific subregion of the gene's map, we can correlate position within a gene with a specific function of the protein product.

The second two classes of intragenic complementation include cases in which one or both the mutations define transcriptional regulatory elements. For example, if a given gene carried both wing- and leg-specific enhancers, a mutant in the leg-specific enhancer (which still allows the gene to function in the wing) might well complement a mutant in the wing-specific enhancer (which still allows the gene to function in the leg). The two classes of this type of interaction are differentiated by their sensitivity to chromosome aberrations that disrupt somatic pairing of the two alleles. In the case just presented, complementation does not require gene pairing, but there are notable examples in which this type of complementation is exquisitely pairing dependent. These cases, which reflect the ability of regulatory elements to act in *trans* on paired homologs, are examples of **transvection**. Cases of transvection usually involve one enhancer mutation paired with a mutation in the gene the enhancer acts on.

The Formal Analysis of Intragenic Complementation

To analyze intragenic complementation maps, we will need to apply a permutation of the complementation test that was not discussed in Chapter 3. Let us suppose that gene *A* encodes a protein with three functionally and spatially discrete functions: X, Y, and Z. We can get missense mutants or in-frame deletions that specifically knock out one of those functions without affecting the other two. Each of these sets of mutants might appear to form a single complementation group, and members of that group would be expected to complement similar function-specific mutants in the other separate elements of the gene. However, we also expect mutants (such as frameshift or early nonsense mutants) that knock out the entire gene to form a large complementation group. The members of this "fully noncomplementing" group are expected to fail to complement all of the function-specific mutants. How then do we combine these data to identify separate functional units within a set of what appear to be overlapping complementation groups?

We need to distinguish between **complementation units** and **complementation groups**. A given set of mutants defines complementation Group A as long as *all* of those mutants fail to complement each other in transheterozygotes. Now suppose that there are two other complementation groups (B and C) such that mutants in B fully complement the mutants in Group A, but the C mutants fail to complement both A and B mutants. We can diagram the complementation pattern as follows:

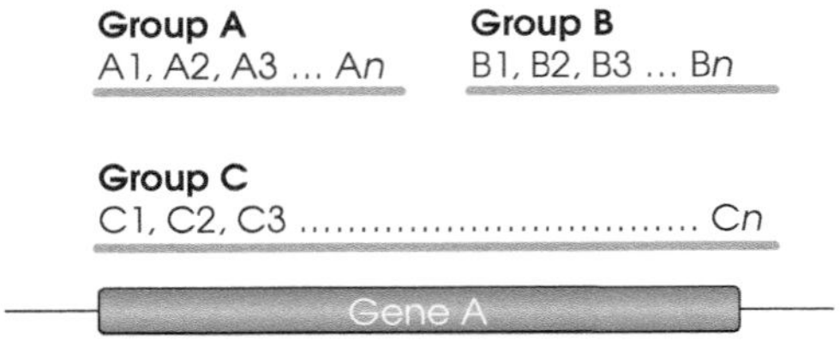

Group A
A1, A2, A3 ... *An*

Group B
B1, B2, B3 ... *Bn*

Group C
C1, C2, C3 .... *Cn*

Gene A

We can now subdivide this map into two complementation units, Unit I and Unit II. The rule here is that those mutants capable of intragenic complementation in Unit I will fail to complement each other, but they may complement mutants in Unit II. Some mutants may obviously fall in both Units I and II. If we keep isolating more mutants in this gene (our reference point for "in this gene" becomes a failure to complement the fully noncomplementing mutants, i.e. the mutants that fail to complement *all* other mutants), we might be able to build a rather more complex map:

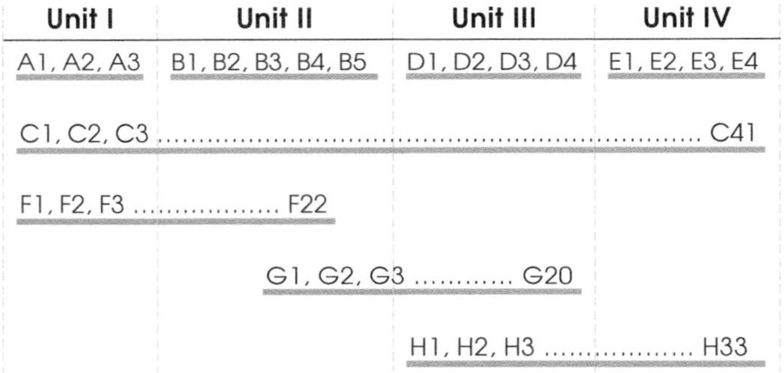

Unit I	Unit II	Unit III	Unit IV
A1, A2, A3	B1, B2, B3, B4, B5	D1, D2, D3, D4	E1, E2, E3, E4
C1, C2, C3 .... C41			
F1, F2, F3 .... F22			
	G1, G2, G3 .... G20		
		H1, H2, H3 .... H33	

To proceed with this analysis, two further conditions must be met:

1) The order of mutants, as revealed by intragenic mapping studies, needs to be consistent with the order of complementation units. (This last sentence is simply a restatement of the idea that there are functionally and spatially separate functional domains. If such domains exist, they can be mapped, and if they can be mapped, then they probably have functional significance.)
2) The mutants in each of the units can be correlated with specific and different functional defects. For example, this gene *B* makes a protein with four enzymatic activities (W, X, Y, Z), such that mutants in Unit I ablate activity W, mutants in Unit II ablate activity X, mutants in Unit III ablate activity Y, and so on.

The ability of this type of genetic fine-structure analysis to identify discrete and separable functional domains within a single gene is now a well-documented and frequently used tool in eukaryotic genetics (Gepner et al. 1996; for additional examples see Tang et al. 1998).

B.3 Fine-Structure Analysis of a Eukaryotic Gene Encoding a Multifunctional Protein

We'll begin with the first class of intragenic complementation: the dissection of genes encoding multifunctional proteins. The first example, describing the *HIS4* gene in *Sachharomyces cerevisiae*, is followed by the historical gory details of a more complex, but conceptually quite similar, analysis of the *rudimentary* gene in *Drosophila*. In both of these cases, the success of the analysis depended on multiple factors:

1) the ability to construct a high-resolution fine-structure map of the gene that positioned a large number of mutants in an unambiguous order,
2) the existence of point mutants that displayed intragenic complementation, and
3) the existence of tools that allowed one to subdivide the function of the gene product so that components of protein function and components of gene structure could be connected.

Genetic and Functional Dissection of the *HIS4* Gene in Yeast

In the 1960s and 1970s, Gerry Fink characterized the gene-enzyme relationships for histidine biosynthesis in yeast. These studies clearly showed that most of the genes involved in this process were widely dispersed in yeast. However, mutants at the *HIS4* locus affected one, two, or three enzymes involved in this process, namely phosphoribosyl-AMP (PR-AMP) cyclohydrolase, phosphoribosyl-ATP (PR-ATP) pyrophosphatase, and histidinol dehydrogenase. PR-ATP pyrophosphatase and PR-AMP cyclohydrolase catalyze the second and third steps in histidine biosynthesis, respectively, and histidinol dehydrogenase catalyzes the last two steps.

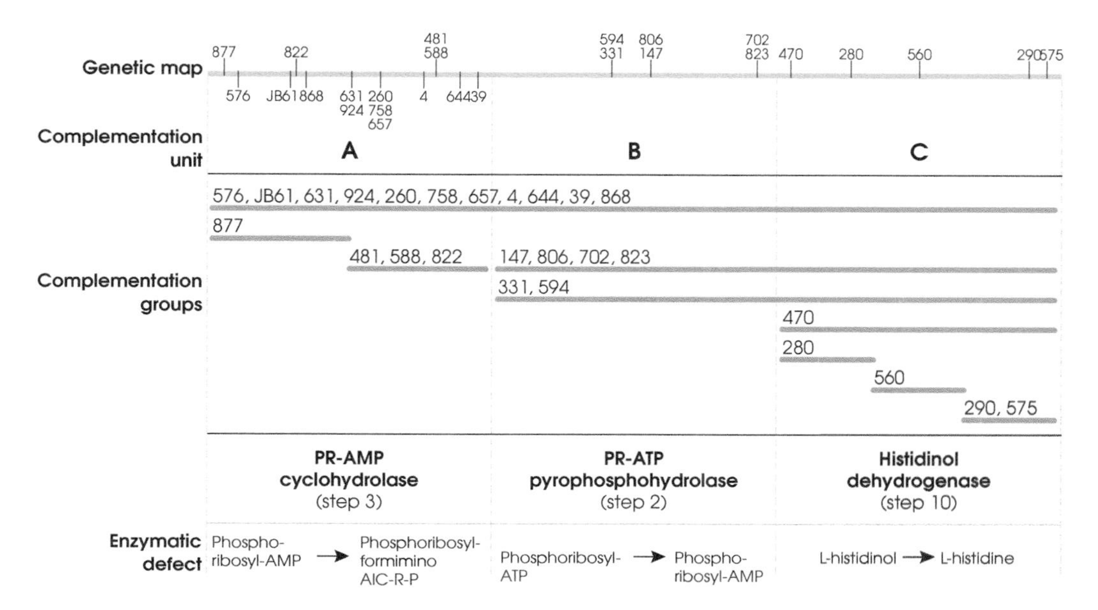

Figure B.2 Fink's complementation map of the *HIS4* gene in yeast shows the map position of each mutation, the complementation group and unit each mutant lies in, the associated enzymes involved in their respective steps in histidine biosynthesis, and the enzymatic defects caused by mutations in each region. *Source:* Adapted from Fink (1966).

To further study this locus, Fink analyzed 58 new *HIS4* mutants in terms of their intragenic complementation patterns and their capacity to produce the various enzymatic activities encoded by *HIS4* (Fink 1966). The positions of the mutants were mapped by intragenic recombination and all 3364 pairwise complementation tests were performed. These tests revealed three separate complementation units, each of which was uniquely associated with one of the three enzymatic defects (Figure B.2). Thus, Fink's data also suggested that the three enzymatic activities were the result of the polar translation of a single mRNA molecule. To prove this, he demonstrated that all of his tRNA mutant-suppressible nonsense alleles fell into the following classes: entirely noncomplementing A–B–C mutants, the B–C mutants, and a C mutant. These nonsense mutants showed a polarized loss of activity, similar to the polar mutants observed for the *lac* operon in *E. coli*. Indeed, Keesey et al. (1979) went on to show that the product of the *his4* gene in yeast is a trifunctional protein of ~95 kD.

Genetic and Functional Dissection of the *rudimentary* Gene in *Drosophila*

A similar, yet more complex, analysis was done for the *rudimentary* (*r*) gene in *Drosophila*. Mutants at the *r* gene produce small, truncated wings, sparse bristles, reduced fertility, and reduced viability. Early attempts to perform complementation tests and intragenic complementation analysis on the *r* locus revealed a complex complementation pattern and three recombinationally separable sites (Green 1964). A detailed study revealed an even more complex pattern (Carlson 1971). Although the 45 alleles could be aligned in a linear array, the mutants defined 16 complementation

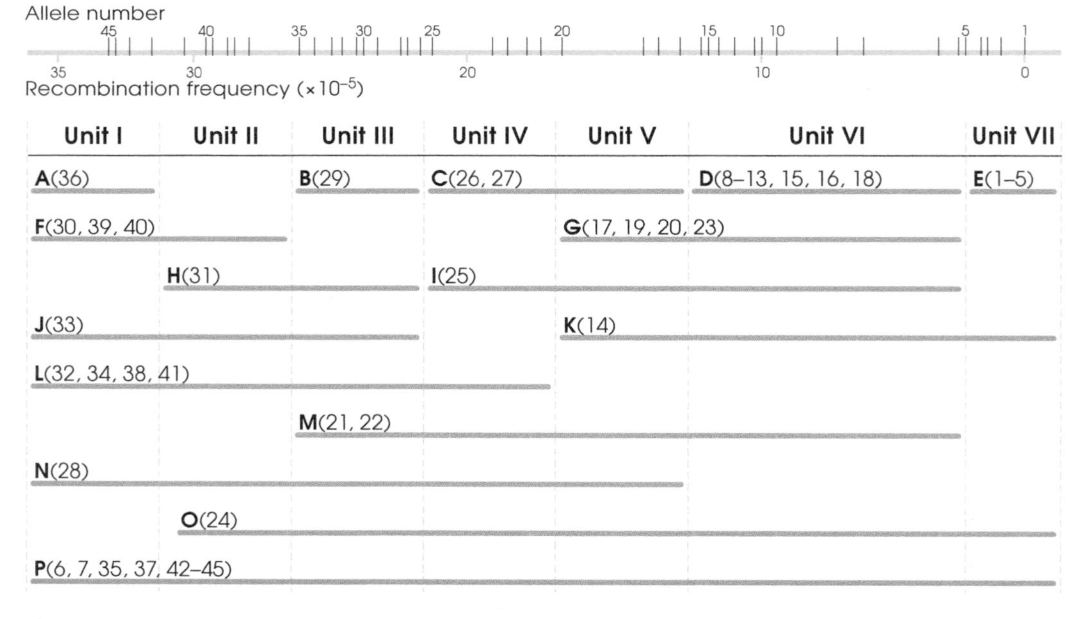

Figure B.3 Recombination and complementation maps of the *rudimentary* gene in *Drosophila*. The 16 complementation groups (A–P) in seven complementation units (I–VII) comprise 45 *r* alleles (1–45). *Source:* Adapted from Carlson (1971).

groups that only roughly correlated with map position (Figure B.3). The complementation groups defined seven complementation units within the *r* gene; the majority spanned more than one complementation unit. Indeed, as pointed out by Judd (1976), "... less than half (18/45) of the mutants studied by Carlson behave as if they belong to only one of the complementation units."

For example, all the mutants in Group L (32, 34, 38, 41) failed to complement each other, and thus defined a group. However, these four alleles also failed to complement the sole mutant (36) in Group A and the mutant (29) in Group B, both of which fully complemented each other. The single mutant (33) defining Group J failed to complement all of the mutants in Units I–III, while the mutant (31) defining Group H failed to complement only the mutants in Units II–III. These examples define five complementation groups and four complementation units. But realize that Unit II is not defined by any mutants unique to that group, as Units I and III are. Rather, its existence was inferred by the ability of Unit I and Unit III mutants to fully complement each other while failing to complement mutants in the overlapping units. The key observation is that the complementation units seemed to be colinear with the position of the mutants, as if the units corresponded to discrete functional domains within the gene. Look at complementation Group P at the bottom of Figure B.3 – the mutants in this group fail to complement all of the other mutants in the *r* gene. Six of these eight noncomplementing alleles map to the left end of the gene, suggesting a critical regulatory region.

One possible explanation for the complexity of these complementation data was the complexity of the collection of *rudimentary* mutants. Carlson used a large number of mutants created by a diverse set of investigators using different methods to produce mutants. When Rawls and Porter (1979) repeated these studies using a new set of mutants, only four complementation units were identified (Figure B.4). These four units were later shown to define discrete functional domains of this gene.[2] Unit D of Rawls and Porter (1979) corresponds to Unit VII of Carlson, Unit C of Rawls and Porter corresponds to Unit VI of Carlson, Unit B corresponds to Unit III of Carlson, and Unit A of Rawls and Porter corresponds Unit I of Carlson. The genetic map and the complementation map are colinear. However, in Rawls and Porter's analysis, noncomplementing alleles map to both the 5′ and 3′ ends of the gene.

Early findings that *r* mutants had a specific nutritional requirement for pyrimidines suggested that this locus might function in some component of the pyrimidine biosynthesis pathway. Indeed, the fully noncomplementing *r* mutants failed to produce three enzymes involved in this pathway: carbamyl phosphate synthase (CPSase), dihydroorotase (DHOase), and aspartate transcarbamylase (ATCase). Further biochemical studies revealed that all three of these enzymatic domains are produced as part of a trifunctional polypeptide translated from a single mRNA. Measurement of these enzyme levels in the presence of the various complementing *r* mutants obtained by Rawls and Porter resulted in the following observations:

- Mutants in complementation Unit A resulted in a significant decrease in the activity of ATCase.
- Mutants in complementation Unit C resulted in a significant decrease in the activity of CPSase.
- Mutants in complementation Unit D resulted in a significant decrease in the activity of DHOase.
- Noncomplementing *r* mutants failed to produce all three enzymes involved in this pathway.

Thus, the genetics alone predicted that there would be a trifunctional protein produced by the *r* message and that the order of domains would be DHOase ⟶ CPSase ⟶ ATCase, and a study of putative regulatory mutations of the *rudimentary* gene confirmed the hypothesis (Tsubota and Fristrom 1981).

2 The *rudimentary* locus does indeed encode a fourth enzymatic activity, GATase, but as this activity was not examined in any of the earlier genetic studies, we have omitted it from consideration here.

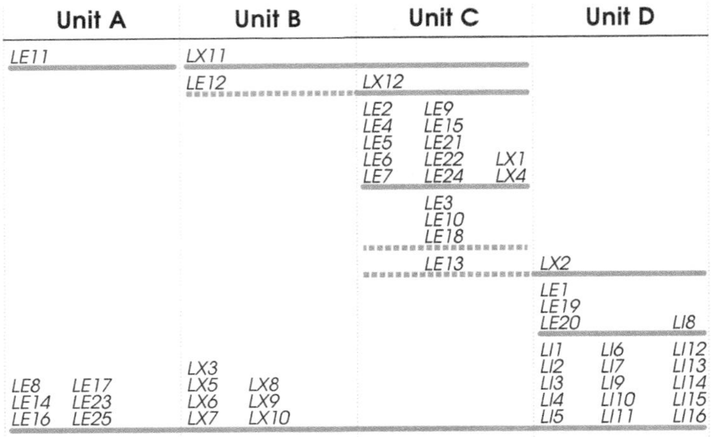

Unit A Unit B Unit C Unit D

LE11 LX11

LE12 LX12

LE2 LE9 LE4 LE15 LE5 LE21 LE6 LE22 LX1 LE7 LE24 LX4

LE3 LE10 LE18

LE13 LX2

LE1 LE19 LE20 LI8

LI1 LI6 LI12 LI2 LI7 LI13 LI3 LI9 LI14 LI4 LI10 LI15 LI5 LI11 LI16

LE8 LE17 LE14 LE23 LE16 LE25

LX3 LX5 LX8 LX6 LX9 LX7 LX10

Figure B.4 Complementation mapping of the *rudimentary* gene. 53 alleles mapped to only four complementation units (A–D), which were later found to correspond to discrete functional domains of this gene. Dashed lines indicate partial complementation. *Source:* Adapted from Rawls and Porter (1979).

Indeed, the *r* locus has been cloned and multiple mutants have been sequenced (Segraves et al. 1984; Freund et al. 1986). The three enzymatic activities are organized on the peptide chain in the following order: NH_2–DHOase–CPSase–ATCase–COOH (Freund and Jarry 1987; Lindsley and Zimm 1992). As predicted, complementation Unit D of Rawls and Porter corresponds to the DHO domain, Unit C identifies the CPS domain, and Unit A identifies the ATC domain. The genetic analysis of *rudimentary* is the epitome of genetic fine-structure analysis in *Drosophila*. Still, the amount of effort required in this analysis can no longer be justified in view of modern molecular tools.

B.4 Fine-Structure Analysis of Genes with Complex Regulatory Elements in Eukaryotes

In the next few examples, we will focus on genes with functionally independent regulatory elements that control gene expression in different tissues or at various times during development. In the previous cases, we considered mutants that failed to complement all other mutants at a locus (i.e. those that do not display intragenic complementation) as mutants that oblate protein production. We considered the intragenic-complementing alleles to be missense mutants that disrupted one of several separable enzymatic activities within the same protein. For genes with complex regulatory elements, our null alleles will usually define the transcription unit while alleles retaining some degree of function will define one of many upstream regulatory elements. Be aware of a critical component of this analysis – the assumption that genetic regulatory elements do not act in *trans*. However, also be aware that *trans*-acting gene activation is well known in *Drosophila* and other organisms, such as humans. We will consider *trans*-acting cases in the second example.

Genetic and Functional Dissection of the *cut* Gene in *Drosophila*

The *cut* gene in *Drosophila* was originally named for alleles that caused the loss of the wing tips, resembling cutting by scissors. However, most of the viable *cut* alleles have other phenotypic defects such as kinked legs (kinked femur), missing or reduced bristles, and aberrations

The allele Ubx^{bx} causes the anterior portion of the metathorax (the posterior of the three segments of the thorax) to develop as the anterior mesothorax (the middle of the three segments of the thorax) normally would. Similarly, the Ubx^{pbx} allele causes the posterior portion of the metathorax to develop as the posterior mesothorax normally would. Combine these two alleles as a double-mutant homozygote and you get a fly with two mesothoracic regions (i.e. a fly with four wings). The allele Ubx^{bxd} affects the development of the prothorax (the first abdominal segment). Ubx^{-} mutants cause all three of these transformations to occur, although the effect is weak in heterozygotes ($Ubx^{-}/+$). Most Ubx^{-}-like alleles are recessive lethal, but fail to complement Ubx^{bx}, Ubx^{pbx}, and Ubx^{bxd} alleles for their respective defects. The dominant Ubx^{Cbx} mutant causes transformations that are the reverse of those caused by the Ubx^{bx} and Ubx^{pbx} mutants (i.e. mesothorax ⟶ metathorax).

The intragenic complementation patterns at this locus are complex. For example, a weak Ubx^{pbx}-like posterior metathorax → posterior mesothorax transformation phenotype is observed in Ubx^{bxd}/Ubx^{pbx} heterozygotes, but none is observed in a Ubx^{bxd} Ubx^{pbx} double mutant ($Ubx^{bxd,pbx}$) heterozygous over wildtype ($Ubx^{bxd,pbx}/+$). Ubx^{-}/Ubx^{bxd} shows an extreme Ubx^{bxd}-like phenotype, but the corresponding $Ubx^{-,bxd}/+$ heterozygotes are nearly wildtype. Similarly, Ubx^{-}/Ubx^{pbx} shows an extreme Ubx^{pbx}-like phenotype, but the corresponding $Ubx^{-,pbx}/+$ heterozygotes are nearly wildtype. The molecular nature of each of the *Ubx* alleles used by Lewis is now known and may help explain some of the intragenic complementation patterns observed at the locus. Ubx^{bx} is a transposon insertion in the 3′ regulatory region of *Ubx*, Ubx^{Cbx} is a duplication within *Ubx*, Ubx^{bxd} is a transposon insertion into the noncoding RNA *bxd* that sits immediately adjacent to *Ubx*, and Ubx^{pbx} is a large deletion adjacent to *bxd*.

Many of these interallelic interactions are sensitive to heterozygosity for chromosome rearrangements. For example, structurally normal Ubx^{bx}/Ubx^{Cbx} heterozygotes show only a weak Ubx^{bx}-like transformation. Yet, in the presence of a large structural aberration on the same chromosome arm that carries the Ubx^{Cbx} allele, a strong intensification of the Ubx^{bx}-like transformation occurs. No such effect is observed in *cis*-double heterozygotes (when the inversion is on the same arm that carries Ubx^{bx}) or in the presence of homozygosity for the rearrangement. It is thus interesting that heterozygosity for a translocation or inversion breakpoint half a chromosome arm away can interfere with intragenic complementation at the BX-C. This effect is so strong that Lewis used it as a method to detect chromosomal aberrations on the third chromosome as a means of biodosimetry for nuclear weapons testing since radiation tends to induce chromosome rearrangements and translocations. Lewis' experiments and subsequent work at BX-C demonstrated that proper somatic pairing plays a critical role in the ability of the BX-C alleles to interact.

Long-range transvection has also been documented at *decapentaplegic* (*dpp*), a gene required for proper dorsal/ventral patterning of the embryo (Gelbart 1982). Rearrangements near the *dpp* locus interfere with transvection in a large region referred to as the critical region. Whether the interaction between *zeste* and *white* (discussed in the previous section) would also be impacted within this region was tested in an elegant experiment performed by Smolik-Utlaut and Gelbart (1987). Typically, *zeste* binds to one of multiple sites near *w* on the X chromosome, resulting in expression of *w*. However, chromosome aberrations in the immediate vicinity of *white* can disrupt the pairing-dependent interactions at *w* that are required for *zeste* to promote *w* expression (Smolik-Utlaut and Gelbart 1987). The question that Smolik-Utlaut and Gelbart set out to ask was whether the zeste-white interaction would be influenced in the same way that *dpp* is by placing paired *white* mini-genes between a heterozygous breakpoint and the *dpp* locus on the left arm of chromosome 2 (Figure B.8). This tested the ability of *zeste* to promote expression of *w* at a distant site in the presence of a breakpoint. The same breakpoint heterozygosity that prevented intragenic

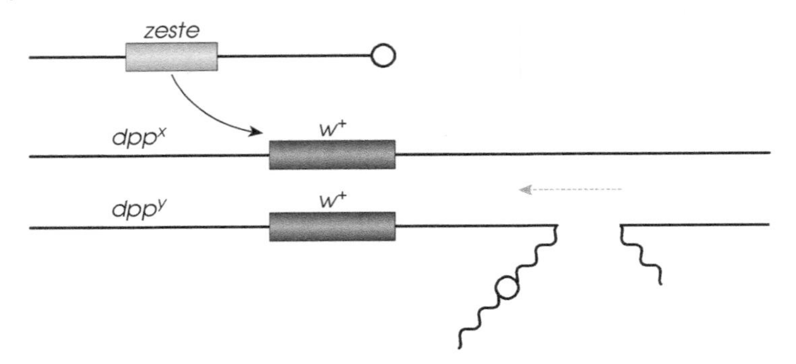

Figure B.8 The Smolik-Utlaut and Gelbart experiment. The ability of *zeste*, which is located on the X chromosome in *Drosophila*, to bind and drive expression of a *mini-white* gene (denoted w^+) that was placed within the critical region of *dpp* was tested in the presence of a heterozygous translocation breakpoint. To perform this experiment, Smolik-Utlaut and Gelbart placed a normal-sequence second chromosome over a translocation chromosome, T(2;3), in which part of the *Drosophila* second chromosome had been transferred to the third chromosome. This created a heterozygous breakpoint within a critical region of *dpp* near the w^+ insertion. The ability of *zeste* to drive the expression of *white* allowed the investigators to determine that the *zeste-white* interaction and transvection are two different proximity-dependent phenomena. *Source:* Adapted from Smolik-Utlaut and Gelbart (1987).

complementation at the *dpp* locus did not disrupt the ability of *zeste* mutants to interact with the paired *white* genes. Based on the proximity-dependent difference in rearrangements, Smolik-Utlaut and Gelbart argued that "the *zeste-white* interaction and transvection are two different proximity-dependent phenomena."

Golic and Golic (1996) have proposed an alternative interpretation of this result. Their view is that differing cell cycle times of relevant tissues may influence the ability of homologous chromosome arms to pair, even in the presence of structural heterozygosity. If that window of time, or "pairing interval," in the cells of the developing embryo (where the BX-C genes act) is shorter than the window for pairing in chromosome eye cells, then one could understand the differences in phenotype and proximity dependence. Quoting from Golic and Golic (1996), "Cells with a longer cell cycle have more time to establish the normal pairing relationships that have been disturbed by rearrangements. . . . *Minute* mutations, which slow the rate of cell division, partially restore a transvection effect that is disrupted by inversion heterozygosity." We favor the Golics' interpretation of this result and thus do not see any need to view the types of transvection observed at BX-C and *white* as different phenomena. Indeed, transvection at both BX-C and *dpp* are sensitive to the genotype of the fly at *zeste* as well as for numerous other loci in the genome.

B.6 Summary

This appendix began with a description of the initial efforts to find structure within the gene. Our interest was in defining the unit of recombination and mutation, the base pair. But with the development of the *cis-trans* test, Benzer opened the door to finding functional structure with a gene. We then asked if we could compare intragenic complementation maps to mutant maps to deduce which regions of a given gene executed which of two or more separable functions of the gene's protein product. As interesting as such methods are, we note that mapping mutants by intragenic complementation is decidedly outdated. One now maps mutants by sequencing, and functional domains of proteins are discovered computationally. Still, out of these studies came the ability to sometimes dissect mutants that defined regulatory elements from those that defined structural genes and some surprising observations about the role of somatic pairing in gene function, at least in *Drosophila*, and perhaps in other organisms as well (Tartof and Henikoff 1991). There are still a few things about gene structure that one can only learn from "doing the genetics."

References

Avery OT, Macleod CM, McCarty M. 1944. Studies on the chemical nature of the substance inducing transformation of pneumococcal types. *J. Exp. Med.* 79:137–158.

Babu P, Bhat S. 1980. Effect of *zeste* on *white* complementation. *Basic Life Sci.* 16:35–40.

Benzer S. 1955. Fine structure of a genetic region in bacteriophage. *Proc. Natl. Acad. Sci. U.S.A.* 41:344–354.

Benzer S. 1962. The fine structure of the gene. *Sci. Am.* 206:70–84.

Carlson PS. 1971. A genetic analysis of the rudimentary locus of *Drosophila melanogaster*. *Genet. Res.* 17:53–81.

Chovnick A. 1989. Intragenic recombination in *Drosophila*: the *rosy* locus. *Genetics* 123:621–624.

Fink GR. 1966. A cluster of genes controlling three enzymes in histidine biosynthesis in *Saccharomyces cerevisiae*. *Genetics* 53:445–459.

Finnerty V. 1976. Gene conversion in *Drosophila*. In: Ashburner M, Novitski E, editor. *The Genetics and Biology of Drosophila*. London: Academic Press. pp. 331–347.

Freund JN, Jarry BP. 1987. The *rudimentary* gene of *Drosophila melanogaster* encodes four enzymic functions. *J. Mol. Biol.* 193:1–13.

Freund JN, Zerges W, Schedl P, Jarry BP, Vergis W. 1986. Molecular organization of the *rudimentary* gene of *Drosophila melanogaster*. *J. Mol. Biol.* 189:25–36.

Gelbart WM. 1982. Synapsis-dependent allelic complementation at the decapentaplegic gene complex in *Drosophila melanogaster*. *Proc. Natl. Acad. Sci. U.S.A.* 79:2636–2640.

Gepner J, Li M, Ludmann S, Kortas C, Boylan K, *et al.* 1996. Cytoplasmic dynein function is essential in *Drosophila melanogaster*. *Genetics* 142:865–878.

Geyer PK, Corces VG 1987. Separate regulatory elements are responsible for the complex pattern of tissue-specific and developmental transcription of the *yellow* locus in *Drosophila melanogaster*. *Genes Dev.* 1:996–1004.

Geyer PK, Spana C, Corces VG. 1986. On the molecular mechanism of *gypsy*-induced mutations at the *yellow* locus of *Drosophila melanogaster*. *EMBO J.* 5:2657–2662.

Geyer PK, Green MM, Corces VG. 1990. Tissue-specific transcriptional enhancers may act in trans on the gene located in the homologous chromosome: the molecular basis of transvection in *Drosophila*. *EMBO J.* 9:2247–2256.

Golic MM, Golic KG. 1996. Engineering the *Drosophila* genome: chromosome rearrangements by design. *Genetics* 144:1693.

Green MM. 1964. Interallelic complementation and recombination at the *rudimentary* wing locus in *Drosophila melanogaster*. *Genetica* 34:242–253.

Green MM. 1990. *The Foundations of Genetic Fine Structure: A Retrospective from Memory*. Genetics Society of America.

Green MM, Green KC. 1949. Crossing-over between alleles at the *lozenge* locus in *Drosophila melanogaster*. *Proc. Natl. Acad. Sci. U.S.A.* 35:586–591.

Harrison DA, Geyer PK, Spana C, Corces VG. 1989. The *gypsy* retrotransposon of *Drosophila melanogaster*: mechanisms of mutagenesis and interaction with the *suppressor of Hairy-wing* locus. *Dev. Genet.* 10:239–248.

Jack JW. 1985. Molecular organization of the *cut* locus of *Drosophila melanogaster. Cell* 42:869–876.

Jack J, Delotto Y. 1995. Structure and regulation of a complex locus: the *cut* gene of *Drosophila. Genetics* 139:1689–1700.

Jack JW, Judd BH. 1979. Allelic pairing and gene regulation: a model for the *zeste-white* interaction in *Drosophila melanogaster. Proc. Natl. Acad. Sci. U.S.A.* 76:1368–1372.

Johnson TK, Judd BH. 1979. Analysis of the *cut* locus of *Drosophila melanogaster. Genetics* 92:485–502.

Judd BH. 1976. Genetic units of *Drosophila* – complex loci. In Novitski E, Ashburner M, editor. *The Genetics and Biology of Drosophila*. London: Academic Press. pp. 767–799.

Keesey JK, Bigelis R, Fink GR. 1979. The product of the his4 gene cluster in *Saccharomyces cerevisiae*. A trifunctional polypeptide. *J. Biol. Chem.* 254:7427–7433.

Lewis EB. 1948. Pseudoallelism in *Drosophila melanogaster. Genetics* 33:113.

Lewis EB. 1952. The pseudoallelism of *white* and *apricot* in *Drosophila melanogaster. Proc. Natl. Acad. Sci. U.S.A.* 38:953–961.

Lewis EB. 1978. A gene complex controlling segmentation in *Drosophila. Nature* 276:565–570.

Lewis EB. 1998. The bithorax complex: the first fifty years. *Int. J. Dev. Biol.* 42:403–415.

Lindsley DL, Zimm GG. 1992. *The Genome of Drosophila melanogaster*. San Diego, CA: Academic Press.

Liu S, McLeod E, Jack J. 1991. Four distinct regulatory regions of the *cut* locus and their effect on cell type specification in *Drosophila. Genetics* 127:151–159.

McKim KS, Heschl MF, Rosenbluth RE, Baillie DL. 1988. Genetic organization of the *unc-60* region in *Caenorhabditis elegans. Genetics* 118:49–59.

Morris JR, Chen JL, Geyer PK, Wu CT. 1998. Two modes of transvection: enhancer action in trans and bypass of a chromatin insulator in cis. *Proc Natl Acad Sci U S A.* 95(18):10740-10745.

Morris JR, Chen J, Filandrinos ST, Dunn RC, Fisk R *et al.* 1999a. An analysis of transvection at the *yellow* locus of *Drosophila melanogaster. Genetics* 151:633–651.

Morris JR, Geyer PK, Wu CT. 1999b. Core promoter elements can regulate transcription on a separate chromosome in trans. *Genes Dev.* 13(3):253–258.

Okagaki RJ, Weil CF. 1997. Analysis of recombination sites within the maize *waxy* locus. *Genetics* 147:815–821.

Oliver CP. 1940. A reversion to wild-type associated with crossing-over in *Drosophila melanogaster. Proc. Natl. Acad. Sci. U.S.A.* 26:452–454.

Oliver CP, Green MM. 1944 Heterosis in compounds of *lozenge* alleles in *Drosophila melanogaster. Genetics* 29:331–347.

Pontecorvo G. 1958. *Trends in Genetic Analysis*. New York: Columbia University Press.

Rand JB. 1989. Genetic analysis of the *cha-1-unc-17* gene complex in *Caenorhabditis. Genetics* 122:73–80.

Rawls JM, Porter LA. 1979. Organization of the *rudimentary* wing locus in *Drosophila melanogaster*. I. The isolation and partial characterization of mutants induced with ethyl methanesulfonate, Icr-170 and X Rays. *Genetics* 93:143–161.

Rosen C, Dorsett D, Jack J. 1998. A proline-rich region in the Zeste protein essential for transvection and *white* repression by *zeste. Genetics* 148:1865–1874.

Ruvkun G, Ambros V, Coulson A, Waterston R, Sulston J *et al.* 1989. Molecular genetics of the *Caenorhabditis elegans* heterochronic gene *lin-14. Genetics* 121:501–516.

Segraves WA, Louis C, Tsubota S, Schedl P, Rawls JM *et al.* 1984. The *rudimentary* locus of *Drosophila melanogaster. J. Mol. Biol.* 175:1–17.

Smolik-Utlaut SM, Gelbart WM. 1987. The effects of chromosomal rearrangements on the *zeste-white* interaction in *Drosophila melanogaster. Genetics* 116:285–298.

Stadler J, Yanofsky C. 1959. Studies on a series of tryptophan-independent strains derived from a tryptophan-requiring mutant of *Escherichia coli*. *Genetics* 44:105–123.

Tang TT, Bickel SE, Young LM, Orr-Weaver TL. 1998. Maintenance of sister-chromatid cohesion at the centromere by the *Drosophila* MEI-S332 protein. *Genes Dev.* 12:3843–3856.

Tartof KD, Henikoff S. 1991. Trans-sensing effects from *Drosophila* to humans. *Cell* 65:201–203.

Tsubota SI, Fristrom JW. 1981. Genetic and biochemical properties of revertants at the *rudimentary* locus in *Drosophila melanogaster*. *Mol. Gen. Genet.* 183:270–276.

Watson, JD, Crick FH. 1953. Molecular structure of nucleic acids; a structure for deoxyribose nucleic acid. *Nature* 171:737–738.

Yanofsky C. 1958. Restoration of tryptophan synthetase activity in *Escherichia coli* by suppressor mutations. *Science* 128:843–844.

Yanofsky C, Crawford IP. 1959. The effects of deletions, point mutations, reversions and suppressor mutations on the two components of the tryptophan synthetase of *Escherichia coli*. *Proc. Natl. Acad. Sci. U.S.A.* 45:1016–1026.

Yanofsky C, Stadler J. 1958. The enzymatic activity associated with the protein immunologically related to tryptophan synthetase. *Proc. Natl. Acad. Scic. U.S.A.* 44:245–253.

Appendix C

Tetrad Analysis

Why do we concern ourselves with converting crossover data into frequencies of tetrads with zero, one, two, etc. exchanges? What information does the latter gives us that is not apparent from the former? The answer lies in the fact that we are interested in the meiotic properties of chromosomes. The question of how chromosomes segregate is closely tied with the nature of conjunction of bivalents, and we have seen that a chiasma can hold the elements of a bivalent in conjunction. When we consider questions of nondisjunction or of the properties of rearrangement heterozygotes, we will have reason to consider the occurrence or nonoccurrence of exchange. Thus, we are often less interested in knowing the crossover frequency than we are in estimating the exchange distribution. The term **exchange distribution** refers to the actual distribution of chiasmata among the bivalents at meiosis I – the fraction of bivalents that underwent no exchange events (E_0), one exchange (E_1), two exchanges (E_2), and so on ($E_3 \ldots E_n$).

Obviously, the best way of measuring the frequency of meiotic recombination (other than directly counting chiasmata) would be one that allowed us to recover all four products of each meiosis and therefore easily determine the exchange classification of the bivalent that produced those crossovers. This process is referred to as **tetrad analysis**. Certain classes of fungi (the ascomycetes, such as *Neurospora crassa*) provide an advantage for such tetrad analysis because all meiotic products are recovered in a linear **ascus**, or sac, according to the sequence of segregations in the two meiotic divisions (Perkins 1953, 1955). Alternatively, there is an algebraic method of estimating the exchange distribution that can be used when only single products of meiosis can be recovered (Weinstein 1936, but see Zwick et al. 1999).

C.1 Tetrad Analysis in Linear asci

In Neurospora, the two meiotic divisions occur in such a way as to create a linear array of four meiotic products. The order of each doublet of cells in the ascus corresponds to the order of chromatids on the metaphase plate. In Figure C.1, the cells at the top arose from one product of meiosis I and the cells at the bottom arose from the other. Each of the four meiotic products then undergoes a complete mitosis to create two daughter cells. The linear order of spores in each ascus makes it possible to map the centromeres in relation to linked markers, according to the frequencies of first- or second-division segregation of markers.

If no exchange occurs between the *A* gene and its centromere, then the order of spores in the ascus will be *AAAAaaaa*. Now, suppose a single exchange occurred proximal to marker

Genetic Theory and Analysis: Finding Meaning in a Genome, Second Edition.
Danny E. Miller, Angela L. Miller, and R. Scott Hawley.

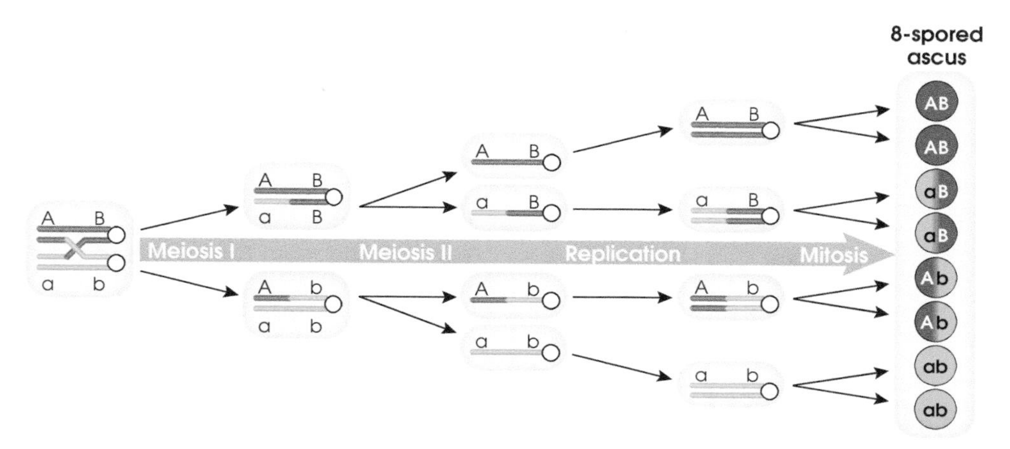

Figure C.1 Tetrad analysis in Neurospora. Because the meiotic products in Neurospora linearly align in an ascus, full tetrad analysis is possible in this organism. This example shows the segregation pattern of a centromere-proximal marker (B) and a distal marker (A) with an exchange between the two genes.

A. We might now expect the order of spores to be *AAaaAAaa*. (Other possible sequences are *aaaaAAAA*, *aaAAaaAA*, *AAaaaaAA*, and *aaAAAAaa*, depending on the orientation of the two chromosomes on their meiosis II spindles.) The critical issue here is that the *A* and *a* alleles did not separate from each other at meiosis I. Rather, the occurrence of an exchange prevents the two different alleles from being separated into separate cells until meiosis II.

Using the terminology of fungal genetics, the absence of an exchange between a marker and its centromere is indicated by first-division segregation and the presence of that exchange is heralded by second-division segregation. (Incidentally, the occurrence of second-division segregation provides strong evidence that crossing over occurs between chromatids at the four-strand stage of meiosis I; crossing over between unreplicated chromosomes would not give rise to second-division segregation.) The frequency of asci showing second-division segregation is a measure of the chiasma frequency – only half the spores in each of those asci will carry crossover chromatids. Thus, we must divide the frequency of asci showing second-division segregation by two to obtain the frequency of crossover chromatids.

Using several linked markers, we can confirm many of the rules of crossing over listed in Section 4.3 and obtain more direct estimates of the exchange distribution. Most importantly, we can directly assess the frequency of the three classes of double-crossover events. By confirming that the ratio of two-strand to three-strand to four-strand double exchanges is indeed 1 : 2 : 1, we can verify the assertion that there is no chromatid interference.

The existence of such double (and presumably higher-order) exchanges creates an oddity in the expected relationship between map distance and the frequency of second-division segregation. Naively, one might expect that as one progresses away from the centromere, the frequency of second-division segregation would gradually increase to a maximum of 100%. Thus, the maximum map length would be 50% – the equivalent of independent assortment. In reality, in Neurospora, the frequency of second-division segregation rises, but not linearly, to a maximum of just below 80% and declines to approach a final value of 66.7% (Figure C.2) (Perkins 1962).

How, then, are we to understand the fact that the relationship of second-division segregation to map length (as determined by summing small intervals) is not linear? And why does it eventually decline from values often as high as 80% (and for some organisms as high as 88%) to a final value

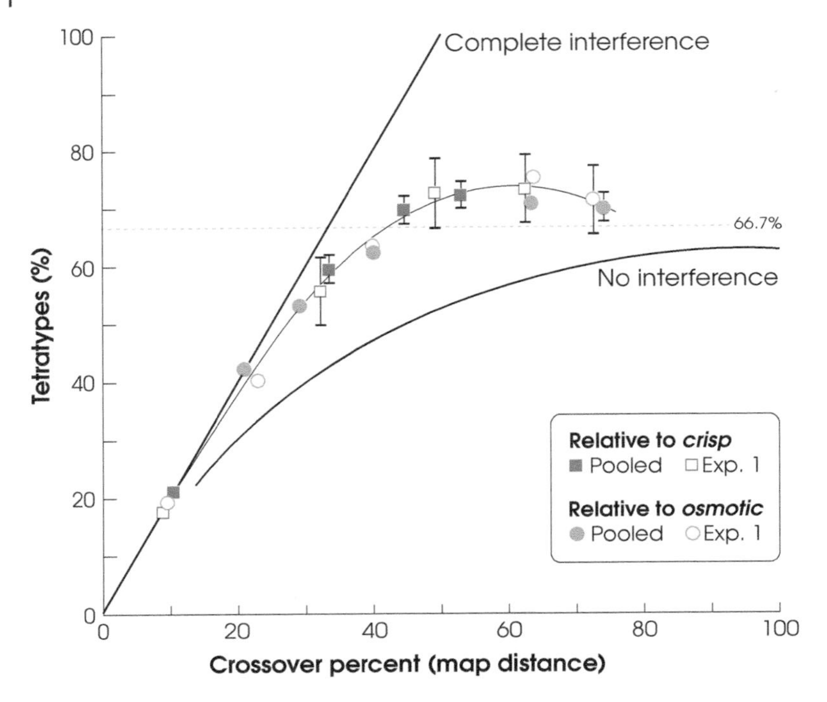

Figure C.2 The Neurospora mapping function. Interference can be observed in the frequency of tetratype (TT) segregations in Neurospora relative to the terminal markers *crisp* and *osmotic*. Theoretical curves are indicated for complete interference and no interference. 95% confidence limits are indicated for critical points. *Source:* Adapted from Perkins 1962.

of 66.7% for all genes studied? The answer to both questions lies in the fact that over long distances the double (and higher level) exchanges will effectively randomize the order of the four copies of each marker (gene) within a tetrad. In this case, the maximum frequency of second-division segregation is 66.7%. If this is confusing, just imagine that you had four spores, *AAaa*, in your hand like marbles and you are now going to drop them one at a time *at random* into a test tube-like ascus. If you drop an *A*-bearing spore in first, then you have a two-thirds chance that the next spore will be an *a* spore (this will look like second-division segregation; however, the next two spores fall) and only a one-third chance that the second spore to fall will be an *A* spore (obviously, this will look like second-division segregation; however, the next two spores, both *a* spores, fall). For this reason, the maximum possible frequency of second-division segregation is going to be 66.7%, and thus the highest possible recombination frequency will be 33.3%.

However, we might also ask why the initial frequency of second-division segregation can rise well above this eventual "maximum" of 66%. The answer lies in the fact that crossover interference is high in Neurospora (indeed, double crossing over is virtually absent in intervals of less than 20 cM). Thus, in the leftward (centromere-proximal) part of the graph in Figure C.2, we are seeing only the effect of increasing the frequency of single crossover events. Based on the tenet that a single exchange may involve any two nonsister crossover chromatids, we expect that the maximum frequency of second-division segregation will approach 100% (the case where all bivalents in the population have one exchange).[1]

1 The alert reader may wonder why this "hump" in the mapping function was not observed in Stahl's idealized graph presented in Figure 4.10. The answer lies in the fact that Stahl's idealized function was written for creatures lacking interference.

The fact that we do see values in excess of 66% is critical because it provides strong evidence that either sister-chromatid exchanges do not occur, or if they do, they do not compete (or interfere) with interhomolog exchanges. If sister exchanges were as likely as nonsister exchanges, then the maximum possible frequency of second-division segregation would be 66.7%, even in that mythical population of single-exchange bivalents. Indeed, Perkins (1955, 1962) tabulated multiple examples of frequencies of second-division segregation that exceeded 66.7% in a variety of fungi, including cases in *Podospora* that exceed 95%.

An excellent review of the uses of ordered tetrad analysis may be found in Zhao and Speed (1998a; see *also* Zhao et al. 1995). This study also contains a critical discussion of the statistical issues inherent in linear tetrad analysis. Although there are a few other fungi that produce ordered asci, the most commonly used fungi (the yeasts) do not.

C.2 Unordered Tetrad Analysis

Two of the most commonly used organisms for genetic analysis, *Saccharomyces cerevisiae* and *Schizosaccharomyces pombe*, also produce asci following meiosis. Each ascus contains four spores, with each spore consisting of a single meiotic product. Unfortunately, however, the spores are not ordered – they are actually arranged in a pyramid-like structure within the ascus. Nonetheless, we can still do genetic analysis. We can no longer use the equator of a linear ascus as a centromere marker, but we can follow the segregation of any two other markers. Let us now extend our analysis to two genes (*A* and *B*), which may be located on the same or different chromosomes. Given that the initial mating consisted of haploid *AB* cells to haploid *ab* cells, the resulting diploid will have the genotype *AB*/*ab*. Following sporulation, one can recover the following three classes of tetrads: **parental ditypes** (PDs), or nonrecombinant tetrads; **nonparental ditypes** (NPDs), or recombinants; and **tetratypes** (TTs), or tetrads with half PDs and half nonparental.

Parental ditype	Nonparental ditype	Tetratype
AB	*Ab*	*AB*
AB	*Ab*	*Ab*
Ab	*aB*	*aB*
Ab	*aB*	*ab*

Imagine that *A* and *B* are on different chromosomes but so near their centromeres that recombination does not occur between either marker and its centromere. In that case, we will observe only PDs and NPDs (Figure C.3). Because the two pairs of centromeres will orient randomly with respect to each other at meiosis I, PDs and NPDs will be equally frequent. You would see a TT *only* if recombination occurred between at least one of the two markers and its centromere.

Now, suppose that the *A* and *B* genes were immediately adjacent to each other on the same chromosome. In the absence of exchange between them, all that would be seen is PDs. Rare exchanges between them could create TTs, but an NPD could only result from a presumably quite rare four-strand double crossover. Thus, in this case, the frequency of PDs will greatly exceed the frequency of NPDs. An excess of PD asci relative to NPDs is evidence for linkage. Actual estimates of map length can be determined by using the formula presented in Box C.1. The methods for mapping centromeres in unordered asci and for mapping genes to either side of the centromere in these organisms are presented in Box C.2.

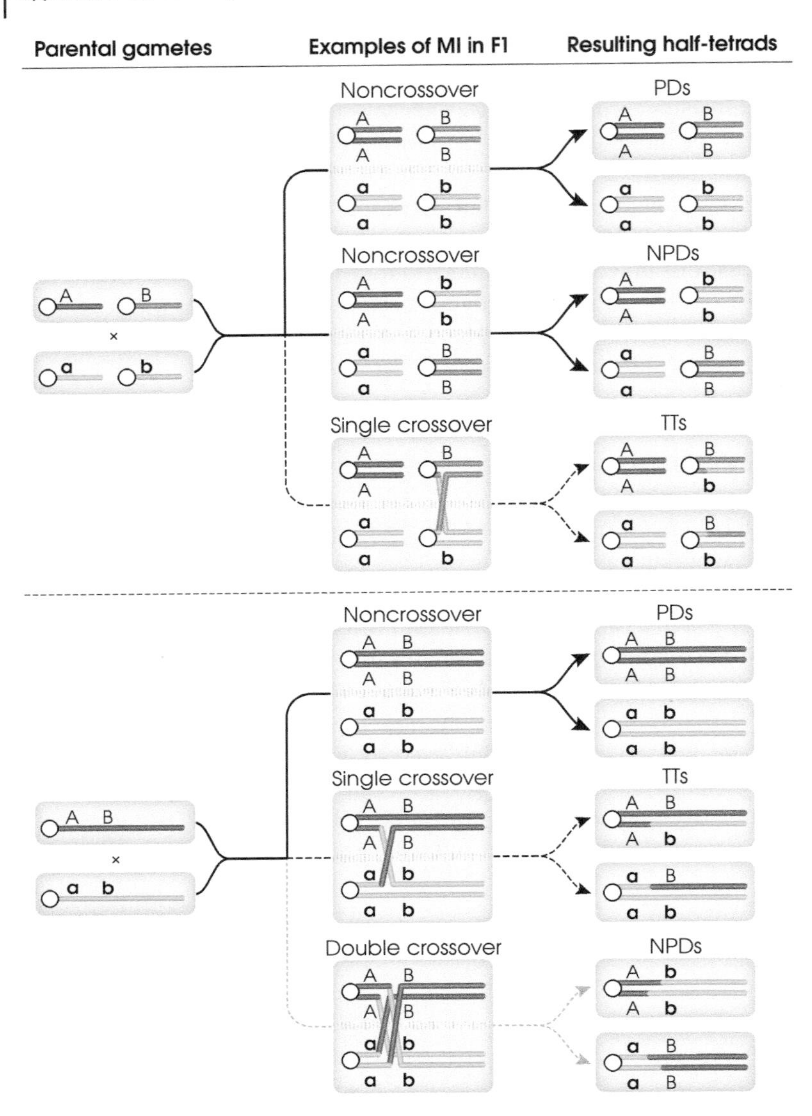

Figure C.3 The effect of linkage on the recovery of parental ditypes (PDs) and nonparental ditypes (NPDs). If centromere-proximal markers A and B are on different chromosomes (top), PDs should equal NPDs. Although infrequent crossing over between B and its centromere will increase the fraction of tetratypes (TTs), it will not alter the PD–NPD ratio. If A and B are linked and on the same chromosome (bottom), many more PDs will be recovered than NPDs because infrequent four-strand double crossovers between the A and B loci are the only way to get NPDs. *Source:* Adapted from Johnson et al. 1995.

Unordered tetrad analysis is also possible in a higher plant, *Arabidopsis thaliana*. The mutation *quartet1* (*qrt1*) causes the four products produced by each pollen meiosis to remain attached to each other, making tetrad analysis possible (Copenhaver et al. 2000, 2002; Sun et al. 2012; Yelina et al. 2013). Studies using these mutants have produced an exceedingly thorough understanding of the control of exchange distribution in this organism. Similarly, tetrad analysis has been described and exploited in the green algae *Chlamydomonas reinhardtii* (Dutcher 1995).

Box C.1 Using Tetrad Analysis to Determine Linkage

We can use the three classes of asci to determine the map length between any pair of markers. In organisms where interference is high, one could simply use the half frequency of tetratypes (TTs) as a good estimate of map length. (Again, the frequency of TTs is really a measure of the chiasma frequency and is equal to twice the frequency of observed crossover chromatids.) However, in most organisms, this metric would underestimate the amount of recombination by failing to consider the double-crossover events. Only a quarter of such doubles will generate nonparental ditype (NPD) asci; three-strand double crossovers will generate TTs, while two-strand doubles are not observable. So, we can estimate the frequency of doubles as 4 × NPD. To account for the fact that each double exchange has *two* exchanges, we will count these twice. Consequently, the number of crossovers resulting from double exchanges is 8 × NPD.

Unfortunately, our TTs can also arise from three-strand doubles. We want our estimate of single exchanges to be free of these events. Fortunately, the frequency of three-strand doubles can be estimated as 2NPD. (Remember the ratio of two-strand to three-strand to four-strand doubles is 1 : 2 : 1 and the frequency of NPDs is the frequency of four-strand doubles.) The true frequency of single exchange is therefore equal to TT − 2NPD.

We can now sum the frequency of exchanges involved both in single- and double-crossover events by adding these two quantities: (TT − 2NPD) + 8NPD = TT + 6NPD. However, since each exchange produces two crossover and two noncrossover chromatids, we need to divide this quantity by two to get the frequency with which we recover crossover chromatids (the map length). Thus:

$$\text{Recombination frequency} = \tfrac{1}{2}\left(\text{TT}+6\text{NPD}\right)/\left(\text{PD}+\text{TT}+\text{NPD}\right)$$
$$= \left(\tfrac{1}{2}\text{TT}+3\text{NPD}\right)/\text{Total tetrads examined}$$

Map length is expressed as 100 times the recombination frequency, and accordingly:

$$\text{Map length} = \left[\left(\text{TT}+3\text{NPD}\right)/\text{Total tetrads examined}\right]\times 100$$

This is the so-called Perkins equation and can be used in organisms with either ordered or unordered tetrads.

Box C.2 Mapping Centromeres in Fungi with Unordered Tetrads

The methods described in Section C.2 allow any two genes defined by mutant alleles to be mapped relative to each other by unordered tetrad analysis, but how does one map the centromeres by this method? The answer is: in pairs. Two markers that both define their centromeres will show an equal frequency of parental ditypes (PDs) and nonparental ditypes (NPDs) without producing tetratypes (TTs). Hence, once you have identified one centromere marker,[2] you can easily find others.

2 This was initially done by Don Hawthorne using a variant of *S. cerevisiae* that produced linear, but oddly ordered, asci.

C.3 Half-Tetrad Analysis

Unfortunately, in most higher eukaryotes, full-tetrad analysis is still not possible. However, it is possible in many organisms to recover two of the four products of meiosis. This was first accomplished using compound (or attached) X chromosomes in Drosophila, as described in Box 4.3. Subsequent workers developed tools for **half-tetrad analysis** in a variety of other organisms, including mice and humans (for review see Johnson et al. 1995).

Perhaps the most useful of these methods has been developed in a vertebrate system, namely zebrafish. Using a technique developed by Streisinger et al. (1986), meiotic half-tetrads (derived from a single product of meiosis I) can be generated from females of virtually any genotype (Johnson et al. 1995). This is accomplished by using high pressure to block the second meiotic division. Blocking meiosis II leads to diploid ova, which when fertilized by UV-inactivated sperm develop as **gynogenetic**, or **matroclinous**, females. Such females will always be homozygous (or reduced) for centromeric markers. Indeed, in the absence of any exchange, they would be homozygous for all markers. But the occurrence of a single exchange in the preceding meiotic prophase will result in heterozygosity for all markers distal to the exchange (Figure C.4).

The distance between any gene and its centromere can then be calculated as half the frequency with which heterozygous gynogenetic offspring are recovered from a cross involving females heterozygous for a recessive allele at that locus.[3] Unfortunately, heterozygotes are often hard to distinguish from wildtype homozygotes. But we can estimate the frequency of wildtype homozygotes by using the easily assayed fraction of mutant homozygotes. Thus, one can do the calculation by subtracting twice the frequency of mutant homozygotes from 1 and dividing that number by two. This calculation, which presumes absolute interference, has been shown to work well for short gene-to-centromere distances. Methods for the statistical analysis of half-tetrad data have been presented by Zhao and Speed (1998b).

Johnson et al. (1995) presented a more general application of this method for linkage analysis that allows the investigator to determine gene–centromere distances, even over distances large enough to allow double crossover events. The method is based on following the segregation of a given marker with respect to the segregation of a Polymerase Chain Reaction (PCR)-based centromere marker. As shown in Figure C.3, if the two markers lie on the same chromosome, then PD > NPD; if they are unlinked, PD = NPD. Once markers have been positioned on a given linkage group, the distance between them can be assessed by determining the fraction of the time the two markers differ in terms of first- versus second-division segregation. (That is, how often are offspring produced that remain

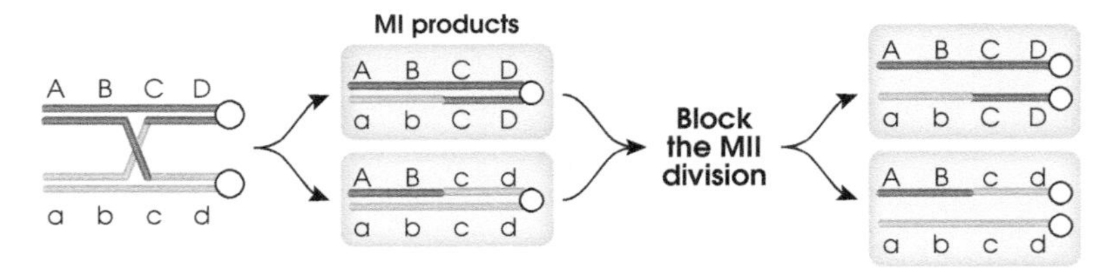

Figure C.4 Half-tetrad analysis in zebrafish. This analysis can be completed by blocking the second meiotic (MII) division and then fertilizing eggs with inactive sperm. The offspring will be heterozygous for all markers distal to the exchange.

3 We hope you understand why we multiplied by 0.5 here. Remember that each animal carries two chromatids, only one of which was crossover-bearing.

Mapping Two Genes that are Very Close to the Same Centromere

How might you map genes to opposite sides of the centromere in an organism like yeast that yields unordered tetrads? Consider the genotype *ABG/abg* where *A* and *B* are tightly linked and where *G* is a very tightly linked centromere marker on another chromosome. Consider the pair *AG* or *BG* that gives the lowest frequency of TTs. Say that pair is *AG*. Among those *AG* TTs, *B* will segregate with *A* if they lie on the same side of the centromere (i.e. all these TTs will be PD with respect to *A* and *B*). But if the *A* and *B* are on opposite sides of the centromere, then these same *AG* tetrads will be TT with respect to *A* and *B*.

References

Carpenter A. 1988. Thoughts on recombination nodules, meiotic recombination, and chiasmata. In: Kucherlapati R, Smith G, editor. *Genetic Recombination*. Washington, D.C.: American Society of Microbiology. pp. 529–548.

Copenhaver GP, Keith KC, Preuss D. 2000. Tetrad analysis in higher plants. A budding technology. *Plant Physiol.* 124:7–16.

Copenhaver GP, Housworth EA, Stahl FW. 2002. Crossover interference in *Arabidopsis*. *Genetics* 160:1631–1639.

Dutcher SK. 1995. Mating and tetrad analysis in *Chlamydomonas reinhardtii*. *Methods Cell Biol.* 47:531–540.

Johnson SL, Africa D, Horne S, Postlethwait JH. 1995. Half-tetrad analysis in zebrafish: mapping the *ros* mutation and the centromere of linkage group I. *Genetics* 139:1727–1735.

Perkins DD. 1953. The detection of linkage in tetrad analysis. *Genetics* 38:187–197.

Perkins DD. 1955. Tetrads and crossing over. *J. Cell. Physiol. Suppl.* 45:119–149.

Perkins DD. 1962. Crossing-over and interference in a multiply marked chromosome arm of Neurospora. *Genetics* 47:1253–1274.

Streisinger G, Singer F, Walker C, Knauber D, Dower N. 1986. Segregation analyses and gene-centromere distances in zebrafish. *Genetics* 112:311–319.

Sun Y, Ambrose JH, Haughey BS, Webster TD, Pierrie SN, *et al.* 2012. Deep genome-wide measurement of meiotic gene conversion using tetrad analysis in *Arabidopsis thaliana*. *PLoS Genet.* 8:e1002968.

Szauter P. 1984. An analysis of regional constraints on exchange in *Drosophila melanogaster* using recombination-defective meiotic mutants. *Genetics* 106:45–71.

Weinstein A. 1936. The theory of multiple-strand crossing over. *Genetics* 21:155–199.

Yelina NE, Ziolkowski PA, Miller N, Zhao X, Kelly KA, *et al.* 2013. High-throughput analysis of meiotic crossover frequency and interference via flow cytometry of fluorescent pollen in *Arabidopsis thaliana*. *Nat. Protoc.* 8:2119–2134.

Zhao H, Speed TP. 1998a. Statistical analysis of ordered tetrads. *Genetics* 150:459–472.

Zhao H, Speed TP. 1998b. Statistical analysis of half-tetrads. *Genetics* 150:473–485.

Zhao H, McPeek MS, Speed TP. 1995. Statistical analysis of chromatid interference. *Genetics* 139:1057–1065.

Zwick ME, Cutler DJ, Langley CH. 1999. Classic Weinstein: tetrad analysis, genetic variation and achiasmate segregation in Drosophila and humans. *Genetics* 152:1615–1629.

Glossary

acentric fragment a chromosome fragment lacking a centromere
achiasmate homologous chromosomes that did not undergo exchange
additive genes two or more genes that contribute to the same trait but act in different pathways
affected haplotype a linked set of alleles characteristic of chromosomes passed to affected individuals
algebraic tetrad analysis a method to determine the number of nonexchange, single-exchange, double-exchange, etc., homologs based on observed data
alignment the process of bringing homologous chromosomes into rough apposition along their lengths; the process of mapping sequencing data to a reference genome
alkylating agent a type of chemical mutagen that introduces an alkyl group onto a DNA base, thus modifying its capacity for complementary base pairing
amorph (*see* nullomorph)
analogs two genes in different organisms that serve similar functions but evolve independently from one another
anaphase I the movement of homologous chromosomes toward opposite spindle poles during meiosis I
anaphase II the separation of sister centromeres and the movement of sister chromatids to opposite poles during meiosis II
antimorph a dominant mutant that produces a protein product that antagonizes, or poisons, the wildtype protein
ascus a sac-like structure containing the ordered row of spores that result from meiosis in ascomycete fungi
asymmetric heteroduplex a region in which the heteroduplex is found on only one of the two chromatids involved in a recombination event
auxotroph a mutant that is unable to produce a particular essential compound and thus will fail to grow without the addition of that compound to the growth medium
auxotrophy when an organism cannot synthesize a nutrient or molecule necessary for growth
balanced stock a stock with a balancer chromosome that permits a deleterious allele to be maintained over time
balancer chromosome a structurally aberrant chromosome that uses multiple, overlapping inversions to suppress either the occurrence or recovery of crossovers
base-pair deletion the removal of one DNA base
base-pair insertion the addition of one DNA base

Genetic Theory and Analysis: Finding Meaning in a Genome, Second Edition.
Danny E. Miller, Angela L. Miller, and R. Scott Hawley.

base-pair substitution mutant (*see* point mutant)
binomial distribution a discrete probability distribution that has two possible outcomes, such as a coin toss
binomial expansion theorem a formula used to expand binomial expressions of the form (a + b)^n, where a and b are any real numbers or variables, and n is a non-negative integer
biosynthetic pathway sequence in which two or more genes function at different steps to synthesize a product
bivalent a pair of homologous chromosomes that are physically connected via chiasmata
BLAST stands for Basic Local Alignment Search Tool and allows you to compare a DNA or protein sequence to a database of sequences
blastn a type of blast that allows you to query a DNA database using a DNA query
blastp a type of blast that allows you to query a protein database using a protein query
blastx used to search a protein database using DNA input sequence, translates the DNA input in all six possible frames
bypass suppression when a second mutation bypasses the phenotype caused by a first mutation
cell autonomy in mosaic studies, this addresses the question of whether the phenotype is determined by the genotype of the cell or influenced by other cells in the organism
centimorgan (cM) a recombination unit used to measure linkage; map unit; 1 cM = 1% recombination
centric heterochromatin tightly packed chromatin directly adjacent to the centromere
centromere the tightly packed heterochromatic region of a chromosome around which the kinetochore is assembled and to which spindle fibers attach to move the chromosome along the spindle
centromere effect describes the observation that crossover frequency for a particular segment is reduced the closer to the centromere a DNA segment is
chemical mutagenesis an efficient method to artificially induce mutations in a genome by exposing an organism to a chemical, particularly an alkylating or crosslinking agent
chiasmata the physical manifestation of meiotic recombination events between nonsister chromatids of homologous chromosomes
chiasma-type hypothesis the hypothesis that chiasmata form as a consequence of crossing over between two nonsister chromatids
chromosomal bouquet (*see* telomere bouquet)
cis-trans test used to determine whether two mutations that affect the same phenotype are located in the same gene (*cis* configuration) or in different genes (*trans* configuration); complementation test
classical hypothesis the hypothesis that chiasmata form without chromatid breakage and repair, and chromatids that have formed chiasmata may or may not be resolved as crossovers
codominance the state in which the phenotypes of two different alleles of a gene are both expressed simultaneously
collochores attachment sites for the spindle fibers that pull the chromosomes apart during cell division
combined haploinsufficiency a situation in which the simultaneous reduction in the dosage of two different genes produces a mutant phenotype; type 3 second-site noncomplementation
complementation groups a set of mutations or genetic variants that fail to complement each other in a genetic assay, indicating that they affect the same gene or functional unit; see complementation test

complementation test a method to determine whether two mutations occur in the same gene or different genes whereby a double heterozygote is made and a mutant phenotype is observed if the two mutations define the same gene (failure to complement) but not if they are in separate genes; *cis-trans* test

complementation units the smallest region of DNA that encodes a single functional product, such as a protein or RNA molecule

compound heterozygote an individual with two different mutant alleles of the same gene

conditional loss-of-function mutant a type of loss-of-function mutant in which the loss of gene or protein activity is observed under one set of conditions (e.g. higher or lower temperature or treatment with a particular drug) and not under another

conformational suppression when a protein–protein interaction results in suppression of a phenotype

conservative mutation a missense mutation that results in the incorporation of an amino acid that is chemically or structurally similar to the wildtype one

constitutive mutation mutation that causes the protein encoded by a gene to be locked into an active (always-on) form

convergent evolution the process by which unrelated organisms evolve similar traits or characteristics independently in response to similar environmental pressures or selective forces

copy number variant (CNV) a deletion or duplication of a large block of DNA

Cre-Lox system can be used to induce site-specific mitotic recombination in a manner similar to FLP/FRT

CRISPR/Cas9 a technique that allows a specific sequence in a genome to be targeted for editing, which could include the removal or alteration of a specific segment of DNA

critical region (*see* genetic inclusion interval)

crossing over the physical interlocking of homologous chromosomes that results from the breakage and exchange of genetic material between the homologs during meiosis I

crosslinking agent a type of chemical mutagen that chemically interlocks two sites on a DNA molecule, which induces small deletions when incorrectly repaired

crossover interference the phenomenon whereby the occurrence of one crossover event within a given chromosomal region decreases (interferes with) the probability that a second exchange will occur within or close to that region

crossover-associated gene conversion a gene conversion that occurs in conjunction with a crossover event

de novo assembly the computational process of constructing a genome from DNA or RNA sequencing reads

de novo mutation a new mutation that is not seen in an organism's parents

deficiency (Df) a chromosomal aberration caused by the deletion of some or all of the DNA that encompasses a particular gene; could also be called a CNV and is a type of nullomorph

degenerate in reference to the amino acid code, the fact that multiple nucleotide combinations can code for the same amino acid

diakinesis the final stage in meiotic prophase during which homologs shorten and condense in preparation for nuclear division

dicentric bridge a chromosome with two centromeres

digenic inheritance mutations in two different genes that result in a disease phenotype only when inherited together; second-site noncomplementation in humans

diplotene stage of meiosis I during which the attractive forces that mediate homologous pairing disappear and the homologs, held together only by their chiasmata, begin to repel each other
discordant reads a pair of sequencing reads in which one read pair aligns to one position in the genome and the other read pair aligns to a different position in the genome
dominant enhancement (*see* second-site noncomplementation)
dominant mutation a mutation that displays a phenotype when heterozygous over a wildtype allele
dominant negative a mutation that produces a poisonous protein product; antimorph
dosage compensation the mechanism by which an organism regulates the amount of gene product produced so that hemizygous and homozygous individuals produce the same amount
double heterozygote an organism heterozygous for mutant alleles at two different genetic loci
double Holliday junction (DHJ) two Holliday junctions formed after a double-strand break as an intermediate step in homologous recombination
double-strand break a break in both strands of a double helix that serves as the initiating event for meiotic recombination
double-strand break repair (DSBR) model model of recombination in which a double-strand break occurs on one chromatid, after which the break is resected to form a gap with two 3′ overhangs that then invade the intact duplex to create a double Holliday junction, which in turn can be resolved as a crossover
duplication a chromosomal aberration consisting of an extra copy of a particular region of the genome; a type of CNV
E-value represents the expected number of random matches between a query sequence and a database sequence that would have a similarity score at least as high as the observed score by chance when performing a BLAST search
enhancer (*see* enhancer mutant; transcriptional enhancer)
enhancer mutant a mutant that allows a normally recessive mutant at another gene to exert a strong phenotypic effect, even when heterozygous with a wildtype allele at that gene
entanglement model proposes that chromosomes are held, or entangled, together by proteinaceous bridges that promote alignment and pairing during meiosis
epigenetic centromere a functional centromere that is established through epigenetic modifications of the chromatin rather than through the presence of a specific DNA sequence
epistasis the interaction between two genes; the effect of one gene on the expression of another
epistatic gene a gene whose action masks the effect of another, hypostatic gene
exchange (*see* crossing over, gene conversion)
exchange distribution describes the distribution of crossover or noncrossover events along a chromosome after meiosis
exchange rank a statistical measure used in tetrad analysis to quantify the degree of crossing over between two chromatids
expect value (*see* E-value)
extragenic suppression when a suppressor mutant lies in a different gene than the original mutant
fate mapping a technique that labels early embryonic cells to determine their developmental trajectory
FLP/FRT system two-component site-specific recombination system where the recombinase enzyme FLP is able to induce mitotic recombination at a specific site, called FRT

four-strand double crossover a set of crossovers that involves all four chromatids in a pair of homologous chromosomes

frameshift mutant the insertion or deletion of one or more base pairs within the coding sequence that alters the reading frame and leads to the creation of a premature stop codon, usually soon after the insertion or deletion; nullomorph

gain-of-function mutant a mutant that causes a gene to be inappropriately regulated or to produce a protein with a new or different function

gametogenesis the formation of egg or sperm cells

gene conversion a double-strand break repair event in which genetic information is copied from one chromosome onto another, resulting in the 3 : 1 segregation of that allele

genetic buffering the ability of one gene to buffer the effects of a loss-of-function mutation at another

genetic inclusion interval the chromosomal region between two visible markers that contains the gene of interest; critical region

gynandromorph an organism that has both male and female characteristics

gynogenetic a type of asexual reproduction in which an egg cell is activated to develop into an embryo without being fertilized by a sperm (also known as parthenogenesis); it can occur naturally in some animals, such as certain species of reptiles, fish, and insects

half-tetrad analysis a genetic technique used to study the segregation and recombination of alleles during meiosis in fungi, particularly in yeast where only two spores are analyzed; *see* tetrad analysis

haplotype a set of linked genes that occur near each other on a chromosome and are inherited together

haplotype analysis a multipoint approach to evaluating co-transmission of human marker information

heat-shock promoter a conditional promoter that is activated by temperature

hemizygote an individual with a single allele at a specific position instead of two, as in 46, XY males who have only one copy of the X chromosome

heterochromatin DNA that is tightly packed and thus typically inaccessible to transcriptional machinery

heterochronic mutant a mutant that causes a gene to be expressed at the wrong time

heteroduplex a region of base-pair mismatch corresponding to the base pair(s) that differ(s) between two alleles

heteromorphic physically or visibly different from one another

high-copy suppression (*see* multicopy suppression)

Holliday intermediate heteroduplex structure formed near the site of a Holliday junction during the resolution of a DSB during meiosis

Holliday junction the site of strand exchange formed from a reciprocal crossover during chromosomal recombination

Holliday model a model of recombination proposed by Robin Holliday to explain the occurrence of both gene conversions and crossovers in which an identical single-strand nick is made on both homologs, followed by strand invasion and ligation, creating a Holliday junction; branch migration, which results in two symmetric heteroduplexes; and resolution as either a crossover or noncrossover gene conversion

homolog a gene related to another gene by common descent, through either speciation or gene duplication (*see also* homologous chromosomes)

homolog recognition region (HRR) a region on each homologous chromosome that plays a role in homolog pairing

homologous chromosomes a maternal and paternal set of chromosomes, each consisting of two sister chromatids, that share a similar size, centromere position, and corresponding genetic loci

homologous segregation the movement of homologous chromosomes to opposite spindle poles and their subsequent separation during the first meiotic division

homology-directed repair (HDR) the process of introducing a new sequence into a genomic segment that has been targeted by CRISPR/Cas9 by providing a complementary, but slightly altered, sequence for the cell to use to repair the break induced by CRISPR/Cas9

horizontal gene transfer the transfer of genetic material between organisms that are not directly related by vertical descent, such as from one species to another, or between different domains of life (e.g. from bacteria to eukaryotes) (*see* xenologs)

hypermorph a mutant that produces either a harmful excess of the normal protein product or a hyperactive one

hypomorph a mutant that produces some degree of residual activity but not enough to show wildtype activity; weak allele

hypostatic gene a gene whose action is masked by another, epistatic gene

imprecise excision the process by which a small deletion is created when a transposable element moves and takes flanking DNA sequence with it

in situ hybridization a technique that detects specific RNA fragments in tissue using labeled complementary DNA fragments

inactivating null a type of nullomorph that produces a protein product that exerts no obvious activity

indel a mutation that involves the insertion or deletion (or both) of DNA sequence, typically smaller than 50 bp

independent assortment Mendel's second law of inheritance, which states that two genes that lie on different chromosomes will be inherited independently from one another

informative meioses those meioses tested that yield information on whether they are recombinant or nonrecombinant

interchromosomal effect the observation that when exchange is suppressed on one pair of homologs, most typically by inversions, the number of recombination events increases on those homologs that are able to undergo exchange

interference (*see* crossover interference)

intergenic suppression (*see* extragenic suppression)

interstrand crosslinks the chemical interlocking of two sites on opposite strands of a DNA molecule caused by a crosslinking agent

intragenic complementation when two alleles of the same gene complement each other, most often because of mutations that occur at different active sites within the protein product

intragenic recombination recombination that occurs within a single gene or locus

intragenic suppression when two interacting mutations lie within the same gene

intrastrand crosslinks the chemical interlocking of two sites on the same strand of a DNA molecule caused by a crosslinking agent

inversion a chromosomal aberration in which a segment of DNA is reversed

inverted meiosis involves the segregation of the sister chromatids during the first meiotic division, followed by pairing and recombination of the resulting haploid chromosomes during the second meiotic division

ionizing radiation a mutagenesis tool (typically X-rays or gamma rays) that induces multiple double-strand breaks in DNA, which when repaired incorrectly result in chromosome aberrations

isochromosome a chromosome in which the two arms, joined by a centromere, are identical

isogenic a large region of DNA, often an entire chromosome, that is genetically identical

joint molecule double Holliday recombination intermediates produced during meiosis

lateral gene transfer (*see* horizontal gene transfer)

leptotene the first subdivision of prophase I of meiosis, which defines the initial phase of chromosome individualization and during which initial homolog recognition and alignments are made

Li–Fraumeni syndrome cancer predisposition syndrome caused by germline variants in the tumor suppressor *TP53* in humans

linkage the tendency for two gene pairs at different positions on homologous chromosomes to segregate together during meiosis

LOD score (*Z*) a ratio that compares the probability that two genes are linked to the probability that they are unlinked (0.5)

loss of heterozygosity when one copy of a gene or the surrounding region is lost, leaving a single copy

loss-of-function mutant a mutant that reduces the level of gene product

low-complexity region a region of DNA where short DNA sequences are repeated multiple times; this may make the region difficult to study using many sequencing methods

map length the calculation of recombination distance

map unit (*see* centimorgan)

matroclinous a pattern of inheritance in which a genetic trait is passed down exclusively through the maternal line, i.e. from the mother to her offspring

meiosis I the first division of meiosis, during which homologous chromosomes pair, recombine, and segregate from each other, while sister chromatids remain bound together

meiosis II the second division of meiosis, often thought of as a haploid mitosis, during which sister chromatids segregate from each other to create haploid gametes

meiotic drive the ability of a particular allele to ensure it is passed on to offspring at a greater than 50% frequency

Meselson-Radding model a model of recombination in which a single-strand nick is made on only one chromatid in a duplex, followed by strand displacement and invasion, D-loop excision, and the formation of a Holliday junction, which results in an asymmetric heteroduplex region and resolution as either a crossover or noncrossover gene conversion

metaphase I the period before the first meiotic division during which bivalents are lined up at the metaphase plate in preparation for homologous chromosomes to segregate to opposite poles

metaphase II the period before the second meiotic division in which chromosomes line up at the metaphase plate in order to segregate sister chromatids to opposite poles

metaphase plate position at the middle of the meiotic spindle where chromosomes line up during metaphase

metaphase-anaphase transition the release of sister chromatid cohesion along meiotic chromosome arms, but *not* in the region surrounding the centromeres, which frees the two chromosomes in each bivalent from their chiasmate attachments, allowing each chromosome to proceed toward the closest spindle pole

MI nondisjunction failure of homologous chromosome to separate properly during the first meiotic division; after sister chromatid segregation during MII the gametes will have an extra copy of the chromosome, but with different polymorphisms

microarray a testing modality that can help determine DNA copy number; can be used with RNA after it is converted to cDNA

microsatellite array a segment of DNA comprising short (<10 bp), tandem repeats that vary in number (generally 5–50 times) among individuals

MII nondisjunction failure of sister chromatids to separate properly during the second meiotic division; the gamete will thus have two identical chromatids

minichromosome a small chromosome with a centromere, telomeres, and replication origins

missense mutant a type of base-pair-substitution mutant that changes the sequence of a given codon, which then directs the incorporation of an amino acid different from the one specified at that position in the wildtype allele

monosome transmission test used to determine the parental origin of a particular chromosome by crossing a chromosome carrying a deletion to an individual who is monosomic for that chromosome

mosaic an organism in which the cells or tissues within that organism have different genotypes and can be distinguished as having a separate origin

mosaic analysis an experiment in which at least some of the cells in an organism are modified to study a genetic change that may be lethal or tissue-specific; or when cells from one organism are injected into another in order to study a specific phenotype

mRNA surveillance the monitoring of mRNA for transcripts that contain premature stop codons so that they can be destroyed

multicopy suppression a situation where increasing the dosage of one gene suppresses the phenotype caused by a mutation in a different gene, perhaps by allowing the cell to bypass the defect

multipoint linkage analysis used to determine linkage of multiple markers in a small region of the genome

mutator element (*see* transposon)

negative complementation a situation in which the phenotype of a compound heterozygote is considerably more extreme than that of either of the two homozygotes alone

neomorph a mutation that causes a gene to be active in an abnormal time or place

nonallelic noncomplementation (*see* second-site noncomplementation)

nonbiosynthetic pathway sequence in which two or more genes function at different steps but do not necessarily result in the synthesis of a product

nonconservative mutation a missense mutation that results in the incorporation of an amino acid that is chemically or structurally different than the wildtype one

noncrossover gene conversion a gene conversion that is not associated with a crossover

nondisjoin describes chromosomes that fail to segregate properly during meiosis; one daughter cell has an extra copy of a chromosome while the other daughter cell will have a missing copy of the same chromosome

nondisjunction the failed segregation, or missegregation, of homologous chromosomes during meiosis

nonhomologous end joining (NHEJ) a cellular process that ligates two nearby broken DNA ends together; typically results in the deletion of one or more base pairs

nonparametric linkage analysis evaluation of an affected population or group to determine allele sharing without prior assumption of a model of inheritance

nonparental ditype a tetrad that has only recombinant chromosomes (*see* parental ditype)
nonsense mutant a type of base-pair substitution mutant that alters a given codon to create one of the three stop codons: UAA, UAG, or UGA
nonsense-mediated decay term used to describe the active decay of mRNA products with premature stop codons
Northern blot a method for studying gene expression; a specific RNA sequence can be detected through the use of gel electrophoresis and a hybridization probe, typically a labeled complementary DNA fragment
nucleotide–nucleotide BLAST (*see* blastn)
null allele (*see* nullomorph)
null mutant (*see* nullomorph)
nullomorph a mutant that, although it may still make part or all of the protein the gene encodes, no longer produces a functional protein product and thus has no remaining gene function
off-target effects the unintended cutting of a region of the genome caused by targeted gene disruption methods such as CRISPR/Cas9 or TALENs
orthologs genes that have evolved through vertical descent from a common ancestral gene and are found in different species due to speciation events; contrast with xenologs
P element a DNA transposon frequently used in Drosophila for mutagenesis
pachytene the third subdivision of prophase I of meiosis during which recombination occurs and that is characterized by full-length synaptonemal complex
pairing the identification and matching up of homologous chromosomes, which occurs during leptotene of prophase I of meiosis
paracentric inversion a chromosomal inversion that does not include the centromere
paralogs two genes created by a duplication event in a species, they may or may not serve similar functions, and one of those genes may no longer function at all
parametric linkage analysis a statistical method used to identify the location of genes that contribute to a trait or complex disease in a family where multiple individuals are affected
parental ditype a tetrad that has only non-recombinant chromosomes (*see* nonparental ditype)
parental ditypes a type of tetrad that contains two non-recombinant spores and two recombinant spores; the two alleles of a linked gene are in the same combination as the original parental combination
partial loss-of-function mutant a type of loss-of-function mutant in which some degree of product activity can be seen; includes mutants that impair the level of product formation and those that create a partially functional product
perdurance of gene products in mosaic studies it is the worry that a gene product from a prior time point persists and will affect the results
pericentric inversion a chromosomal inversion that spans the centromere
petite mutants mutations that are the result of deletions in mtDNA; they slow down the growth of an organism, such as yeast
phage T4 a type of virus that infects the bacterium *Escherichia coli* that has been a well-studied model organism and has been used extensively as a tool for genetic manipulation
piwi-interacting RNAs (piRNAs) noncoding RNA elements that function in the germline to silence transposable elements to protect the genome from *de novo* insertions
plaque a visible area of bacterial lysis on a bacterial lawn or agar plate, indicating the presence of infectious phage particles
plasmid rescue (*see* transformation rescue)

pleiotropic affecting more than one function

point mutant a mutant that results from the change of one base in the DNA sequence to another

poisonous interactions a situation in which two different mutant proteins interact to make a product that poisons the cell; type I second-site noncomplementation

Poisson distribution discrete frequency distribution of the probability of independent events occurring during a defined time interval

poly-A selection the selection prior to RNA sequencing of only those transcripts that contain a poly-A tail

prometaphase the period during which bivalents attach to or create the meiotic spindle and congress to the center of that spindle

prophase I the period after the last cycle of DNA replication during which homologous chromosomes pair and recombine, the end of which is signaled by the breakdown of the nuclear envelope

prophase II a brief period, in those organisms in which a true telophase I occurs, during which the nuclear envelope once again breaks down

protein null a type of nullomorph that fails to produce a protein product at all

protein–protein BLAST (*see* blastp)

pseudoautosomal region (PAR) a small region of the X and Y chromosomes that share homologous DNA sequences and are able to pair and recombine during meiosis

pseudoreversion (*see* intragenic suppression)

pseudorevertant an intragenic suppressor mutant that converts a dominant mutation into a recessive or loss-of-function mutation

recessive mutation a mutation that displays a phenotype only when homozygous

reciprocal shift experiment an experiment using heat-sensitive and cold-sensitive alleles of two genes that involves a shift from normal temperature to high or low temperature followed by a shift in the opposite direction; provides information about the order in which the genes function in the cell division cycle

recombination (*see* crossing over)

recombination nodule a spherical structure that marks the site of recombination along a meiotic chromosome during pachytene

regional tetrad analysis a method to determine the frequency and distribution of recombination events between two linked genes on a chromosome

repetitive region a region of DNA in which short DNA sequences are repeated multiple times

restriction fragment length polymorphism (RFLP) a genetic variant that can be used as a marker for genetic mapping

ribosomal RNA depletion the selective removal of rRNA transcripts prior to RNA sequencing

RNA interference (RNAi) the introduction into an organism of double-stranded RNA that binds complementary mRNA sequences to inhibit translation or to degrade that mRNA

RNA sequencing (RNA-seq) the sequencing of mRNA transcripts from a particular tissue or during a particular time during development to identify which genes and isoforms are expressed

RT-PCR reverse transcription polymerase chain reaction; a form of PCR used to detect gene expression in which RNA is first reverse transcribed into cDNA and then amplified by traditional PCR methods

second-site noncomplementation (SSNC) a situation in which mutations in two different genes produce a mutant phenotype, while a mutation in either gene alone does not

second-site revertant (*see* pseudorevertant)
segregation the movement of chromosomes toward opposite spindle poles and their subsequent disjunction into separate cells during cell division
semidominant mutation a mutation that displays a phenotype as a heterozygote that is more severe than the wildtype phenotype, yet less severe than the homozygous phenotype
separation-of-function mutation in genes with multiple biochemical activities, a mutation that disrupts one function but not the other(s)
sequestration a situation in which the mutant form of one protein sequesters the wildtype form of the other protein into an inactive complex; type II second-site noncomplementation
silent mutation (*see* silent substitution)
silent substitution a mutant in coding sequence that does not change the amino acid directed by that codon or a mutant in a noncoding region that does not affect gene expression
single-nucleotide polymorphism (SNP) an insertion, deletion, or substitution of a single DNA base pair; an SNP or indel
single-nucleotide variant (SNV) substitution of a single DNA base pair
sister chromatids the two identical DNA strands that result from chromosome replication and are joined at their centromeres
split reads in DNA sequencing, two halves of one read that can be aligned to multiple locations in the genome
sporulation the process in yeast whereby a diploid individual is induced to undergo meiosis, yielding four haploid spores
strand breakage breakage of the sugar-phosphate backbone of a DNA strand
structure-function study a study that is designed to determine the structure or function of a gene of interest
suppression masking of a phenotype by a second mutation
suppressor mutant the mutant that is masking, or suppressing, the first mutation (see suppression)
symmetric heteroduplexes created during the branch migration step of DNA repair
synapsis the formation of a full-length synaptonemal complex between homologs to connect them together along their lengths
synaptonemal complex a macromolecular protein structure that forms between paired homologs and links them together during prophase I of meiosis
synthesis-dependent strand annealing the double-strand break repair pathway by which noncrossover gene conversions are formed during meiosis
synthetic antimorph when two mutant products produce a poisonous gene product
synthetic lethality two mutants that are viable in a wildtype background, but lethal when combined
TALENs transcription activator-like effector nucleases; a form of a targeted gene or sequence disruption
tandem duplication a type of duplication in which the duplicated segments sit right next to each other
targeted gene disruption a method of mutagenesis in which one specific gene is mutated or its function is disrupted, while the rest of the genome remains unchanged
tblastn searches a nucleotide database for all possible translations in all six reading frames, and then compares the resulting protein sequences to a query protein sequence, which allows for the detection of homologous sequences in nucleotide databases that may not have been annotated as protein-coding genes

d

e

u

v

w

x

y

z